MATERIALS SCIENCE AND TECHNOLOGIES

X-RAY DIFFRACTION

STRUCTURE, PRINCIPLES AND APPLICATIONS

Materials Science and Technologies

MATERIALS SCIENCE AND TECHNOLOGIES

X-RAY DIFFRACTION

STRUCTURE, PRINCIPLES AND APPLICATIONS

KAIMIN SHIH
EDITOR

nova publishers
New York

For permission to use material from this book please contact us:
Telephone 631-231-7269; Fax 631-231-8175
Web Site: http://www.novapublishers.com

Additional color graphics may be available in the e-book version of this book.

Library of Congress Cataloging-in-Publication Data

X-ray diffraction : structure, principles, and applications / editor, Kaimin Shih (Department of Civil Engineering, The University of Hong Kong, Hong Kong).
pages cm
Includes index.
ISBN: 978-1-62808-591-4 (hardcover)
1. Materials--Microscopy. 2. X-rays--Diffraction. 3. X-ray crystallography. 4. Transmission electron microscopy. I. Shih, Kaimin.
TA417.23.X155 2013
548.028--dc23

2013028248

Published by Nova Science Publishers, Inc. † New York

CONTENTS

PREFACE

"Why XRD?" After deciding to edit this book, I still often found myself questioning why I had picked a technique to come up for a book that had been developed more than 100 years. Since I have been using the quantitative X-ray diffraction (QXRD) technique for nearly 15 years, I certainly understand how powerful X-ray diffraction (XRD) can be in evaluating materials for a wide variety of new applications. However, what about others? Do other researchers still find anything new and exciting in their work with XRD? The answer to my question is a resounding "yes", which gradually became clear during my preparation of this book as the chapter contributions were collected.

First, a little historical background: In 1895, Wilhelm Conrad Röntgen discovered X-rays. X-rays have very high frequencies, and their electromagnetic spectra are situated between those of gamma rays and ultraviolet radiation. Since crystals are regular arrays of atoms, the atoms in a crystalline structure can scatter X-ray waves, acting as a diffraction grating, primarily through the elastic scattering of electrons in atoms. These waves may cancel one another out through destructive interference, or add constructively in a few specific directions as determined by Bragg's law. Those specific directions with constructive reflections generate the diffraction pattern on the recording device. In 1912, the diffraction of X-rays was observed by Max von Laue, together with his two assistants, Walter Friedrich and Paul Knipping. The potential of X-ray crystallography for determining the structures of molecules and minerals was immediately recognized after Von Laue's pioneering work, particularly by physicists William Lawrence Bragg and his father William Henry Bragg. During 1912-1913, the younger Bragg developed Bragg's law, which connects the observed scattering with reflections from evenly spaced planes within the crystal. In 1914, von Laue was awarded the Nobel Prize in Physics for his contribution in connecting the scattering angles and the size and orientation of the unit-cell spacings in the crystal. In 1915, the Braggs, father and son, shared the Nobel Prize in Physics for their contribution in crystallography.

In the last 100 years, X-ray diffraction studies have revealed highly valuable information about many ordered atomic structures presenting in a variety of common materials, such as metals, rocks, ceramics, sediments, and even ice. Through the understanding of material structures, the applications of these materials began to be reliable, and the scientific discussions about material properties and structural features became possible. The diffraction method - still one of the most essential tools for the structural characterization of materials today - has been tested for over 100 years, with successful uses in both academia and industry. Today, a broad range of general-purpose and specialized X-ray diffractometers are commonly available, and just about every research project that involves crystalline solids inevitably begins with the collection of material diffraction patterns, such as one-dimensional powder X-ray diffraction patterns.

In addition to its useful role in probing material structures, the popularity of the X-ray diffraction technique is also due in large part to its fundamental simplicity and the affordability of quality diffractometers today. Although some very special scanning schemes may require specific diffractometer configurations and therefore more specialized hardware, most structural identification, and even structural analysis tasks, can be accurately carried out by commercial X-ray diffractometers currently available. However, this advantage should not be taken to mean that X-ray diffraction today is standardized phase identification work that can be easily carried out by non-experts. It is a misconception that anyone can just grab a "freshly produced" diffraction pattern from the XRD room and start to answer their scientific questions, just because he/she sees some diffraction peaks matching the expected phase in the sample. In my research and teaching career, I have heard and seen too many cases of misinterpreting X-ray diffraction data in research results. For example, a crystalline phase observed in the XRD pattern of a new catalyst can be erroneously used to explain its catalytic performance, if an amorphous hump is in the pattern. Another mistake that many students often make is to simply present the XRD phase identification results and start to develop the chemical reaction pathways in the system. They forget that the XRD data provide crystal structure information, not chemical composition information.

The affordability of quality diffractometers for the research community is of course due to the success of the diffractometer manufacturers in improving the precision of the manufacturing work and the quality of the materials used for their products. Moreover, the innovative modularized design has made every laboratory X-ray diffractometer not only a phase identification tool, but also a versatile XRD "experimental platform", capable of a wide variety of new experimental designs. Unfortunately, this information has not yet been well-disseminated to different research communities. In addition, the principles of analyzing X-ray diffraction data have also experienced a great leap, due to improvements in hardware performance capabilities and active method development. Today, it is still common to hear researchers claiming that XRD is not a quantitative tool. With a careful model design, the Rietveld refinement with the fundamental parameters approach and the internal standard for amorphous content quantification has made the quantitative X-ray diffraction technique a very powerful, straightforward, and reliable method to provide phase composition information for solid samples.

In selecting chapters for this book, I felt strongly obligated to report on current advancements in the XRD technique and to demonstrate the wide variety of its applications today. After 100 years, this simple technique not only played a crucial role in the history of scientific development but also has successfully transformed itself into a method in the forefront of extending many of our knowledge boundaries. In this book, the authors first provide a review of the development and applications of the single-crystal XRD technique in Chapter 1. Chapter 2 extends diffraction studies to include coordination polymers for a new direction of exploration. The strength of the Rietveld refinement is demonstrated in solving crystallographic structures for polycrystalline materials, beginning with magnetic shape memory alloys in Chapter 3 and the new rare earth intermetallic compounds in Chapter 4. Continuing on with the wide applications of the Rietveld refinement, the quantitative X-ray diffraction technique has been successfully used on cement materials in the geoenvironmental application described in Chapter 5 and on ceramic materials in the hazardous metal stabilization study described in Chapter 6. In Chapter 7, the powder X-ray diffraction technique is applied to help delineate the core-shell structure information of nanoparticles, and in Chapter 8 the use of the Rietveld refinement for intermetallic materials is aimed at hydrogen technology applications. Chapters 9 and 10 both point out on-going methodological developments in another important function of the X-ray diffraction technique - quantifying the state of residual stress in materials. In Chapter 11, the authors conclude this book by bringing you to witness a state-of-the-art development in how X-ray diffraction principles can combine with electron crystallography to create a powerful method for structure determination.

The accomplishment of this book is owed to unselfish contributions from chapter authors across the world (China, Brazil, Spain, USA, Hong Kong, South Africa, Argentina, UK, and France). I am very grateful to each and every one of the authors, and what readers do not see are the conscientious efforts of the authors in responding to the comments from the peer-reviews. The authors, of course, also deeply thank the reviewers of these chapters for their help shaping this final outcome. My research interest in XRD initiated this book, but this interest could not have been pursued, if it were not for the crucial support from Professor Herbert H. P. Fang, Professor Alfonso H. W. Ngan, and Professor Weng Cho Chew to establish the XRD laboratory in the Faculty of Engineering at the University of Hong Kong. The origin of my XRD research interest should be fully credited to Professor Tim White at Nanyang Technological University (Singapore), who first introduced me to the powerful QXRD technique that still strongly fascinates me to this day. Learning about the Rietveld refinement from Professor Zhili Dong and Dr. Sou Hon Lim in Professor White's group was a critical point leading to my career path today. After my return to Asia, Professor Lingmin Zeng, a very respected powder diffraction scientist, and his colleague Professor Wei He, at Guangxi University (China) have continually helped me develop a very capable XRD team within my research group. My deep appreciation also extends to Ms. Xiaoyin Luo in Bruker's Beijing Office, who greatly facilitated my personnel and facility setups for the XRD experiments at the University of Hong Kong. Finally, I am pleased to acknowledge the key roles played by Ms. Xingwen Lu, Mr. Zhengyuan Zhou, Mr. Changzhong Liao, Ms. Hanlu Yan, and Mr. Minhua Su in the HKU Environmental Materials Group. Their tireless efforts in handling the chapter submissions, peer-review processes, manuscript formatting, and figure processing work made it possible to have this book here in front of you.

Kaimin Shih
The University of Hong Kong

In: X-Ray Diffraction: Structure, Principles and Applications ISBN: 978-1-62808-591-4
Editor: Kaimin Shih

Chapter 1

DEVELOPMENT AND APPLICATION OF X-RAY DIFFRACTION TECHNIQUE FOR SINGLE CRYSTALS

Xiao-Ming Lin, Xu-Jia Hong and Yue-Peng Cai*
School of Chemistry and Environment, South China
Normal University, Guangzhou, People's Republic of China

ABSTRACT

Single crystal X-ray diffraction is a powerful technique for determining structures of inorganic compounds or small organic molecules, as well as biological molecules, including proteins and DNA. This chapter provides a summary regarding various developments in the field of single crystal X-ray diffraction technique during the past decades; moreover it also provides practical tips on crystal growth and selection, diffraction data collection, data reduction and correction, crystal structure solution and refinement, as well as structure analysis and presentation. Further, it works as a curtain-raiser exploring the applications of this technique in various fields such as material science, life science and supramolecular chemistry as well as crystal engineering. This would definitely promote the various prospects of single crystal X-ray diffraction for future research endeavours.

1. INTRODUCTION

The determination of crystal structures by X-ray diffraction technique with its exciting avenues has shown valuable progress and has become a burgeoning field of interest to the scientific community in the past decade.

Single crystal X-ray diffraction analysis has become one of the most important pathways and authoritative methods as it facilitates direct visualization of the internal structure, which helps improve and expand our knowledge regarding the microstructures and properties of material to a great extent.

* Corresponding author: Yue-Peng Cai. Tel: +86-20-3931-0383; Fax: +86-20-3931-0187. E-mail address: caiyp@ scnu.edu.cn.

By determining the crystal structure of a compound, we can understand the atomic arrangement in three-dimensional space and obtain detailed information about the internal lattice of any crystalline substance, such as unit cell dimension, bond length, bond angle, torsion angle, charge distribution, atomic thermal vibration in the vicinity of the equilibrium position, molecular configuration and conformation, intermolecular interaction, packing mode in the crystal and the other structural data. On the other hand, single crystal X-ray diffraction also allows the determination of the three-dimensional structure of alloys, minerals and biological macromolecules, such as proteins, viruses, DNA segments and ribosomes. These aforesaid information could not only help understand the mechanisms involved in important biological processes, but also could be useful to understand the nature of chemical bonding and the questions related to nanotechnology. This chapter is therefore dedicated to other developments and applications of single crystal X-ray diffraction techniques.

2. Development of Single Crystal X-Ray Diffraction Technique

After Max Von Laue's pioneering research [1], the field of X-ray crystallography had a remarkable start. In 1912-1913, the physicist William Lawrence Bragg determined the first atomic-resolution structure and outlined Bragg's law [2, 3]. He along with his father was awarded the Nobel Prize in physics in the year 1915 for their excellent contribution in crystallography. Thereafter, X-ray diffraction has been considered as the principal method for determining the arrangement of atoms in minerals, metals, semiconductors, as well as all kinds of inorganic, organic and biological molecules. It has become a technique popularly used by most of scientists of numerous application in variegated scientific fields.

In the 1960s, X-ray radiography was the only tool used for the measurement of crystal diffraction intensity, for example, Laue method, oscillation method, Weissenburg method as well as procession method [4-7]. However, these methods proved to be cumbersome and had relatively low data accuracy. It took several months or even years to complete the data collection for a small molecule crystal, but had the advantage of accurate Laue group and unit cell. Herein, looking to the drawbacks, these traditional methods are rarely used at present and have not been described in details.

In 1970s, with the advent of the most common automated diffraction techniques, an outstanding instrument, the four-circle diffractometer was invented (Figure 1), which enabled accurate measurement of the intensities of all the accessible diffraction reflections and automatically determined the shape and symmetry of the unit cell of a crystal [8]. The four-circle diffractometer aided in the determination of small molecule crystal structure, and also helped in studies of biological molecules such as proteins and other macromolecules. However, the need to record diffraction data using a scintillation counter point by point proved to be a drawback, and this limitation caused a slow data collection speed, time-consuming and low sensitivity.

Especially, it did not meet the fast recording requirements of those crystal samples with large unit cell volume, weak diffraction capacity, instability or decay of diffraction capability when exposed to X-ray for a long time, for example, many supramolecular systems or biological macromolecules.

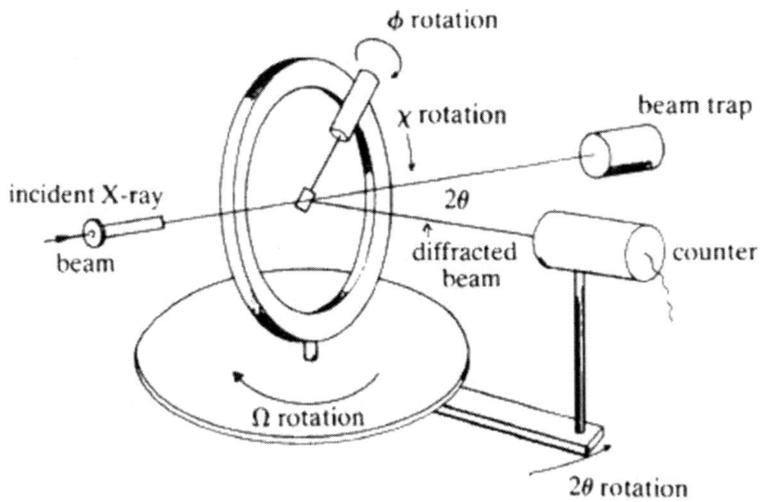

Figure 1. Schematic diagram of four-circle diffractometer.

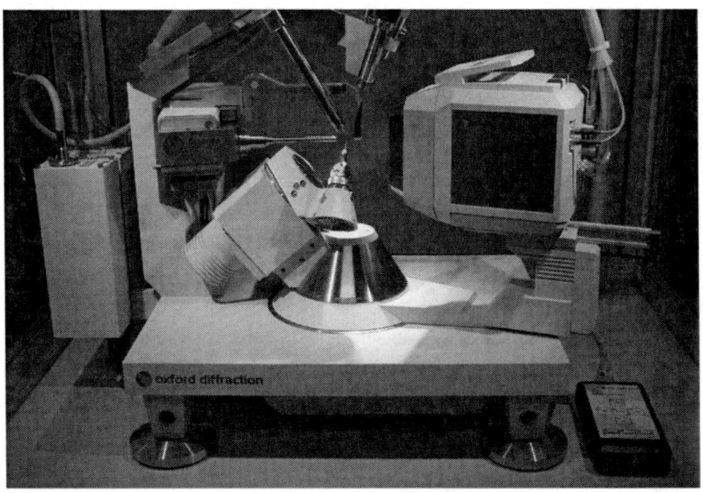

Figure 2. Configurations of Xcalibur 3 Diffractometer (Oxford Diffraction).

In the late 1980s, the applications of two new technologies brought a great breakthrough for the single crystal structural analysis [9, 10].

One is the synchrotron radiation with high intensity photon beam, which provides high brilliant X-ray light source for X-ray crystallography with tunable wavelength required for their experiment; the other is the two-dimensional electron X-ray detector (area-detector), image plate (IP) and charge coupled device (CCD), which was able to significantly improve the speed and sensitivity of the diffraction data collection (Figure 2). Area-detector diffractometer with a high sensitivity could collect high-quality diffraction data even for crystals with small size or weak diffraction ability.

IP and CCD were certainly in the forefront of intensity data collection so that they have become the primary means of X-ray crystal structural analysis. Nowadays, they are frequently used in the collection of intensity data from protein crystals and quite popular in other applications.

As compared to the four-circle diffractometer popularly used two decades ago, the development of single-crystal diffraction technique after 1990s proved to be a positive investment especially the application of the area-detector diffractometer. The area-detector diffractometer has many advantages such as fast collection speed, more capacity, less collection time as well as large numbers of diffraction data collection at the same time. On the other hand, it requires advanced computer with large capacity and more time for the complicated data processing; however, both of these modern X-ray diffraction systems commonly consist of several parts: (a) X-ray, (b) Goniometer, (c) Detector, (d) Computer equipped with the software required to run the diffractometer, (e) other auxiliary equipments, such as the cooling system, high-and low-temperature device, vacuum device, high-voltage device and X-ray security facilities (Figure 3).

3. Crystal Growth and Selection

3.1. Crystal Growth

In order to obtain a crystal suitable for X-ray diffraction experiment, it is important to prepare the perfect crystals with good quality and suitable size. In case of a new compound, it becomes difficult to predict what kind of crystallization a compound would form, i.e., it is difficult to predict and avoid the occurrence of crystallites and reunions; however, there are some basic rules to be followed. The authors would like to present a few effective methods that would help have a vivid idea regarding this.

3.1.1. Evaporation Method

Crystallization from the solution is the most common method used for crystal growth. Here, the solution of the compound to be tested in a suitable solvent is placed in a beaker or a test tube covered with filter paper or a perforated cork to prevent the dust (Figure 4); then, the solution is allowed to evaporate slowly by controlling the evaporation rate and kept stationary.

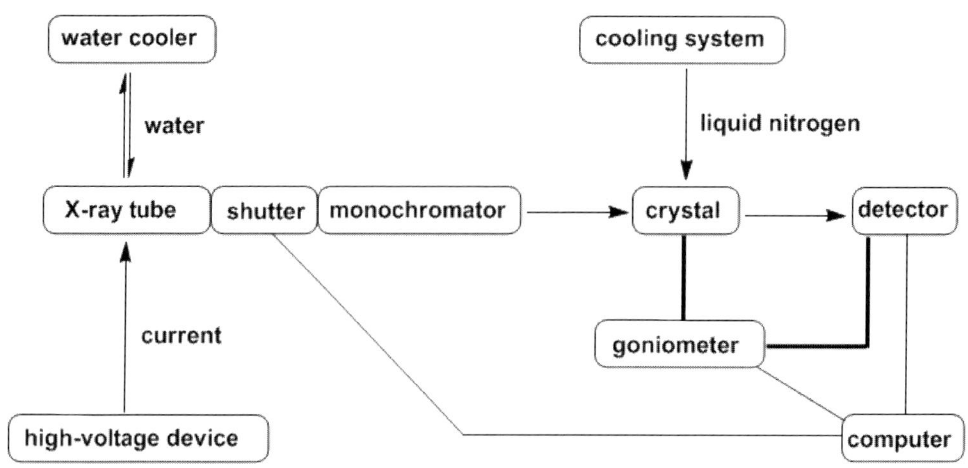

Figure 3. Schematic diagram of diffraction.

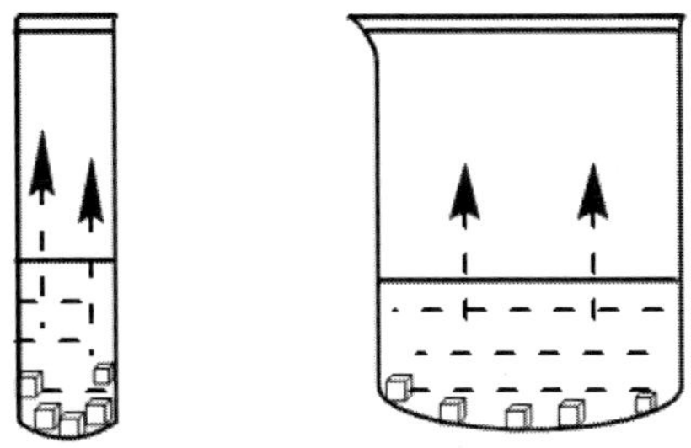

Figure 4. Schematic diagram of solvent evaporation method.

Crystals with appropriate size automatically crystallize in the supersaturated solution. This method is suitable for a variety of organic solvents, water or a mixed solvent, and is also applicable to solution resulting from the direct reaction; moreover, the recrystallization of the product is also possible.

If the aqueous solution can't be crystallized in air due to the excellent solubility, it can be placed in a desiccator with desiccant such as anhydrous $CaCl_2$, concentrated sulfuric acid, silica gel or molecular sieve to absorb the water vapor of the solvent.

Thereafter, the desiccators can be placed in the refrigerator or other incubator to obtain the desired temperature.

3.1.2. Diffusion Method

Liquid Diffusion Method (Layer) This method is usually used in cases where the compound is generated by two reactants. The first reactant dissolved in one solvent is put at the bottom of the test tube, and the second reactant dissolved in another solvent is carefully layered over the above liquid level with a dropper during the experiment (Figure 5a). A few experiments also involve the use of a buffer layer layered between these two solvents to slow down the reaction rate and avoid the generation of precipitates. Particularly, the density of the bottom solvent should be greater than the upper solvent.

Vapor Diffusion The principle of vapor diffusion is as follows: two miscible organic solvents are selected with different solubility of targeted compound, e.g., high boiling point (DMF, DMSO, etc.) and low boiling point solvents (diethyl ether, *etc.*, also called anti-solvent). In a sealed container, the anti-solvent volatilizes into the high boiling point to reduce the solubility of the compounds and promote the formation of nucleation and enhance the crystal growth (Figure 5b).

The operation of a vapour diffusion system is simple: the two reactants dissolved at the high boiling solvent in a small bottle are placed into a large bottle filled with volatile solvent. Followed by this, the large bottle is sealed and allowed to stand until crystals appear.

Gel Diffusion This method is also commonly used to obtain crystal suitable for X-ray diffraction [11]. Many gels such as sodium silicate gel, silica gel, gelatin and agar can be used for crystal growth; here, either a test tube or U-shaped tube could be used for the preparation of crystals (Figure 5c). The operation includes the following steps: the solution A is mixed with the gel, and after the gelling, the solution B is carefully poured above the gel. As the diffusion proceeds, A and B crystallize in the interface or the gel. A simple U-shaped tube apparatus are shown in Figure 5d, and it is particularly suitable for this situation that precipitation is formed immediately when reactants A and B are mixed together.

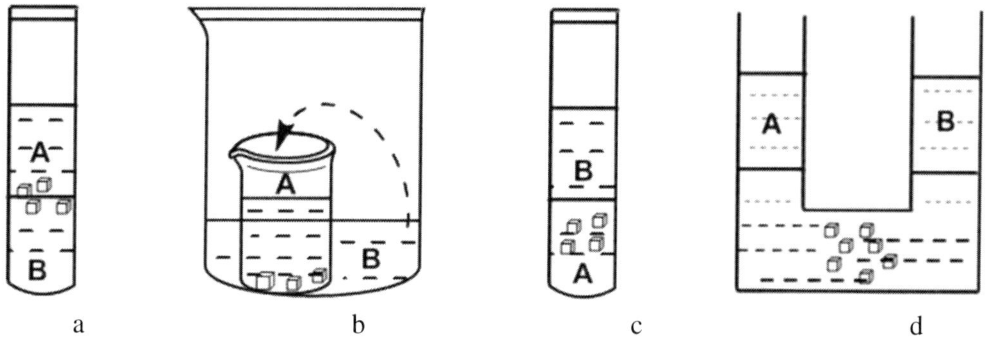

Figure 5. Diffusion method involving (a) liquid diffusion (layer), (b) vapor diffusion, (c) and (d) gel diffusion.

These conditions are relatively mild and easy to obtain high quality single crystal suitable for structural analysis; however, this process is time-consuming and the reactants must be dissolved at room temperature.

3.1.3. Hydrothermal and Solvothermal Method

Hydrothermal method has been widely used in recent years due to its simple equipment and has been proved to be an effective way to obtain particular substances or high quality crystals in the presence of water under high temperature and high pressure condition (Figure 6a). The general procedure of hydrothermal method is as follows: the mixture of the reactants insoluble or poorly soluble under room temperature and deionized water was sealed in a Teflon-lined autoclave and heated under autogenous pressure. In general, the most important skill for crystal growth depends on the temperature crystallization and the temperature will be controlled in the range of 80 to 600 °C. As the temperature and pressure rise, many reactants will gradually be dissolved under hydrothermal condition and would crystallize during the slow cooling process.

Nowadays, people have expanded the scope of its application and use organic solvents instead of water as reaction medium to the similar reaction, so-called solvothermal method. Nonetheless, both these methods have the same operation process and principle except that the solvent used. Solvothermal (hydrothermal) method requires a short reaction time and can work well for many reactants dissolved at room temperature. The solvents used in the synthesis, especially the organic solvents with different functional groups, polarity and dielectric constant, boiling point and viscosity and so on, can greatly enhance synthetic routes and increase the structural diversity of products.

In recent years, solvothermal technique has been widely used in the synthesis of inorganic materials, inorganic-organic hybrid composites and complexes due to its advantages of simple equipment, energy saving, perfect crystal growth [12].

On the other hand, it is rather difficult to understand the reaction process because of its sealing system.

In addition, we also can use a special reactor, such as glass tube (Pyrex glass tube) (Figure 6b), to observe the reaction process and study the reaction mechanism, but this glass tube should be put in a safety place and it is likely to explode if the temperature is higher than 120 °C.

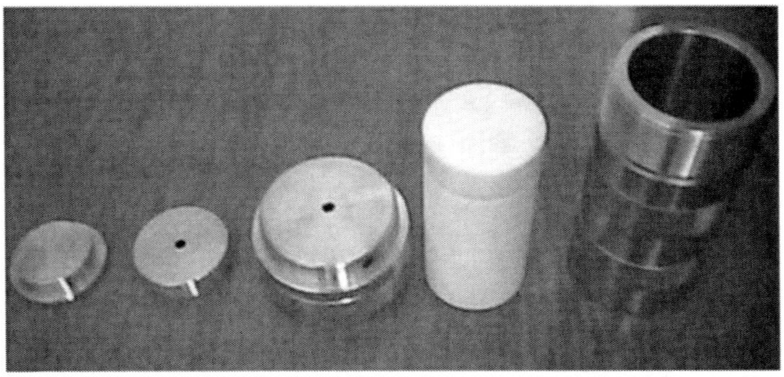

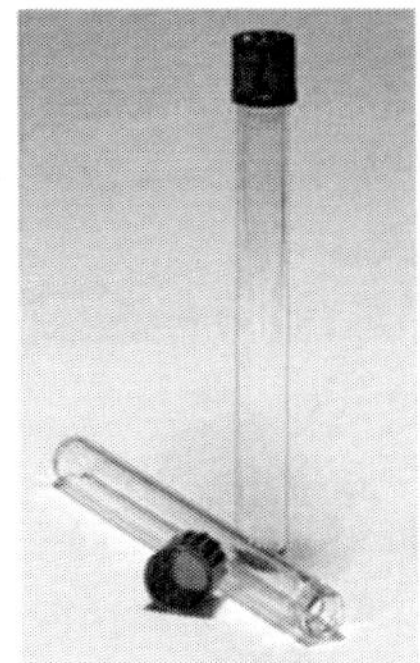

Figure 6. (a) Teflon-lined stainless steel autoclave and (b) Pyrex glass tube 1. Bottom plate; 2. Upper pressure plate; 3. Stainless steel cap; 4. Teflon liner and Teflon lid; 5. Autoclave body.

3.1.4. Other Methods

In addition to the traditional diffusion method and hydrothermal (solvothermal) method, microwave and ultrasonic methods have been developed in recent years. Microwave is a common method used to accelerate the chemical reaction rate. The temperature of the solvent can increase quickly in the microwave environment [13]. The crystal with higher specific surface can be obtained by adjusting different temperature and reaction time. But it is essential to consider that the crystals will be damaged if the reaction exceeds the optimum reaction time.

The ultrasonic method can be used to prepare small crystals in a short time, which can greatly reduce reaction time and temperature [14]; however, this is still in its infancy and needs to establish its universality.

Sublimation is another method used to obtain crystals with high quality [15]. Theoretically, solid substance with a large vapor pressure below its decomposition temperature can be used to prepare the crystal growth. However, this method is rarely used due to the few substances which can meet the sublimation condition.

3.2. Common Conditions for Crystal Growth

In addition to the above discussed synthetic methods, the crystal quality and yield are influenced by many factors, even small changes can lead to large differences in crystallization behaviour. In fact, impurities, microcrystalline, crystal reunion and defects, amorphous powders are inevitable. The growth and quality of the crystal is mainly depended on the formation and the growth rate of nuclei. If the nucleation rate is faster than the growth rate, it is inclined to form a large number of crystallites. Conversely, too fast growth rate may lead to crystal defects. Therefore, we have to study the techniques and methods of crystal growth and pay more attentions to the details of the operation, so that we can obtain prefect single crystals suitable for X-ray diffraction.

Supersaturated Solution Crystallization and dissolution are two opposite process. The basic condition for a compound to crystal is to achieve supersaturation.

One could slowly change the solution components to increase the saturation for encouraging crystallization by vapor diffusion method.

Purity The crystallization process is an extremely important method generally used to purify a substance, however, the crystalline materials need to possess a degree of purity. The purified product can be obtained by the combined way of recrystallization and crystal growth. Impurities are usually inimical to crystallization.

Solvents and Temperature. It is very difficult to predict favorable conditions for nucleation and the growth of crystals. In practice, suitable and successful crystallization conditions are identified by screening during the numerous experiments, for example, some may change the solvents. Appropriate solubility is conducive to the crystal growth and the different solvents of solution component will influence on the growth habit of the crystal shape. For instance, NaCl crystals with cube shape are often obtained in aqueous solution, while octahedral crystalline form can be generated when urea is added to the solution. In addition, some may try several temperatures for encouraging crystallization. Different temperatures can lead to the formation of crystals with distinct shapes and structures, which is evident by witness of a large number of examples [16-18].

Ironically, it is more unwilling to assemble into crystals for a molecule subjected to different conformations due to entropy. Here, the higher the concentration is, the greater the viscosity. Changing a solvent or temperature allows the molecules to tend to a dominant conformation for encouraging crystallization. Meanwhile, it is less likely to make crystallization for those molecules that tend to self-assemble into regular helices.

3.3. Crystal Selection and Mounting

Crystal Selection Obtaining a diffraction-quality crystal is the basic step for the technique of single crystal X-ray diffraction. After crystallization from a solution, several crystals can be laid on a filter paper in order to remove the mother liquor and other small crystals. An ideal crystal should be between 50 and 600 microns in size, as close as possible in all directions. If the crystal is too small, the diffraction efficiency is too weak and the number of observable diffraction points is not enough. While the crystal is too large, the diffraction intensity will decline due to the diffraction absorption of the crystal. Most of the crystal should be regular in shape, transparent, unfractured and optically clear, without significant internal crack of twinning. Although the quality of the crystal can be largely identified under a microscope by careful observation as well as a bit of experience, the final quality of the crystal suitable for experiment requirement depends on the diffraction test [19, 20].

Crystal Mounting. The single crystal suitable for X-ray diffraction is typically mounted on the tip of a thin glass fibre or a loop with epoxy or cement under the microscope, avoiding the data errors due to the glass fibre on the X-ray diffraction. The high quality crystal also should be stable enough in air for several hours, without obvious weathering or decomposition. Inadequate stable crystal should be protected under the liquid nitrogen environment or wrapped by adhesive to separate the air during the collection data process. In addition, the unstable crystal can also be placed inside the glass capillary with the appropriate internal diameter, and then sealed both sides of the capillary with adhesive (Figure 7). Sometimes, the mother liquor can be added into the glass capillary to protect the crystal before sealing.

The fibre or the capillary is attached to a brass mounting pin by modelling clay, and inserted into the goniometer head, then affixed to the diffractometer with a screwdriver (Figure 8).

4. DATA COLLECTION, ANALYSIS AND PRESENTATION

X-ray crystal structure analysis includes crystal growth, crystal selection and mounting, reflection intensity measurement, crystal structure solution and refinement with a variety of structure analysis and data fitting method, as well as the presentation of the structure data and graphics. The X-ray crystal structure analysis process is summarized in Figure 9.

Data Collection A crystal is centered and exposed to an intense beam of X-ray, usually monochromatic X-ray ($Mo_{K\alpha}$, $\lambda = 0.71073$ Å or $Cu_{K\alpha}$, $\lambda = 1.54178$ Å), generating a diffraction pattern of spots know as reflections. A preliminary set of rotational frames are collected by an automatic collection routine. Reflection from these images are auto-indexed by a computer to determinate the primitive unit cell and the orientation matrix, which are further refined using least-square to obtain the final unit cell parameter and orientation matrix.

Then, a full set reflection intensities data are recorded at different orientations of the crystal. Generally, data collection could be done by a charge coupled device (CCD) or image plate (IP) detector using an incremental scan method. The data collection time depends on the crystal sample and the instrument.

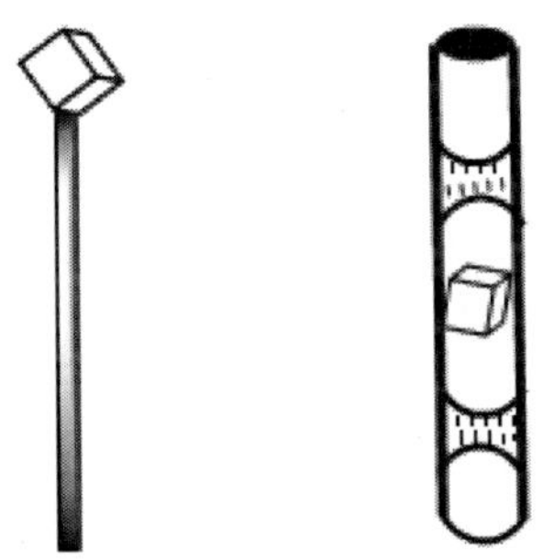

a b

Figure 7. Crystal mounting (a) correct operation: on the tip of a glass fibre (b) sealing in the glass capillary.

Figure 8. Goniometer head and screwdriver.

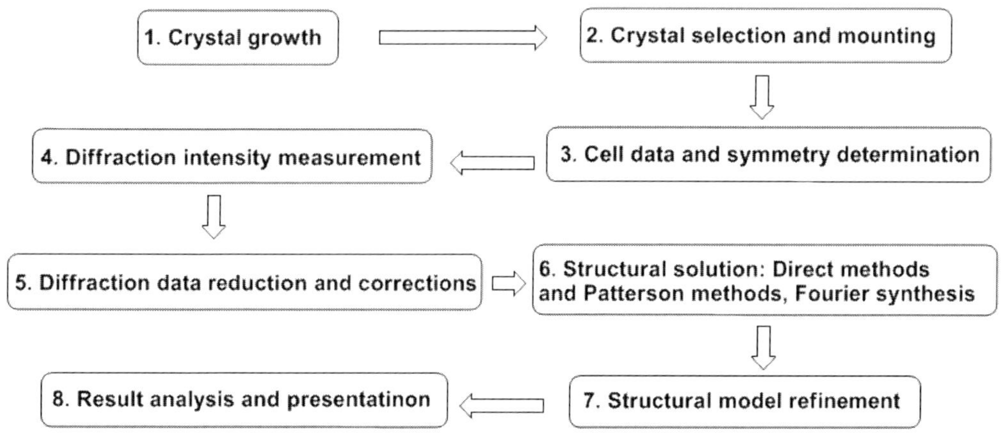

Figure 9. Process of crystal structure analysis.

In addition, some variable parameters such as rotation angle, scan angle, exposure time, the range of data collection as well as scan method, should be taken into consideration [21].

Data Reduction and Corrections. After the data collections are completed, the data reduction and correction should be carried out in order to process the entire data set. Data were processed for reduction and cell refinement, and absorption correction could be done using numeric absorption correction, empirical absorption based on scans or DIFABS method [22, 23]. These processing procedures and correction are typically finished by the software package of program.

Phase Problem. Based on the crystal diffraction investigations, one can only obtain the cell parameters, space group and intensity data, but it does not provide information regarding the precise position of the atoms in the unit cell, i.e., single crystal X-ray diffraction can only obtain the intensity I_o value, without phase information. This I_o value can be converted to the absolute value of the structure factor, namely structure factor amplitude $|F_o|$. The phase problem and structure factor amplitude must be solved to obtain the electron density map, which can be used to build a starting model of the molecule. Initial phase problem can be obtained using a direct method or Patterson method, which is most prevalent currently [24-26].

Structure Solution. After obtaining a preliminary set of phase, combined with the experimental structure factor amplitude, a new set of electron density map is generated through Fourier transforms (FT), from which we can obtain more atomic coordinate information. An initial crystal structure model could be obtained after the atom assignment has been completed, and it could be used to refine the phases to obtain a better model. This understanding and investigation is repeated for several times until the model is optimized.

Structure Refinement. This procedure may be more mystifying and difficult to the beginner, as it is often done through a "trial and error" which finally leads to resolving the correct structure model.

Once a better structure model is determined, several runs of refinement are necessary for the atomic coordinates and isotropic displacement to improve the atomic coordinated parameters. The best possible fit of the structure model are generally be assessed with R_1, R_2 and GOOF values, which can interpret the quality of a solution; several reports have detailed these values [27].

Deposition of the Structure. Once the correct model of a crystal structure has been finished, the resulting crystallographic information file (CIF) in a format ready for publication is often deposited in a crystallographic database such as the Cambridge Structural Database (CCDC, for organic compounds, metal-organic compounds and coordination compounds), the Inorganic Crystal Structure Database (ICSD, for inorganic compounds), the Protein Data Bank (PDB, for protein structures) or Joint Committee on Powder Diffraction Standards-International Center for Diffraction (JCPDS-ICDD, for substances identification). The original data of the crystal structures could be deciphered either from these databases or based on former reports available that deal with a related crystal structure, especially the CIF format [28].

Structure Analysis and Presentation. The crystal structure analysis is generally conducted using the SHELXTL program suite, which is probably the most widely used software package that includes XPREP, XP and XL. In addition, programs such as PLATON, WinGX, Olex 2 and APEX II are also used in the crystal structure solving and analysis [29, 30].

Additionally, single crystal X-ray structure analysis also provides various forms of visualization of a structure, i.e., stick, ball-and-stick, ORTEP, space-filling, polyhedral, stereo, *etc.* [31-35]. Fortunately, many types of freely downloadable drawing software are available online along with other related programs, for instance, XP, Mercury, Diamond, TOPOS, Olex, Material Studio, *etc.*

In general, the basic structural details is provided in the form of a brief summary report including atomic coordinates, thermal parameters along with an ellipsoid plot and further diagrams appropriate; moreover, electronic material such as cif, res, pdb files *etc.* can also be provided on request.

5. Applications of the Single Crystal X-Ray Diffraction Technique

Single-crystal X-ray diffraction is the most unambiguous method for determining structures of small molecules and macromolecules, providing accurate and precise measurements of molecular dimensions as well as unit cell parameters, bond lengths and angles directly related to the atomic positions in a way that no other science can begin to approach. Keeping this in mind, the specific applications of crystal structure data obtained from the single crystal X-ray diffraction have been detailed herein.

5.1. Unit Cell Parameter

Unit cell parameter is the basic information during the crystal diffraction measurement process. Number of formula units per unit cell *Z*, crystal density *d*, relative molecular mass M_r, cell volume V and Avogadro constant N_A has the following relationship.

$$d = \frac{M_r \cdot Z}{V \cdot N_A}$$

Obviously, as long as one knows any of the four parameters, the fifth parameter can be calculated, for instance, the crystal density measured by X-ray diffraction method can be obtained according to the unit cell parameters as long as you know the M_r value. Conversely, if we know the crystal density determined by the other physical measurements, we can also calculate the Z value according to the M_r value.

5.2. Visualization of the Crystal Structure

Due to the weak forces, various forms, little difference in energy state between the molecules, the general spectroscopy analysis is difficult to obtain the structural information. However, single crystal X-ray diffraction is a unique technique that not only characterizes the intermolecular interaction and packing modes of inorganic compounds, small organic molecules, and biological molecules (including proteins and DNA), but also provides important structural details for exploring the performance of the material.

Inorganic Molecule Figure 10 elucidates the crystal structure of zincite, an inorganic mineral that has the hexagonal wurtzite structure, where each zinc cation is located in tetrahedral coordination with oxygen, moreover, each oxygen atom also shows tetrahedral coordination to four zinc atoms. The results after the structure analysis are as follows: hexagonal crystal system with space group $P6_3mc$, $a = b = 3.2494$ Å, $c = 5.238$ Å, $\alpha = \beta = 90°$, $\gamma = 120°$, $V = 47.58$ Å^3, Z = 2, $R_1 = 0.0102$.

Organic Molecules (Metal-Organic Compounds and Coordination Compounds) Assembly of $Zn(NO_3)_2$ and a tripodal ligand of H_3TCPB under solvothermal condition results into a colorless porous metal-organic framework, $[Zn_3(TCPB)_2 \cdot 2H_2O] \cdot 2H_2O \cdot 4DMF$ (H_3TCPB = 1,3,5-tri(4-carboxyphenoxy)benzene) [36]. A full set of diffraction data collection is completed on an Oxford Gemini S Ultra diffractometer equipped with $Cu_{K\alpha}$ radiation ($\lambda = 1.54178$ Å) at 150 K (Figure 11). Approximate 1.2 thousands of diffraction intensity was collected with a 99.9% completeness and a resolution of 0.88 Å.

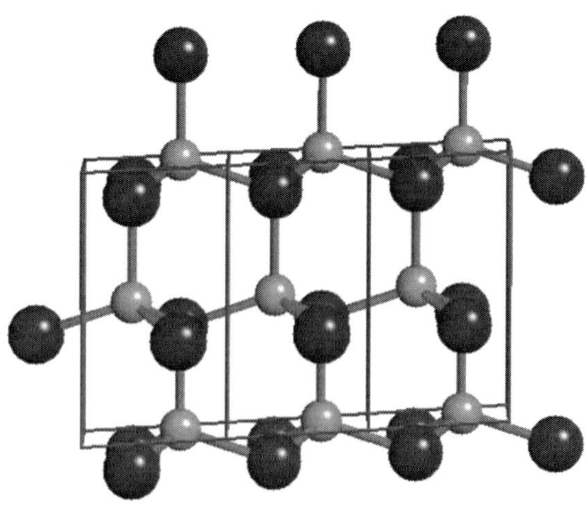

Figure 10. The crystal structure of zincite (ZnO).

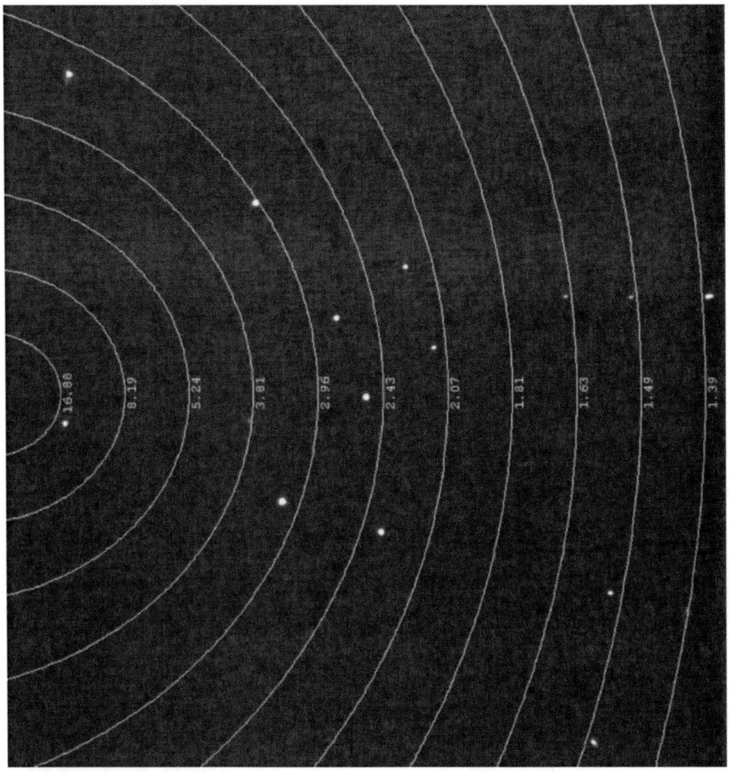

Figure 11. The X-ray diffraction pattern of $[Zn_3(TCPB)_2 \cdot 2H_2O] \cdot 2H_2O \cdot 4DMF$.

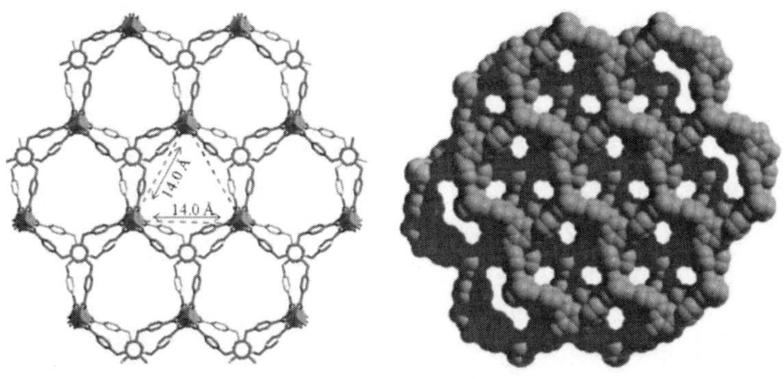

Figure 12. Polyhedral view of 2D network (left) and the offset packing nodes in the crystal structure (right).

The data reduction and structural analysis results are as follow: trigonal system with space group $P\bar{3}1c$, $a = b = 16.7057(4)$ Å, $c = 14.3418(5)$ Å, $\alpha = \beta = 90°$, $\gamma = 120°$, $V = 3466.28(17)$ Å^3, $R_1 = 0.0376$, $wR_2 = 0.1137$. Single crystal X-ray diffraction analysis reveals that it possesses offset packing nodes network containing 1D channels (Figure 12).

Biological Molecules. The diffraction data collection of a biological macromolecule, Saccharomyces cerevisiac RR1-5-NITP (5-NITP=5-nitro-indolyl-2'-deoxyribose triphosphate) (Figure 13), is completed on a synchrotron diffractometer equipped with CCD detector at 100 K [37].

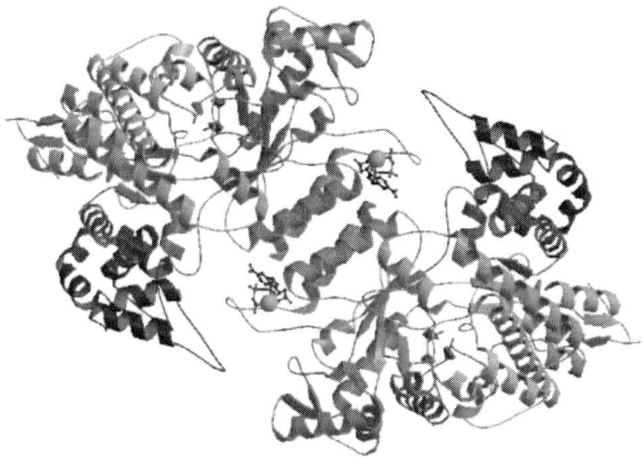

Figure 13. Crystal structure of biological macromolecules saccharomyces cerevisiae.

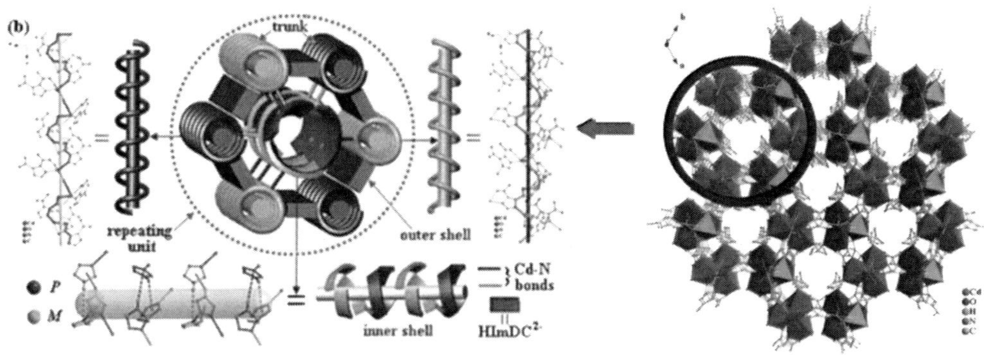

Figure 14. Crystal structure of $[Cd(HImDC)(Py)]_n$.

About 37126 reflections were observed with a resolution of 2.3 Å. The X-ray crystal structure reveals that it crystallizes in orthorhombic system with space group $P2_12_12$, a = 108.35 Å, b = 118.21 Å, c = 63.92 Å, $\alpha = \beta = \gamma = 90°$

5.3. Determination of Bond-Lengths, Bond-Angles and Site-Ordering

In addition to the unit cell parameters and the visualization of structure discussed above, X-ray single crystal analysis also provides other significant information for chemists. It has been known that the distances and angles between atoms can be calculated according to the atomic coordinates, usually expressed in fractional coordinates, i.e. x, y and z coordinates; this information could be easily obtained using single crystal X-ray structure analysis. Furthermore, different chemical bonds having different bond lengths are also easily resolved using this technique; several reports summarize the statistical data of bond lengths of organic compounds, inorganic compounds, as well as coordination compounds [38, 39]. For example, a robust porous MOF containing achiral helical nanotubular channels $[Cd\ (HImDC)(Py)]_n$ [40], derived from cadmium nitrate as the source of cadmium, multidentate H_3ImDC as a bridging ligand, and Im as the auxiliary ligands under hydrothermal conditions (Figure 14).

X-ray diffraction data were collected on a Bruker Apex II diffractometer with graphite monochromated MoK_{α} radiation (λ = 0.71073 Å) at 298 K. Absorption corrections were applied by using multiscan program SADABS.

All the structures were solved by direct methods and refined with full-matrix least-squares technique using SHELXTL. Single crystal X-ray diffraction reveals the formula is composed of $C_8H_6N_4O_4Cd$. The crystal structure data were shown in Table 1 and the selected bond distances and angles were listed in Table 2.

5.4. Intermolecular Interaction

The intermolecular interaction contains hydrogen bond, the π…π stacking interaction between aromatic rings, as well as the van der Waals interaction.

Table 1. Crystal structure data of $[Cd(HImDC)(Py)]_n$

Atomic coordinates parameters							
atom	*x*	*y*	*z*	atom	*x*	*y*	*z*
Cd1	.02348(12)	.03246(14)	.03847(13)	C1	.0331(19)	.051(2)	.036(2)
O1	.0381(17)	.0576(18)	.0447(16)	C2	.0307(18)	.0348(19)	.0300(18)
O2	.0544(19)	.083(3)	.0433(16)	C3	.040(2)	.041(2)	.042(2)
O3	.0475(17)	.075(2)	.0447(17)	C4	.033(2)	.0354(19)	.0283(17)
O4	.0324(15)	.0546(19)	.0533(18)	C5	.048(3)	.039(2)	.036(2)
N1	.0251(15)	.0394(17)	.0339(16)	C6	.098(4)	.038(3)	.047(2)
N2	.0239(16)	.0429(18)	.0353(17)	C7	.089(4)	.040(3)	.067(3)
N3	.0400(18)	.0396(19)	.0367(19)	C8	.044(3)	.054(3)	.042(3)
N4	.055(2)	.060(3)	.051(2)				

Formula: $C_8H_6N_4O_4Cd$ Formula weight: Mr = 334.57.
Crystal system: trigonal Space group: *R*3*c*.
Unit cell parameters: a = 20.6556(6), b = 20.6556(6), c = 13.3915(7) Å, α = 90°, β = 90°, γ = 120°, V = 4948.1(3) Å^3, Z = 18, D_c = 2.021 g cm^{-3}, GOOF = 1.048.

Table 2. Selected bond distances (Å) and angles (°) for complex

Cd(1)-N(2)#1	2.180(3)	N(1)-Cd(1)-N(3)	102.63(13)
Cd(1)-N(1)	2.191(3)	N(2)#1-Cd(1)-O(4)#2	101.66(12)
Cd(1)-N(3)	2.295(4)	N(1)-Cd(1)-O(4)#2	99.28(12)
Cd(1)-O(4)#2	2.404(3)	N(3)-Cd(1)-O(4)#2	95.21(12)
Cd(1)-O(3)#2	2.553(3)	N(2)#1-Cd(1)-O(3)#2	97.26(12)
Cd(1)-O(1)	2.576(3)	N(1)-Cd(1)-O(3)#2	80.99(12)
N(2)#1-Cd(1)-N(1)	151.89(12)	N(3)-Cd(1)-O(3)#2	147.14(12)
N(2)#1-Cd(1)-N(3	93.96(14)	O(4)#2-Cd(1)-O(3)#2	52.29(11)
N(2)#1-Cd(1)-O(1)	88.58(11)	N(1)-Cd(1)-O(1)	68.87(11)
N(3)-Cd(1)-O(1)	90.86(12)	O(4)#2-Cd(1)-O(1)	167.68(10)
O(3)#2-Cd(1)-O(1)	120.12(11)		

Symmetry transformations used to generate equivalent atoms: #1 x-1/3, x-y+1/3, z-1/6; #2-y+4/3, -x+5/3, z+1/6; #3 –y+1/3, -x+2/3, z+1/6; #4 x-1/3, x-y-2/3, z-1/6.

In addition to the strong hydrogen bond, intermolecular interactions are obviously weaker than the covalent bond and the coordination bond. Generally, hydrogen bond contains several types: linear, V, and tricentered hydrogen bond (Figure 15) [41].

On the other hand, $\pi...\pi$ stacking interaction between aromatic rings originates from the attraction between different symbols of electron clouds. It has been divided two kinds: offset face-to face and edge-to face (Figure 16) [42]. It has been known that supramolecular interactions exit in crystal samples are evident by witness of a large number of examples [43-45]. By using the tripodal ligand "ntb" (tris(benzimidazole-2-ylmethyl)amine) and lanthanide nitrate, three isomorphous series of coordination frameworks of the general formula $[Ln(ntb)(NO_3)_3]$·solvents have been assembled [46]. The single-crystal analysis revealed that the polymorphism was related to different fashions of hydrogen bonding interactions, resulting in the formation of different 3D hydrogen-bonded frameworks showing distinct porous and topological structures (Figure 17).

A 2D 4d–4f coordination polymers based on elliptical triple-deck cylinders, namely, $\{[Ag_2Eu(IN)_4(H_2O)_2]\cdot(NO_3)\cdot(H_2O)\}_n$ (IN = isonicotinate), interacts with two neighboring ones via interlayer Ag...O(H_2O) contacts and face-to-face $\pi...\pi$ weak stacking, leading to the formation of a 3D supramolecular network (Figure 18) [47].

5.5. Determination of Molecular Absolute Configuration and Conformation

An absolute configuration in stereochemistry comprises the spatial arrangement of the atoms of a chiral molecular entity (or group) and its stereochemical description (Figure 19). It is absolutely essential to determine the absolute configuration of chiral molecule in order to know its function in biological systems.

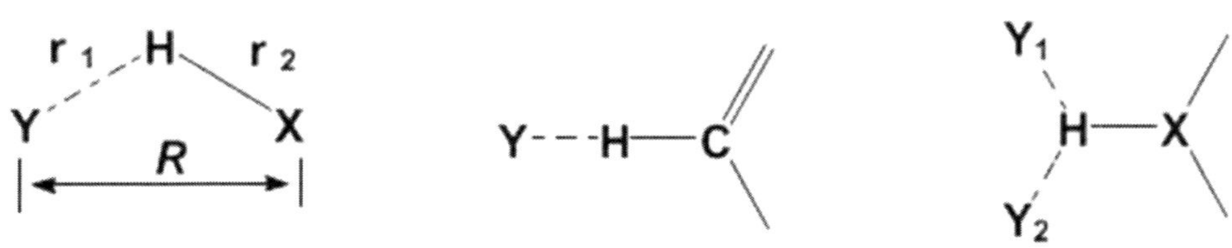

a b c

Figure 15. Structures of hydrogen bonds (a) V, (b) linear, (c) tricentered.

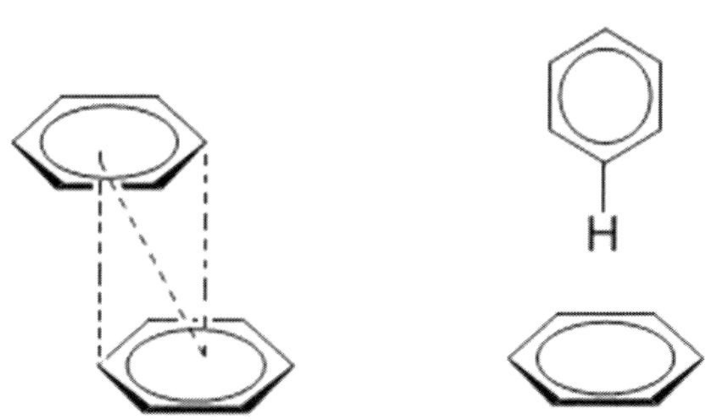

a b

Figure 16. Schematic diagrams of $\pi...\pi$ stacking interaction (a) offset face-to face and (b) edge-to face.

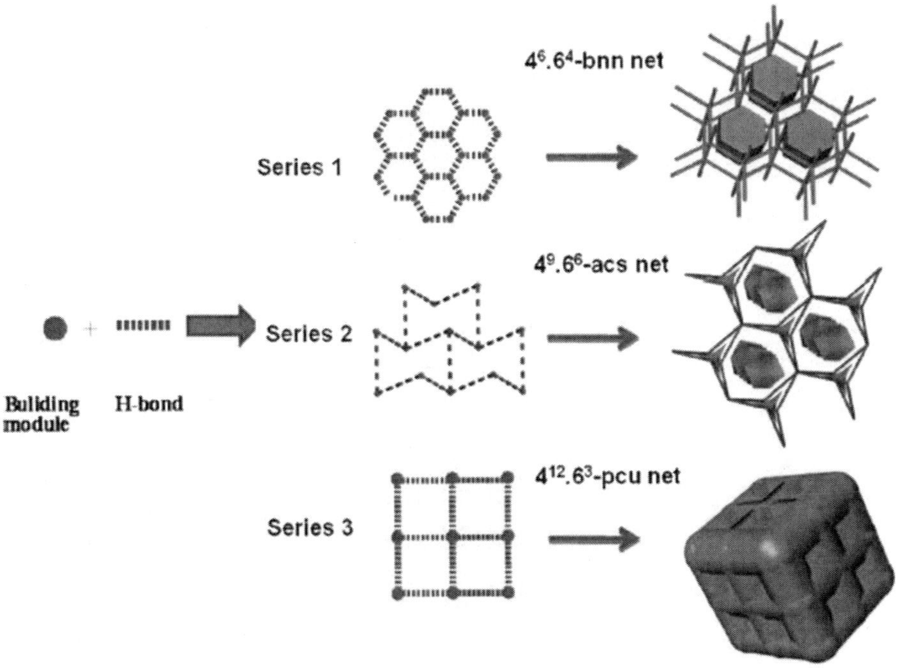

Figure 17. 0D f 3D hydrogen-bonding assembly from the [Ln(ntb)$(NO_3)_3$] building modules.

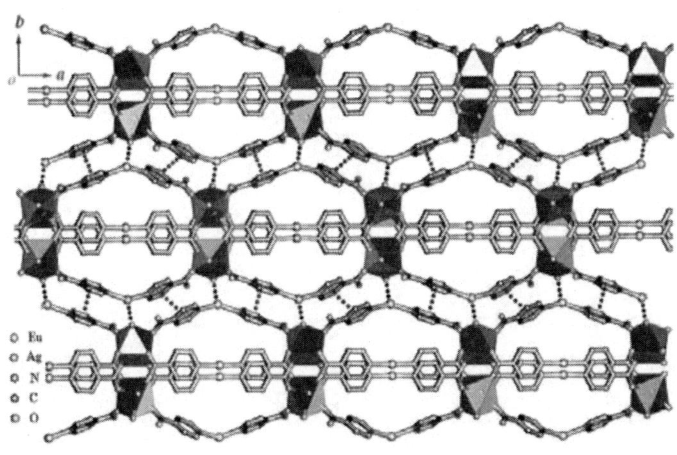

Figure 18. 3D Supramolecular network formed through weak interlayer Ag…O and π…π stacking interactions.

There are a variety of experimental techniques that can be used to determine the absolute configuration of the drug molecules and X-ray crystallography is the method of choice in this aspect.

When the molecules contain heavy atoms, the absolute configuration of the molecule can be accurately determined using the anomalous scattering effect of the heavy atoms. Occasionally, the structure of molecules without heavy atoms could also be confirmed by $Cu_{K\alpha}$ radiation.

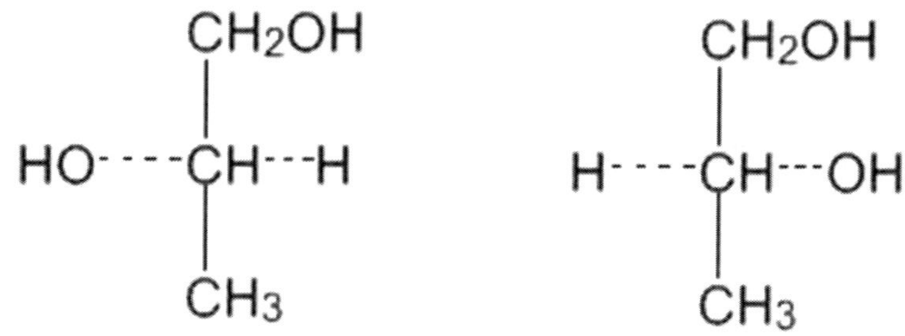

a b

Figure 19. Two different absolute configurations of chiral molecules.

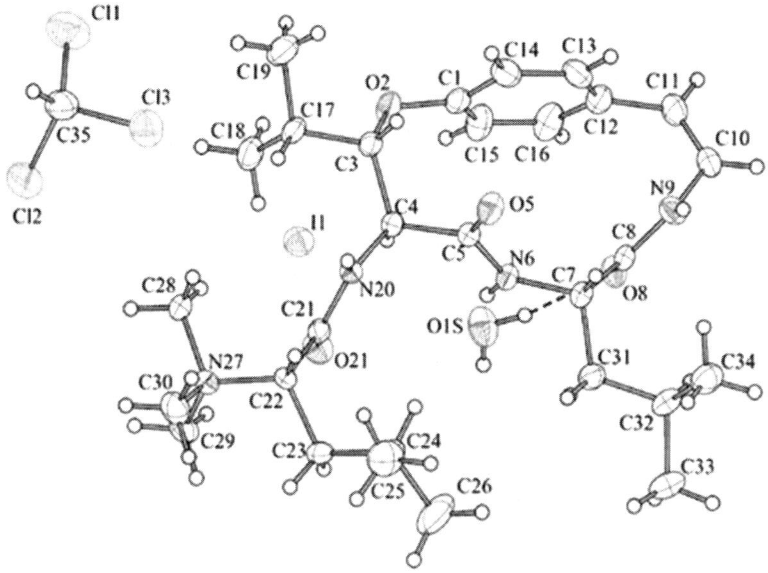

Figure 20. Crystal structure of Tri-N-methylfranganine methiodide.

For example, the compound $C_{30}H_{50}Cl_3IN_4O_5$, namely Tri-N-methylfranganine methiodide (Figure 20), has been studied using single crystal X-ray diffraction at room temperature on a Bruker X8 Kappa APEX II CCD diffractometer using MoK_α graphite momochromatized radiation to determine its absolute configuration using the anomalous scattering of the heavy iodine/chlorine atoms, here, the Flack *x* parameter was refined, giving a value of 0.02(2), which indicates an enantiomorphically pure crystal of the correct handedness [48].

Another example is the 19-norbufenolide A, composed of C, H, N and O atoms, without any heavy atoms (Figure 21). The data collection was performed on a SMART CCD by $Cu_{K\alpha}$ radiation under low temperature and the absolute configuration was unequivocally certified by the result of single crystal X-ray analysis [49]. In addition, much conformational information such as boat or chair conformation, the cis-trans connection mode, twist angle, the relative orientation to the side chain position can been obtained from single crystal X-ray diffraction analysis. Five transition metal complexes were synthesized from a N,N'-bis(3-pyridylmethyl)naphthalene diimide ligand [50]. The structural differences among these complexes are attributed to the different conformations adopted by the ligands (Figure 22).

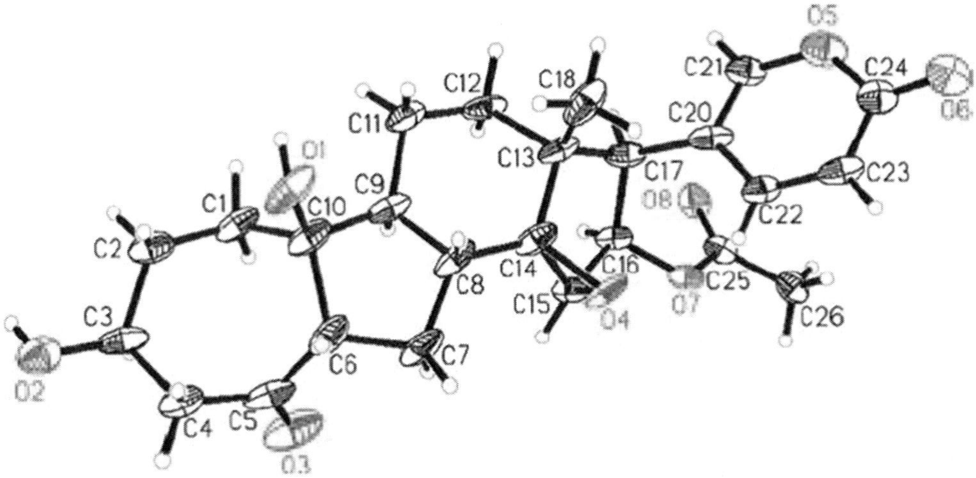

Figure 21. X-ray crystal structure of 19-norbufenolide A.

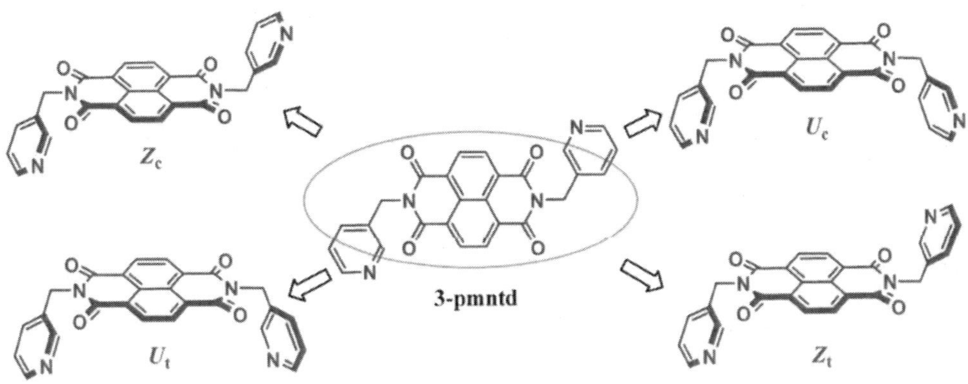

Figure 22. Possible conformational modes of ligand 3-pmntd.

5.6. Simulated Powder X-Ray Diffraction Patterns

In order to study the property of the solid material, first of all, it is absolutely necessary that the substances have a sufficiently high purity. We usually determinate the powder X-ray diffraction (PXRD) patterns of the samples and compare with the standard powder diffraction patterns to measure the phase purity.

However, there is no a standard powder diffraction patterns for the new compounds, especially organic compounds or coordination compounds. If a single crystal structure of this compound could be obtained, it is possible to convert the structural information into a powder X-ray pattern by using a computer program. This structural information generally contains the unit cell parameters, space group, all atomic coordinates, sometimes also including intensity data and reflection indices.

The simulated powder X-ray diffraction patterns can be regarded as a standard powder diffraction spectrum of a specific compounds.

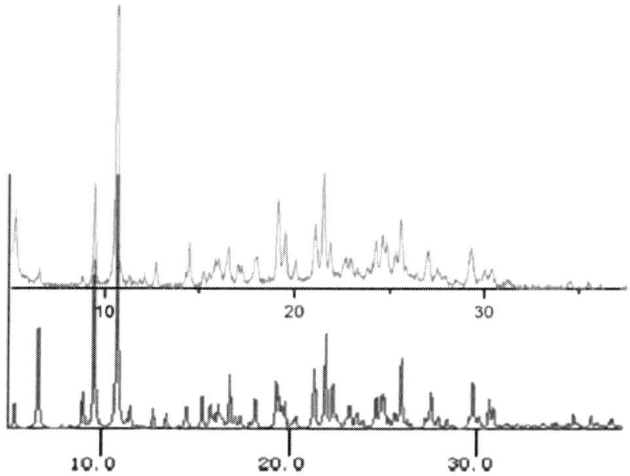

Figure 23. The powder XRD patterns of complex $[Ag_3(tppt)_2](ClO_4)_3{\cdot}8DMSO$ (top) and the simulated one from the single-crystal diffraction data (bottom).

It helps to understand the phase purity by comparing the PXRD pattern of the crystal samples in the experiment with the simulated PXRD pattern. For example, a highly undulating (6,3) honeycomb networks $[Ag_3(tppt)_2](ClO_4)_3{\cdot}8DMSO$ was obtained by the reaction of $AgClO_4$ with a tripodal ligand 2,4,6-tris(4-((pyridin-4-ylthio)methyl)phenyl)-1,3,5-triazine (tppt) [51]. The PXRD patterns of this complex are identical to those simulated from the single-crystal diffraction analysis, indicative of the phase purity of the bulk samples (Figure 23).

CONCLUSION

Over a short period of time, single crystal X-ray diffraction technique has made great progress (from Laue method to the synchrotron diffraction) and become a powerful tool for investigating the structure of solids (inorganic, organic and metallorganic). Prior to the data collection, it is crucial to obtain perfect crystal suitable for X-ray diffraction.

This chapter particularly elucidates the conditions and strategies generally used for the crystal growth.

However, the basic structural details could be conducted by data collection, and structure solution and refinement are necessary to determine the final crystal structure. From a fundamental viewpoint, research on the applications of single crystal X-ray diffraction technique would definitely contribute towards understanding the nature of a substance, which could in turn provide a basis for the development of chemistry, physics, materials science, life sciences and other basic subjects.

The single crystal X-ray diffraction technique will continue to provide researchers with the structure of a substance on the atomic and molecular level that bridge the gap between the molecular and macroscopic worlds.

REFERENCES

[1] Friedrich, W., Knipping, P., von Laue, M., *Sitzungber(kgl.)* Bayerische Akad. Wiss., 1982, 303-322.

[2] Bragg, W. L., Bragg W. L., *Nature*, 1913, 91, 557.

[3] Bragg, W. L., *Proc. Roy. Soc.*, 1913, 89A: 248.

[4] Zhou, G. D., Guo, K. X., *Diffraction of Crystal and Quasicrystal*, Beijing: Peking University, 1999.

[5] Glusker, J. P., Lewis, M., Rossi, M., *Crystal Structure Analysis for Chemists and Biologists*. New York: VCH Publisher Inc., 1995. 406.

[6] Buerger, M. J., *X-Ray Crystallography,* Wiley, New York, 1942.

[7] Buerger, M. J. *The Precession Method in X-Ray Crystallography*, Wiley, New York, 1964.

[8] Zhou, G. D., *Crystal Structure Determination.* Beijing: Science Press, 1981.

[9] Helliwell, J. R., In: A. J. C.Wilson (ed.). *International Tables for Crystallography*, Vol. C, Mathematical, Physical and Chemical Tables, Kluwer, Dordrecht. 1992, Section 2.1 and 2.2.

[10] Mosset, A., Galy, J., *Topics in Current Chemistry*, 1988, 145: 1.

[11] Henisch, H. K. *Crystals in Gels and Liesegang Rings*, Cambridge University Press, Cambridge, 1988.

[12] Zhang, X. M., Tong, M. L., Chen, X. M. *Angew. Chem. Int. Ed.*, 2002, 41, 1029.

[13] Kappe, C. O. *Angew. Chem. Int. Ed.*, 2004, 43, 6250-6284.

[14] Li, Z. Q., Qiu, L. G., Xu, T. *Mater. Lett.*, 2009, 63, 78-90.

[15] Hulliger, J. *Angew. Chem., Int. Ed.*, 1994, 106, 151.

[16] Lin, X. M., Fang, H. C., Zhou, Z. Y., Chen, L., Zhao, J. W., Zhu, S. Z., Cai, Y. P. *CrystEngComm*, 2009, 11, 847-854.

[17] Cui, P., Ren, L. J., Chen, Z., Hu, H. C., Zhao, B., Shi, W., Cheng, P. *Inorg. Chem.*, 2012, 51, 2303–2310.

[18] Kanoo, P., Gurunatha, K. L., Maji, T. K. *Cryst. Growth Des.*, 2009, 9, 4147-4156.

[19] Glusker, J. P., Lewis, M., Rossi, M. *Crystal Structure Analysis for Chemists and Biologists,* VCH, New York, 1994.

[20] Lindley, P. F. In: A. J. C. Wilson (ed.), *International Tables for Crystallography*, Vol. C, Mathematical, Physical and Chemical Tables, Kluwer, Dordrecht, 1992, Secition 3.4, pp. 145-149.

[21] Rühl, S., Bolte, M. Z., *Kristallogr.*, 2000, 215, 499.

[22] Massa, W., *Crystal Structure Determiantion*. Berlin: Springer, 2000.

[23] Walker, N., Stuart, S. *Acta Crystallogr.*, 1988, 145, 1-6.

[24] Patterson, A. L., *Phys. Rev.*, 1934, 46, 372-387.

[25] Egert, E., PATSEE, *Adjunction to SHELXS for Patterson Search Methods*. Institute of Org. Chem., Marie-Curie St. 11, D-60439 Frankfurt am Main, Germany.

[26] Giacovazzo, C. SIR97, *Integrated Program for Crystal Structure Solution.* Dipartimento Geomineralalogico, Campus University, via Obrbona 4, 70125 Bari, Italy.

[27] Sheldrick, G. M., *SHELXS-97, Program for Crystal Strucutre Solution*. Göttingen University, Germany, 1997.

[28] Hall, R. S., Allen, F. H., Brown, I. D. *Acta Cryst.,* 1991, A47, 655-658.

[29] Spek, A. L., PLATON 99, *Program for the Geometric Interplanation of Structural Data*. Lab. Voor Kristal-en Structuurchemie, U. Utrechtm Padualaan 8, 3584 CH Utrecht, Netherland.

[30] Farrugia, L. J., Win, G. X., *A Window program for crystal structure analysis*. Department of Chemistry, University of Glasgow, G12 8QQ, Scotland.

[31] Johnson, C. K. *ORTEP III, Fortran Thermal Ellipsoid Plot Program*. Oak Ridge National Laboratory, Oak Ridge, TN 37831-6197, US.

[32] Lin, X. M., Ying, Y., Chen, L., Fang, H. C., Zhou, Z. Y., Zhan, Q. G., Cai, Y. P. *Inorg. Chem. Commun.*, 2009, 12, 316–320.

[33] Cai, Y. P., Zhou, X. X., Zhou, Z. Y., Zhu, S. Z., Thallapally, P. K., Liu, J., *Inorg. Chem.*, 2009, 48, 6341–6343.

[34] Zhou, X. X., Cai, Y. P., Zhu, S. Z., Zhan, Q. G., Liu, M. S., Zhou, Z. Y., Chen, L.; *Cryst. Growth Des.*, 2008, 8, 2076–2079.

[35] Lin, X. M., Li, T. T., Chen, L. F., Zhang, L., Su, C. Y. *Dalton Trans.*, 2012,41, 10422-10429.

[36] Lin, X. M., Li, T. T., Wang, Y. W., Zhang, L., Su, C. Y. *Chem. Asian J.* 2012. 7, 2796-2804.

[37] Ahmad, M. F., Wan, Q., Jha, S., Motea, E., Berdis, A., Dealwis, C., *Mol. Cancer Ther.*, 2012, 11, 2077-2086.

[38] Allen, F. H., Kennard, O., Watson, D. G., Brammer, L., Orpen, A. G., *J. Chem. Soc. Perkin Trans.*, 1987, S1-S19.

[39] Orpen, A. G., Brammer, L., Allen, F. H., Kennard, O., Watson, D. G., Taylor, R., *J. Chem. Soc. Dalton Trans.*, 1989, S1-S89.

[40] Gu, Z. G., Cai, Y. P., Fang, H. C., Zhou, Z. Y., Thallapally, P. K., Tian, J., Exarhos, G. J. *Chem. Commun.,* 2010, 46, 5373-5375.

[41] Desiraju, G. R. *Acc. Chem. Res.*, 1996, 29, 441-473.

[42] Steed, J. W., Atwood, J. L., *Supramolecular Chemistry*. Chichester: John Wilry and Som, 2000, 26.

[43] Chen, J. Q., Cai, Y. P., Fang, H. C., Zhou, Z. Y., Zhan, X. L., Zhao, G., Zhang, Z. *Cryst. Growth Des.*, 2009, 9,1605–1613.

[44] Liu, M. S., Yu, Q. Y., Cai, Y. P., Su, C. Y., Lin. X. M., Zhou, X. X., Cai, J. W. *Cryst. Growth Des.*, 2008, 8, 4083–4091.

[45] Li, G. B., Fang, H. C., Cai, Y. P., Zhou, Z. Y., Thallapally, P. K., Tian, J. A. *Inorg. Chem.*, 2010, 49 (16), 7241–7243.

[46] Jiang, J. J., Pan, M., Liu, J. M., Wang, W., Su, C. Y. *Inorg. Chem.*, 2010, 49 (21), 10166–10173.

[47] Cai, Y. P., Yu, Q. Y., Zhou, Z. Y., Hu, Z. J., Fang, H. C., Wang, N., Zhan, Q. G., Chen, L., Su, C. Y. *CrystEngComm*, 2009, 11, 1006-1013.

[48] Caro, Miguel S. B.; de Oliveira, Leonardo H.; Ilha,Vinicius; Burrow, Robert A.; Dalcol, Ionara, I.; Morel, Ademir. F. *J. Nat. Prod.*, 2012, 75, 1220-1222.

[49] Tian, H. Yan., Wang, L., Zhang, X. Q., Wang, Y., Zhang, D. M., Jiang, R. W., Liu, Z., Liu, J. S., Li, Y. L., Ye, W. C. *Chem. Eur. J.*, 2010, 16, 10989-10993.

[50] Deng, H. Y., He, H. R., Pan, M., Li, L., Su, C. Y. *CrystEngComm*, 2009, 11, 909-917.

[51] Yang, Q. Y., Zheng, S. R., Yang, R., Pan, M., Cao, R., Su, C. Y. *CrystEngComm*, 2009, 11, 680-685.

In: X-Ray Diffraction: Structure, Principles and Applications ISBN: 978-1-62808-591-4
Editor: Kaimin Shih

Chapter 2

STRUCTURE CHARACTERIZATION OF COORDINATION POLYMERS BY X-RAY DIFFRACTION DATA

Filipe B. de Almeida[1], Maria C. R. Freitas[2], Charlane C. Corrêa[1,3], Luciano H. Chagas[1], Márcia C. de Souza[1], Weberton R. do Carmo[1] and Renata Diniz[1*]

[1]Departamento de Química, Universidade Federal de Juiz de Fora, Juiz de Fora, Minas Gerais, Brazil

[2]Instituto de Química, Universidade Federal Fluminense, Niterói, Rio de Janeiro, Brazil.

[3]Departamento de Física, Universidade Federal de Minas Gerais, Belo Horizonte, Minas Gerais, Brazil

ABSTRACT

Supramolecular chemistry, which can be described as the chemistry beyond the covalent bonds, has become a very interesting focus of investigation in the last years. The main purpose of this kind of study is the strategic construction of specific arrangements, and the complete comprehension of the connection between structure and physical–chemical properties. In recent years, the design and synthesis of metal–organic hybrid compounds have attracted increasing attention due to the facility of well-defined organic molecular building blocks be assembled into intriguing frameworks. Organic aromatic polycarboxylate ligands have been extensively employed in the preparation of such metal–organic compounds with multidimensional networks. It could be used to create a variety of coordination polymers in the form of 1D polymeric chains, 2D sheets, and 3D networks. This chapter will first include a general description about supramolecular chemistry, coordination polymers and solid state characterization using X-ray diffraction. Then, the specific description of crystal structures of coordination polymers and theirs topological analysis will be introduced.

* Corresponding author: Tel: +55 32-2102-3310. E-mail address: renata.diniz@ufjf.edu.br.

1. X-Ray Diffraction

Wilhelm Conrad Röntgen received the physical Nobel Prize in 1901 by his discovery of the X-ray radiation. Initially the nature of this discovery was unknown, but the hypothesis that this radiation can have wavelenght of angstroms provided studies from Max Von Laue in 1912 with the use of crystals as a diffraction grating for the X-ray [1]. The experience using crystals proved the undulatory proprieties hypothesis of the X-rays. One year after the Bragg's (Willian Lawrence Bragg and his father William Henry Bragg) exploited the phenomena to solve the first structure and determined the law that governs these materials. They concluded that materials having internal organization, act like grid of diffraction, since the distance of the scattering planes are in the same order of the wavelength.

The studies of Röntgen, M. v. Laue, Bragg's and others, for not list all, allows the development of an important tool for the characterization of compounds in the solid state, the X-ray diffraction (XRD). This can be used to characterize crystals or polycrystalline samples that have long range organization of the atoms [2]. All this kind of materials produces diffraction patterns when they interact with the X-ray source. Crystallography is an important science which uses XRD as a powerful tool for crystalline studies.

Powder diffraction can characterize material known as polymorphous. Polymorphism means "having many forms" and the definition is a crystalline solid that exist in two or more different forms where the constituent units are arranged in different ways [3]. The existence of polymorphous is frequently observed in pharmaceutical solids, and it can exhibit different physical proprieties.

Studies of supramolecular and metal organic frameworks (MOFs) defined later in topics **2** and **3,** respectively, can be examined by XRD, since the technique allows the determination of bond distances, bond angles and the atomic positions. XRD allows the rationalization of the solid array of coordination polymers. The arrangement is extended by covalent metal-ligands linkage forming the structures of one, two, or three dimensions. These materials are of great interest for the scientific community, since they have a wide range of applications. A large number of publications with the XRD technique can be found due to development of hardware and software that facilities the application of XRD. These data will favor the intermolecular interactions studies, such as hydrogen bonds (H-bond) and π-stacking favorable in the supramolecular array.

These studies can elucidate the topologies from H-bond networks or coordination polymers. In the study of intermolecular interactions, it is possible to correlate geometric factor, as distance and angle, with the strength of interaction. In case of H-bond possibly to classified, the strength and type based on the distance between the donor and acceptor atoms can be provided. Studies of Novak [4] illustrated the correlation values of distance in H-bond, obtained from XRD, with infrared spectrum. The π-stacking is caused by electrostatic interactions, between aromatic compounds present in adjacent planes. For existence of π-stacking among different rings, the plane-to-plane separation (interplanar distance), as offset distance of centroids, should have a maximum value. In the case of the six-ring of carbons, it should be less than 4 Å and the offset distance of centroids (horizontal shift) is less than 2,5 Å [5]. These two no-covalent interactions are known as weaker than covalent bond, but are in the main cases responsible by the stabilization of supramolecular structure. The topological parameters of π-stacking are displayed in Scheme 1.

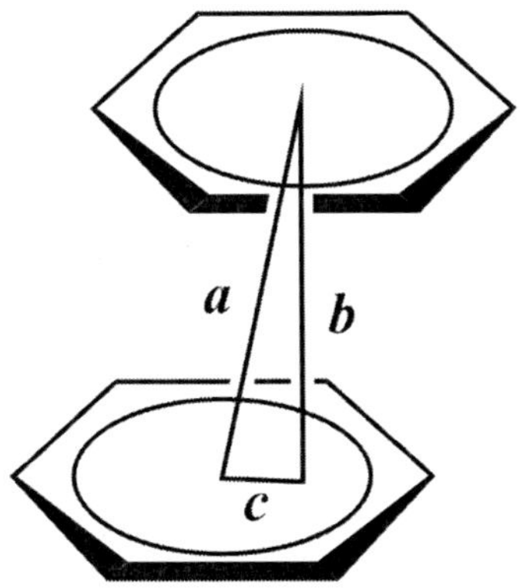

Scheme 1. Topological parameters of π-stacking: a = centroid-centroid distance, b = interplanar distance and c = horizontal shift.

2. Supramolecular Chemistry

Undoubtedly, the X-ray crystallography of single-crystals has represented an active and interesting research area shortly after the discovery of X-rays. Nevertheless, the subjects of crystal design and crystal engineering only developed fast in recent years. This was due to several factors such as the development of powerful computers and programs which enable structural determination and visualization. Furthermore, databases development is also a very important factor. Single-crystals are formed by the translation of the unit cell into three spatial directions. Analyzing more deeply into the intermolecular interactions and the geometrical parameters within a crystalline material require one to recognize patterns leading to molecular networks. Molecular networks, which possess translational symmetry, are defined as chemical assemblies for which specific interaction patterns or assembling cores are repeated through the space [6].

The modular self-assembly approach is equally well applied to all levels of dimensionality, since this is often determined directly by the nodes of the network. The dimension of a molecular network depends on the number of translations operating at the level of the assembling cores. Thus, 1D networks are formed when a single translation occurs. Similarly, 2D and 3D networks are defined when two or three translations of the same or different assembling cores are present respectively. The growth of crystal engineering has also coincided with advances in our understanding of the intermolecular interactions and the solid-state chemistry behavior. Supramolecular chemistry is inherently linked to the concepts of crystal engineering [7].

Supramolecular chemistry is a highly interdisciplinary field covering the chemical, physical, and biological characteristics. Chemical species of higher complexity are held together and organized by means of intermolecular (non-covalent) interactions. The term supramolecular chemistry was defined by Jean-Marie Lehn, who won the Nobel Prize for his work in 1987, as the 'chemistry of molecular assemblies and of the intermolecular bond'. More colloquially this may be expressed as 'chemistry beyond the molecule' [8]. Other definitions include phrases such as 'the chemistry of the non-covalent bond' and 'non-molecular chemistry'. Originally, supramolecular chemistry was defined in terms of the non-covalent interaction between a 'host' and a 'guest' molecule, comprising the design and studies of intermolecular interactions which are responsible for the structure and behavior of the various chemical and biological systems. The work in modern supramolecular chemistry encompasses not just host-guest systems, but also molecular devices and machines for molecular recognition, so called 'self-processes' such as self-assembly and self-organization.

In fact, it is known that weak attractive forces are important in deciding the conformation of organic compounds and the 3D structure of biomacromolecules. Among molecular interactions, the van der Waals force, electrostatic interactions and hydrogen bonds are the most important. Particularly, the hydrogen bond (H-bond) is of great importance in natural sciences, and is directly related to biological aspects such as the molecular recognition that could be a basis for the creation of life. Thus, the approach used in supramolecular chemistry is centered on the association of molecular species and to obtain a particular property or functionality. In this sense, supermolecules have been designed with aim to achieve specific properties such as self-assembly, transport and molecular recognition [9].

Inside the supramolecular approach, not enough to simply connect various molecular species to obtain supermolecules, it is necessary to find a particular purpose or functionality. Furthermore, the degree of electronic coupling between the subunits is also a very important parameter. Supermolecules are classified as species in which only the interactions between the subunits can disrupt the electronic properties of the same. Thus, it is possible to design supermolecules in function of the characteristics of their components. Conventionally, a supermolecule is distinguished from a large molecule from three aspects. Firstly, the molecular components of a supermolecule are non-covalently linked; secondly, the intrinsic properties of these molecular components persist and; thirdly, the properties of a supermolecule are not a simple superposition of the constituent molecular components properties, i.e., there is a well-defined supramolecular function [10].

A very interesting focus of investigation in the supramolecular chemistry is the self-organization of the supramolecular structures. The concept of self-organization or self-assembly, which is based on the spontaneous formation of supramolecular structure from simple molecular building blocks, is strictly related to molecular interactions. Among these interactions, the H-bonds represent a fundamental role in the DNA structure as well as the secondary and tertiary structure of proteins besides several other biological systems. Nowadays, the main challenges in supramolecular chemistry have been the strategic construction of specific supramolecular arrangements, and the comprehension of the relationship between these structures and their physical-chemical properties. In this sense, the coordination polymers construction, for example, can provide an important contribution in supramolecular research [3]. Thus, from the scientific and technological view, several applications are possible such as optic, electronic and catalytic materials.

The molecular organization can be understood as any combination of characteristics that confer certain order and specific properties to a system. Several structural features may exhibit behaviors that fall within this general definition of molecular organization [9]. In the literature, various phenomena related to the molecular organization have been studied including the chelate, the macrocyclic, the template, and the neighboring group or anchiomeric effects [11]. It is suggested that, in coordination and inclusion chemistry, several varying manifestations of molecular organization result in various effects, which are derived from the structural factors like topological (or linking), metric (or relative size), shape (or geometric), and rigidity added to the complementarity between interacting species [11].

Several recent studies have focused on the development of self-assembling synthetic large molecules or molecular arrays [7]. These systems use a variety of interactions, some of which are clearly non-covalent (such as H-bonds) and others have a significant covalent component (such as metal-ligand interactions). Ultimately, these self-assembly reactions and the resulting self-organization of the system rely solely on the intrinsic information contained in the molecular components structure. Hence, there is an increasing trend towards the study and the manipulation of intrinsic 'molecular information' [9]. Particularly, non-covalent bonds are responsible for the construction and stability of the several chemical and biological systems [12]. Overall, the study of these intermolecular interactions, which is named supramolecular chemistry, is of paramount importance in the search for new drugs and new materials, as well as for understanding the behavior of many systems of interest to mankind.

3. Coordination Polymers

The term coordination polymer was first used by Y. Shibata in 1916 to describe dimers and trimers of several cobalt(II) ammine nitrates and has been in continuous use throughout the scientific community [13].

The expansion of the coordination polymer search was increased with the support of two areas: crystal engineering and supramolecular chemistry. In both areas, the use of different types of ligands can influence the formation of coordination polymers, for example, the use of a convergent ligand will result in a metallosupramolecule whereas a divergent ligand will form a coordination polymer. Figure 1 shows the use of convergent and divergent ligands in the formation the metallosupramolecule and coordination polymer, respectively. In this case, the design and the chemistry of a compound may be similar, except that the use of different ligands may or may not form a coordination polymer [14].

Coordination polymers are compounds containing metal ions linked by coordinated ligands that extent infinitely into one, two or three dimensions, via metal-ligand bonding [15, 16]. A coordination polymer consists of metal and ligands and in some cases of guest solvents and counter-ions. The porous properties of some of these materials are promising for several applications, from gas storage to catalysis [17].

The main metals used are transition metals, and the first-row elements mainly due to their kinetic reliability and stability. The lanthanides have drawn considerable attention due to their higher connectivity conducting to interesting topologies, in addition to the other properties as luminescence. All metals are used as their halide, nitrate and perchlorate among other salts. The non-coordinating anions should always be used. The choice of the ligand is important for constructing the desired coordination polymers. The ligand should be divergent and have at least two divergent coordination sites. The preferred ones are pyridyl, imidazole, nitrile or carboxylate functional groups, and these ligands can from rigid to flexible molecules [17].

The infinite network must be formed by coordination bonds, but molecules linked by hydrogen bonding are not coordination polymers. For example, a structure linked by coordination bonds in two directions and hydrogen bonding in another direction is a 2D coordination polymer. The structural determination of these ordered polymeric materials through X-ray crystallography can contribute to the important information about the structure-property correlations.

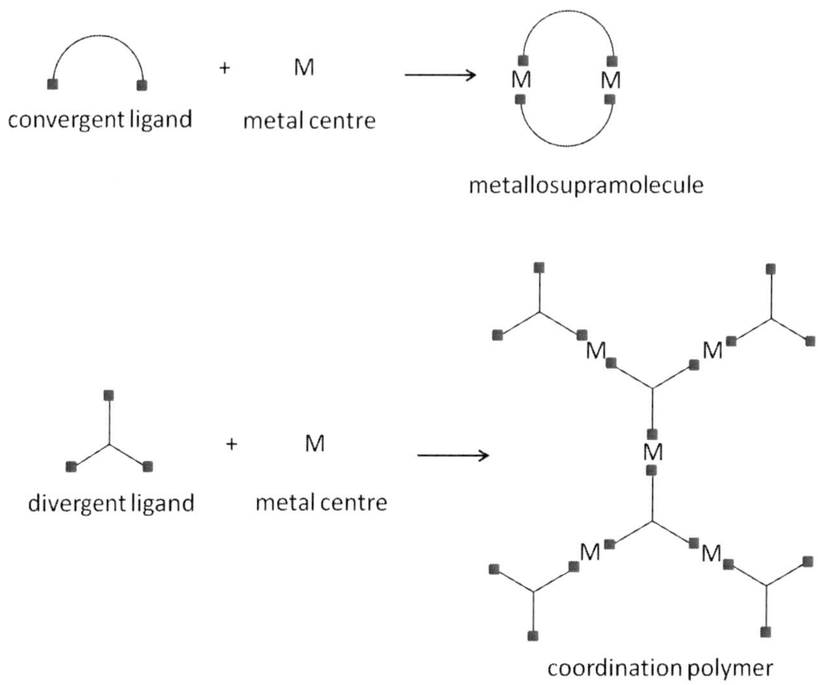

Figure 1. The convergent and divergent ligands in the formation of metallosupramolecule and coordination polymers, respectively. Adaptated for XXX.

One of the most important techniques in crystal engineering is to reduce the crystal structure of compounds to networks. Networks can help to describe the knowledge of difficult structures, since the net must also have a repeating pattern, a finite number of unique nodes, and links. A polymeric network is a set of interlinked nodes, each link connects two nodes and each node is linked to three or more other nodes. The metals and ligands can act as either nodes or links. A net is also a topological description, but not a geometric description. For example, two nets can be topologically identical, although they are geometrically very different. By this reasoning, there is no topological difference between square-planar nodes and tetrahedral nodes, and both are 4-connecting nodes. Hence, it is the connectivity of a network that defines it, not its geometry.

There is a great number of mathematical properties that can be defined in a network. One of the most well-known is to determine the lowest circuits around each unique node. The lowest circuit is the number of nodes in the smallest loop in the net that can be constructed containing two given links from a given node. If all nodes have the same connectivity, then according to Wells the net can be represented by the symbol (n,p), where n is the size of the lowest circuit and p is the connectivity of the nodes.

Some types of polymeric nets can be found: 1D, 2D, 3D and interpenetrated nets [17]. The Figure 2 shows 1D polymeric nets. Most of the 1D polymeric nets have the simple topology of a linear chain extending in a unique direction. The 2D polymeric nets may be formed by a succession of nodes connected by links. The most commonly encountered (6,3) [18] and (4,4) [19] 2D polymeric nets are presented in Figure 3. There are many 3D polymeric nets [20, 21], but most of these networks are complicated to describe. A small part of these polymers can be described by simple 3D polymeric nets.

(a) (b)

Figure 2. Some 1D polymeric nets: (a) linear chain and (b) ladder-like.

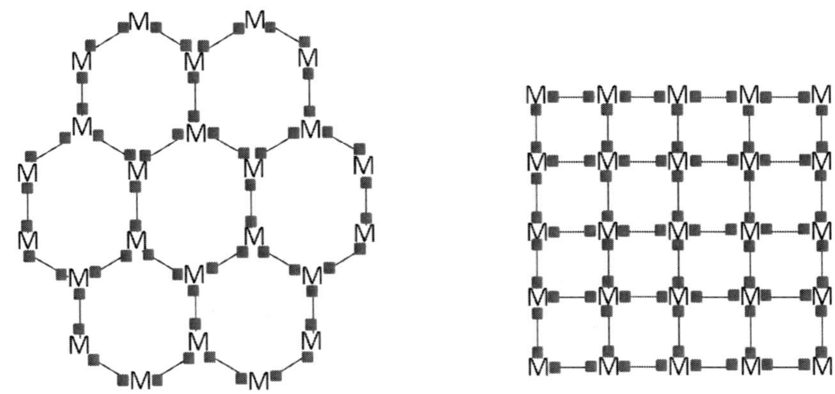

(a) (b)

Figure 3. A collection of 2D nets: (a) (6,3) and (b) (4,4).

One of the main ways in which a crystal can maximize its packing efficiency is the interpenetration. The interpenetration occurs when some polymeric nets are not connected but cannot be separated topologically without breaking the links. There are a great number of consequences to the crystal engineers for interpenetration [22]. The main one is the reduction in the porosity of the solid. However, the interpenetration can favor the porous material by structurally stabilizing frameworks that would collapse upon solvent removal, if they were a single net. There are 1D and 2D interpenetrations and, in some cases 3D interpenetration also exists [23, 24].

3.1. Structural Topology

Once knowing the polymeric networks, one can define the network topology by mathematics and a network is initially reduced to a series of nodes and connectors. A series of comparisons between networks can now be performed. Figure 4 shows two geometrically different networks, but they are topologically identical. The first has an arrangement in "bee hive" form, while the second one is of a brick wall. However, both have the same topology and can be initially identified by tri-connected nodes.

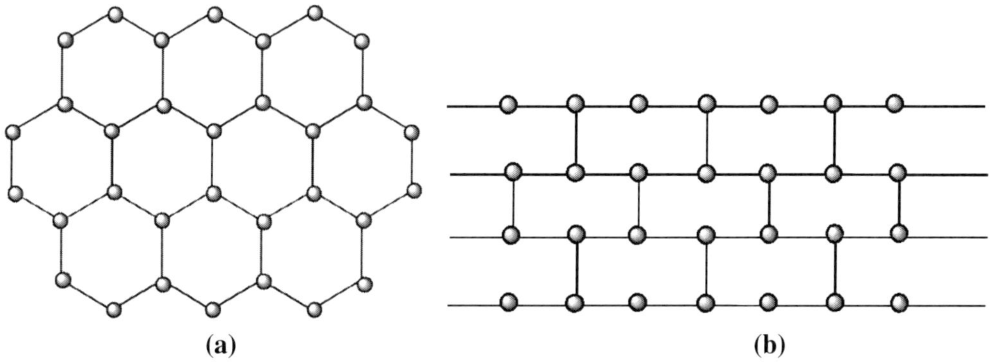

Figure 4. Polymeric network with tri-connected nodes with the arrangements in form: (a) bee hive and (b) brick walls.

The networks shown in Figure 4 are topologically identical, because they can be converted each other by distortion, without breaking the connection. Similarly, there is no topological difference between plane quadratic and tetrahedral nodes, because they are both tetra-connected. Although geometry is not a strictly topological quality, it can be an important factor in the design and the selection of polymeric networks. If the node is a metal ion, the connectivity may be different from its local geometry, in other words, the connectivity number may not equal to the coordination number. A good example of this case is shown in Figure 5. In this example, the coordination number (CN) of metal ion is six and the metal ion acts as a tri-connected node.

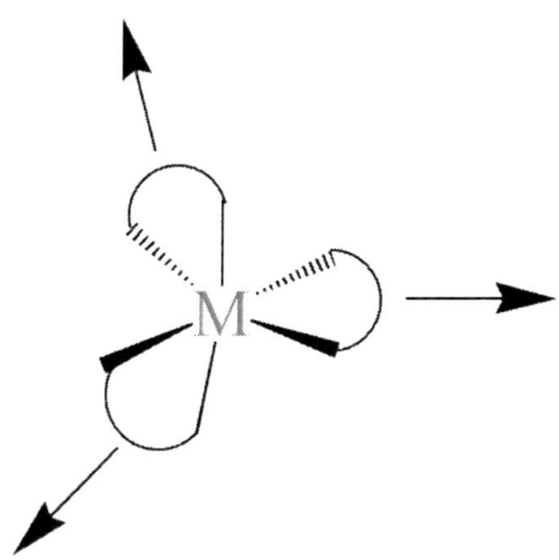

Figure 5. Coordinating environment of a metal with CN = 6 (octahedral), whose ligands are bidentate, resulting in a tri-connected node.

The major challenge to a polymer network is to decide how it will be defined and described, i.e. where are the nodes and connectors and what are the important interactions. For example, the metal-ligand clusters are commonly seen in the coordination polymers (Figure 6). In order to simplify the description of the polymer network, these metal-ligand clusters are often considered as nodes, also known as SBU (secondary building unit), instead of metal atoms and ligands individually.

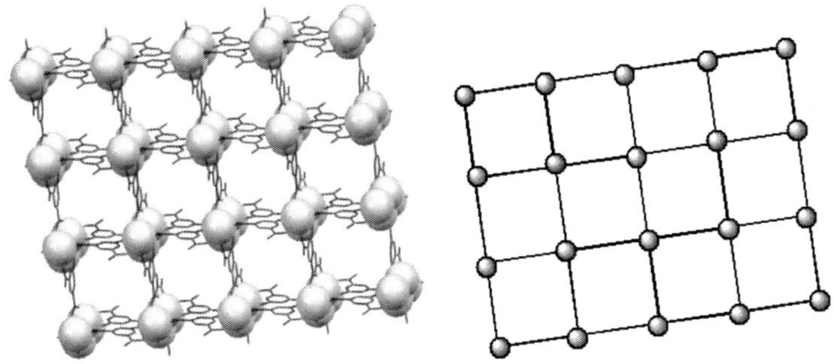

(a) (b)

Figure 6. 3D coordination polymer containing metal-ligand clusters: (a) polymer chains formed by Mn^{2+} atoms and carboxylate ligands and (b) simplification of clusters.

Once the network is defined, it must be identified and a symbol must be assigned to it. At first, the obvious information is the number of the different nodes and their connectivity in topology. One should be aware that the number of topological nodes can be much smaller than the number of crystallographic nodes in a real crystal structure. The crystal structure is the result of the connectivity, geometry, shape and position of all the elements, not only the nodes and connectors. Knowing the crystal structure for the 2D and 3D networks that have closed paths, the network topology can be calculated and represented by a mathematical description of the crystal lattice, and therefore four steps must be taken, which are:

1) simplification process
2) identification and separation of individual patterns,
3) topological analysis of patterns, and
4) topological analysis of the whole network.

Initially, the simplification process is to remove all elements that do not present topological importance in order to remain the essential elements: nodes and connectors/ spacers (vertices and sides), as can be seen from Figure 6. Subsequently, the identification and separation of individual patterns and its topological analysis should be carried out. In 2D and 3D networks, such notation is based on the analysis of the "circuits" or "closed paths" and the shortest closed path must contain a pair of connectors of a given node. According to *Wells* [25], if circuits generated by all possible pairs of connectors are in the same size, the network is called uniform. If all nodes have the same connectivity, the network is called the platonic uniform network. Therefore, for a simple uninodal 2D network, the mathematical notation *(n, p)* can be used. For example, the 2D polymer network of Figure 6 shows a structural topology of type (4, 4). The closed path contains four nodes *(n=4)* represented by metal clusters (circles) and each node is tetra-connected to neighborhoods *(p=4)*. Note that not all networks can be described by *Wells's notation*. Figure 7 shows a uninodal network with tri-connected nodes, in which three pairs of connectors at each node generate closed paths of different sizes.

For the description of these 2D or 3D uninodais networks with different closed paths, the topological notation can be given by *Schläfli's symbols* [26,27]. *Schläfli's symbols* simply list the closed paths of different sizes formed from a node, the connectivity of the nodes is not shown explicitly. In this description, when closed paths of the same size are formed, a superscript is used to indicate the number.

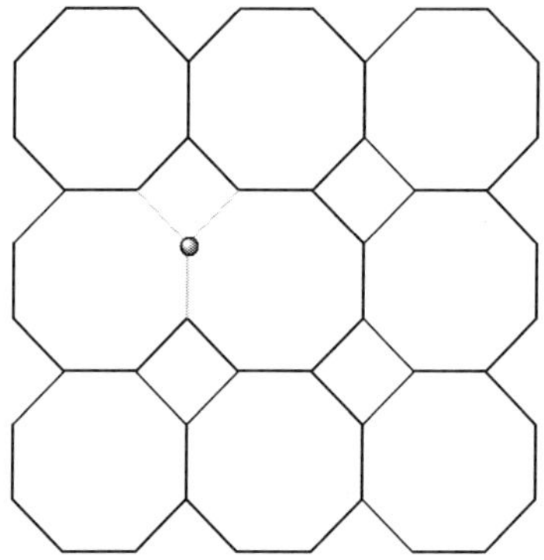

Figure 7. Uninodal network containing three pairs of connectors that generate closed paths of different sizes.

Thus, *Schläfli's symbols* are characterized by the number of sides *(q)* in the closed path. For a p-connected node, there are *[p(p-1)/2]* angles, because each two sides of a node define an angle. This notation is given by symbols $(q_1^a \cdot q_2^b \cdot q_3^c \ldots q_x^n)$, q_1, q_2, $q_3,\ldots,q_n$ are the number of sides in the closed paths and *a, b, c,...,n* are the numbers of closed paths of q_1, q_2, $q_3,\ldots,q_n$ sides generated by the angles from one node.

The Figure 7 shows an example of a polymer network in which the topology is calculated by *Schläfli's description*. Initially, two different closed paths can be observed, one is four sides and another is eight sides, so, $q_1 = 4$ and $q_2 = 8$. This tri-connected node generates three angles with one path of four sides ($q_1 = 4$), and two paths of eight sides ($q_2 = 8$). Therefore, *Schläfli's topological notation* for this network is $(4^1{\cdot}8^2)$, where the sub index 1(one) can be omitted (rewritten as $(4{\cdot}8^2)$).

2D uninodal networks will show a conflict between *Wells's notation* and *Schläfli's symbols*, if $p \geq 4$. Considering the quasi-square grid in Figure 6b, it can be observed that according to *Wells's notation*, this is a net (4,4). In the literatures, many considered *Wells's notation* (4,4) equivalent to *Schläfli's symbol* (4^4), which was incorrect. The true *Schläfli's symbol* for that network is $(4^4{\cdot}6^2)$, since the nodes are tetra-connected *(p=4)* with *[p(p-1)/2]* angles. In this example, six angles generate four paths of four sides and two paths of six sides, because the angles between each two different connectors are all analyzed for *Schläfli's symbol*. Therefore, it can be observed that in *Wells's notation* the paths of six sides formed by connectors in the *trans* position are ignored.

Finally, *Schläfli's symbols* $(q_1^a \cdot q_2^b \cdot q_3^c \ldots q_x^n)$ for each node can be used to describe multinodal network and uninodal network with different number of paths. A simple way to view this case could be done through the topology of the network in Figure 8.

In this case, we have two different nodes (tri-connected and tetra-connected) and two different closed paths (six sides and four sides). Therefore, the tri-connected node forms three angles with two paths of six sides and one path of four sides, and this is marked as $(4{\cdot}6^2)$. The tetra-connected node forms six angles with two paths of four sides, six sides and eight sides, and it is marked as $(4^2{\cdot}6^2{\cdot}8^2)$. Thus, the final topology of the network is $(4{\cdot}6^2)$ $(4^2{\cdot}6^2{\cdot}8^2)$, and each bracket is equivalent to a node. In this case, the nodes are tri- and tetra-connected, respectively.

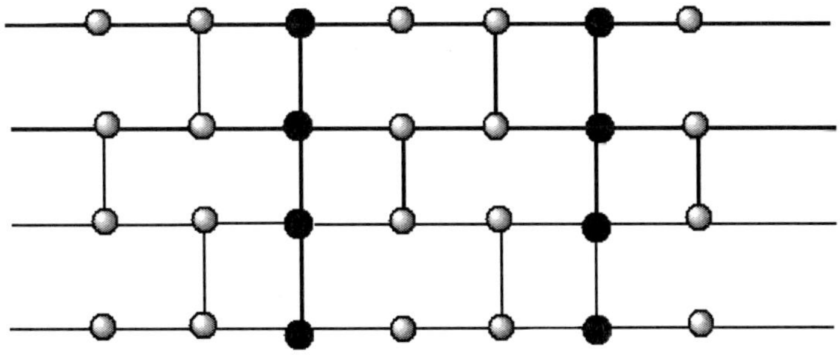

Figure 8. Multinodal polymer network in the form of long and short brick walls.

The analysis procedures of 3D networks and 2D networks are the same, but the visualization of closed paths is much more complicated for 3D networks. Furthermore, the high connectivity of the nodes will generate many closed paths, it cause the difficulty for the topology analysis. To minimize this difficulty, a software package called TOPOS [28] was developed to calculate the structural topology of the networks. TOPOS package as well as its manual can be obtained through the website: http://www.topos.ssu.samara.ru/. By the operation of TOPOS program, files generated by the SHELX [29] refinement program with the extensions of (*.cif) (Crystallographic Information File)* or (*.res)* are converted into the entry file of (*.cmp) (Chemical Formulae of Compounds)*. Particularly, it has been used in the analysis of complex networks with multinodes.

3.2. Benzenecarboxylate Ligands

For several years, scientists have investigated the physical and chemical proprieties of polymers with benzenecarboxylate ligands, because these ligands have been proven to be good candidates for the construction of coordination polymers and metal-organic framework (MOF). These benzenecarboxylates ligands are commonly referred to: terephthalic acid (1,4-H_2BDC), trimesic acid (1,3,5-H_3BTC), pyromellitic acid (1,2,4,5-H_4BTC), hemimellitic acid (1,2,3-H_3BTC) and trimellitic acid (1,2,4-H_3BTC), which are shown in the Figure 9. Commonly, these ligands are divided into centrosymmetric molecules and noncentrosymmetric molecules.

The terephthalic acid, a rigid and divergent molecule, can be used to synthesize the polymer with higher apparent area and pore volume, comparing to zeolites [31]. Normally, trimesic acid is applied to the construction of MOFs because of its thermal stability [32]. Additionally, due to the high molecular symmetry and the hydrogen-bond capability, pyromellitic acid is also selected as building blocks of coordination polymers [33]. In Habib and co-workers' research [34], asymmetrical ligands construct intriguing structure that cannot be achieved by symmetrical ligands, because of the strong steric hindrance and special orientation of carboxylic groups. In the literatures mentioned above [31, 32, 34-46], the carboxylate anions adopt diverse coordination modes (monodentate, bidentate, chelate and different bridging conformations) to the metallic ion. Notably, the changes of position and the number for the carboxylic group show great effects on the construction of different kinds of crystal structures with various topologies, due to their versatile coordination modes.

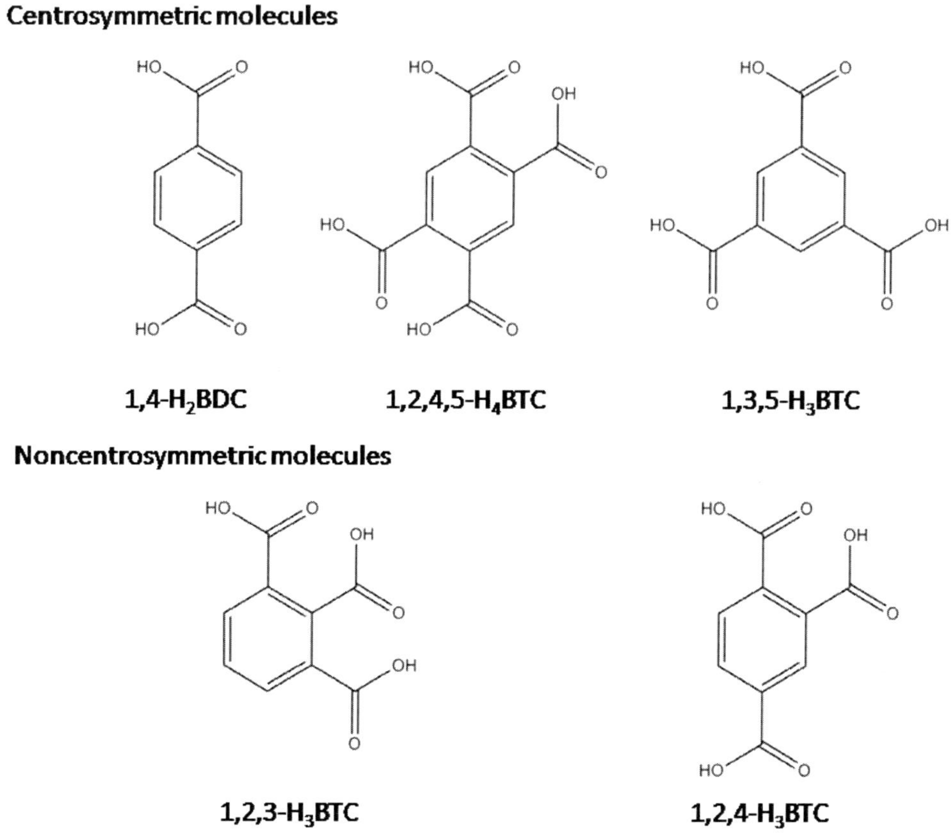

Figure 9. Examples of the main carboxylates ligands.

In some cases, N-donor ligands are normally used together with the benzenecarboxylate ligands for the construction of polymeric network and MOFs. In Luo and coworkers' research [47], the N4N4'-di(piridin-4-yl)biphenyl-4,4'-dicarboxamide (Figure 10a) and the terephthalate were used as mixed ligands to the construction of polymer with a three-fold interpenetrating net. In Liu and coworkers' study [48], the 1,3,5-benzenetricarboxylate and bis(imidazole) ligands were used in the construction of the 3D polymer, and the same network was observed. It is notable that porous materials, constructed by these mixed ligands, can be applied to not only gas storage, such as CO_2 and H_2, but also the liquid chromatographic separation [47]. In the other case, the fluorescent proprieties of terephthalic acid and dinitratobis(phen) cadmium complex were studied [49]. A new family of compounds containing Cu(II) and a mixture of aromatic multi-carboxylate 1,2,4,5-benzenetetracarboxylic acid and three isomeric dipyridyl bridging ligands was presented by Hung and co-works [50]. They observed that the changing of the nitrogen atom in the isomeric dipyridyl was sufficient for to form 3D frameworks. The mixture of 1,2,3-benzenetricarboxylic acid, 4,4'-bipyridyl and zinc ion was used for the construction of chiral self-penetration network 3D with chiral helical characteristic [51]. The Zn(II) is connected by the polycarboxylate ligand to form a 2D layer, and the bypiridil spacer was responsible for extending these 2D layers to the 3D network. Six new coordination polymers were synthesized by Ren *et al.* [52], and these complexes were constructed from the mixtures of 1,2,4,5-benzenetetracarboxylate, 1,3,5-benzenetricarboxylate, 1,2,4- benzenetricarboxylate and 4,4'-bis(1,2,4-triazol-1-ylmethyl)biphenyl) (Figure 10b). The magnetic susceptibility measurement of the complex formed from 1,2,4-benzenetricarboxylate indicated the presence of weak ferromagnetic interaction between the metal ions bridged by carboxylate groups.

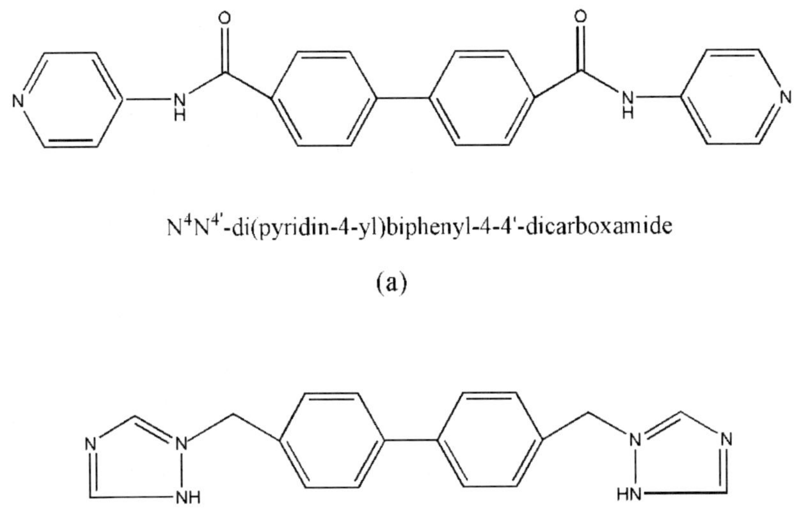

$N^4N^{4'}$-di(pyridin-4-yl)biphenyl-4-4'-dicarboxamide

(a)

4,4'-bis(1,2,4-triazol-1-ylmethyl)biphenyl

(b)

Figure 10. Molecular structure of N-donors a) $N^4N^{4'}$-di(piridin-4-yl)biphenyl-4,4'-dicarboxamide, and b) 4,4'-bis(1,2,4-triazol-1-ylmethyl)biphenyl.

With the examples mentioned above, we observe the importance of selecting the benzecarboxylate ligand and the N-donor ligands in the design of new interpenetration network for MOFs and coordination polymers.

3.2.1. Structures and Applications

The coordination polymers were mainly based on the assembly of coordination bonds between various types of organic ligands (carboxylate ligands, nitrogen ligands, etc.) and metal ions. In coordination polymers, different types of ligands usually present in different coordination sites, even a kind of ligand showing different coordination modes in a coordination polymer. The large structure variety of coordination polymers has been published and some interesting structures with their possible applications will be described below.

An interesting 2D coordination polymer, formed by Isonicotinohydrazide (INH) and Zn^{2+}, was reported by *Freitas et al.* in 2011. The crystal structure of this polymer belong to monoclinic crystal system with space group $P2_1$ and cell parameters a = 8.1190(2) Å, b = 17.977(4) Å, c = 9.1051(2) Å, β = 100.87(3)° and V=1305.1(4) Å^3 [53]. The coordination bonds between INH and Zn^{2+} ion, together with the coordination geometry of Zn^{2+} ion, could be seen in Figure 11.

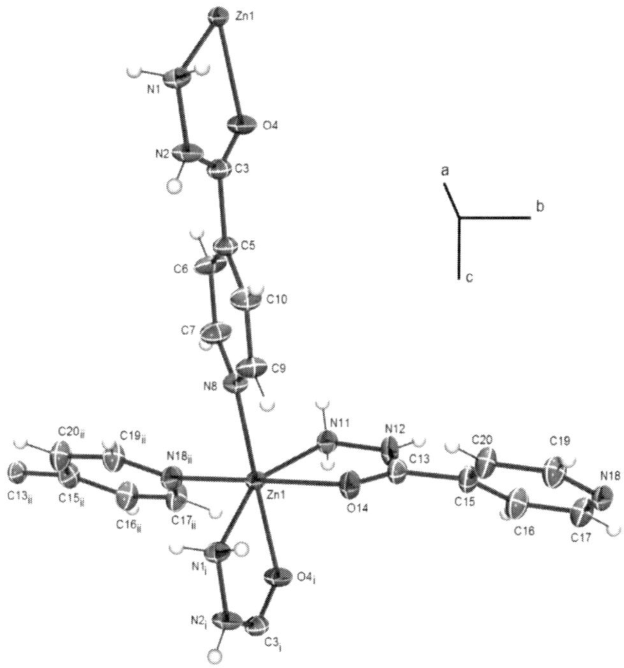

Figure 11. Crystalline structure formed by the coordination of INH with Zn^{2+}(i: x, y, 1 + z; ii: 1 - x, -½ + y, 2 - z).

The formation of this 2D coordination polymer depends on the coordination of INH and Zn^{2+} with the extension along the crystallographic axis *b* and *c*. The fact that INH is a divergent ligand is also important to the formation and the expansion of polymers. These properties, together with the hydrogen bond interactions between solvent and polymer structure and the planarity of INH molecules, give rise to the formation of pores in the structure. In Figure 12, we can see this formation along the crystallographic axis *a*. This structure is expanded in 3D arrangement by interlayer hydrogen bonds and the pores present a volume around 677 $Å^3$, filled by water and perchlorate molecules. In relation to this simple arrangement, this structure presents the topology with Schlafli symbol of $(4^4{\cdot}6^2)$ by considering zinc atom as a node and INH as a connector (Figure 12).

Additionally, INH is often used in tuberculosis treatment when it interacts with metal ions [54]. In this work, toxicity assays of the INH and the coordination polymer against *A. salina* were compared. The resulting assays showed that the coordination polymer is 8.5 times more active than the pure INH, in the same experimental conditions, making this compound a candidate to cancer treatment. This application does not have direct relation with the coordination polymer, but with the coordination mode to enhance the INH activity.

In the work of do Carmo et al. [45], the 3D coordination polymer, with the formula of $Na_5[Cu_3(C_9H_3O_6)_3(C_9H_4O_6)(H_2O)_{10})]{\cdot}(H_2O)_3$ containing the 1,2,3 - benzenetricarboxylic acid (3BTC) ligand, was reported.

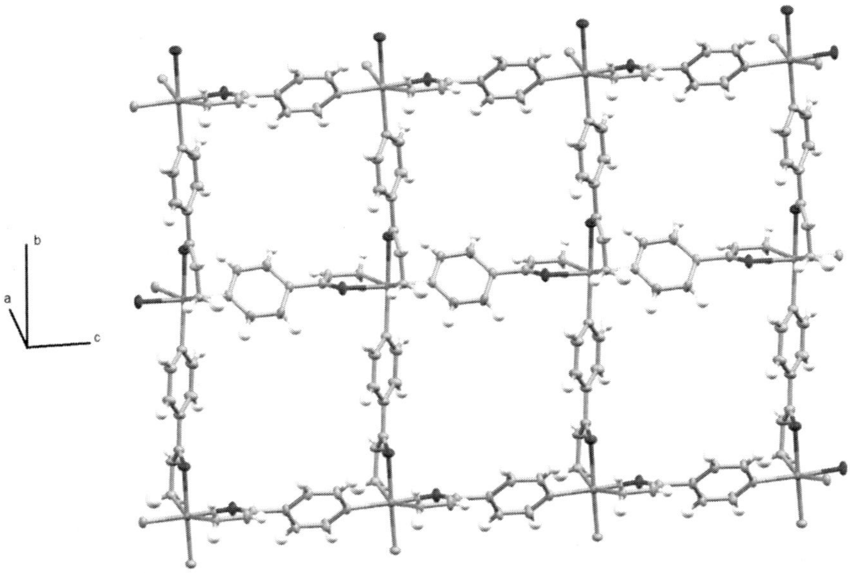

Figure 12. Disposal of extended coordination for polymer 2D.

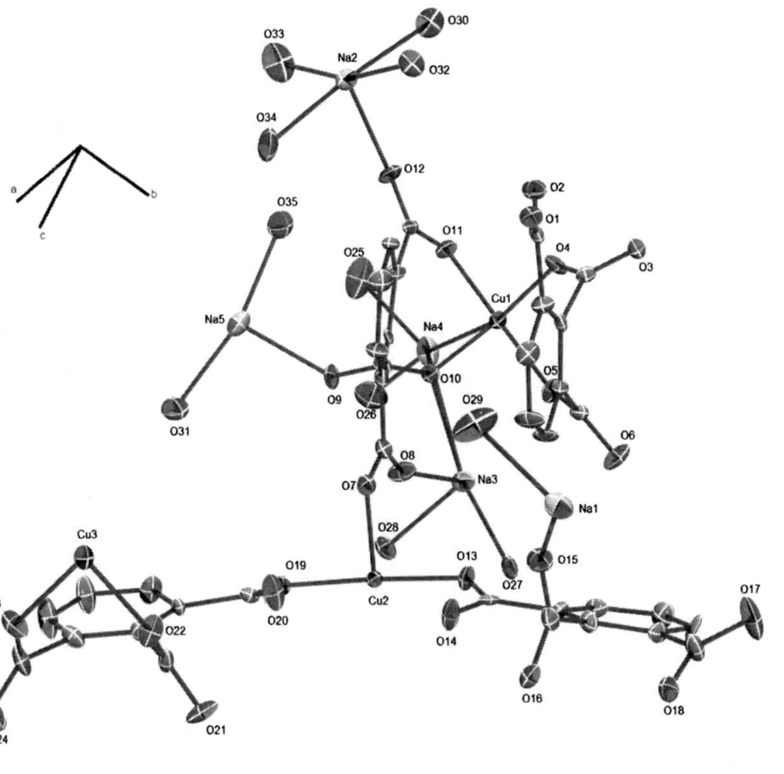

Figure 13. Asymmetric unit of the compound $Na_5[Cu_3(C_9H_3O_6)_3(C_9H_4O_6)(H_2O)_{10})]·(H_2O)_3$ (ellipsoid Ortep representation at 35% probability).

This compound was crystallized in the triclinic crystalline system, with a space group *P1* and cell parameters: a = 10.205(1) Å, b = 10.820(1) Å, c = 12.306(1) Å, α = 77.524(5)°, β =

73.743(5)°, γ = 89.201(5)° and V=1272.1(2) Å^3. In its asymmetric unit, there are four molecules of 3BTC, five sodium cation, three cooper ions and thirteen water molecules (Figure 13).

The dimensionality of this polymer can be analyzed from the expansion in two different crystallographic planes: *ab* and *bc*. This structure expands in *bc* plane through of the coordination between O10-Cu1, O4-Cu1, O-Cu2 and O7-Cu2. By the coordination bonds of O1-Cu2, O7-Cu2 and the connection between Cu3 and 3BTC, this structure expands in direction of the *ab* plane (Figure 14). With the extension along *ab* and *bc* planes, 3D coordination polymer was formed with a topology of Schlafli symbol ($4^{24}\cdot6^{4}$). This coordination polymer is a good candidate for application in non-linear optics, because its crystalline structure is non-centrosymmetric [55].

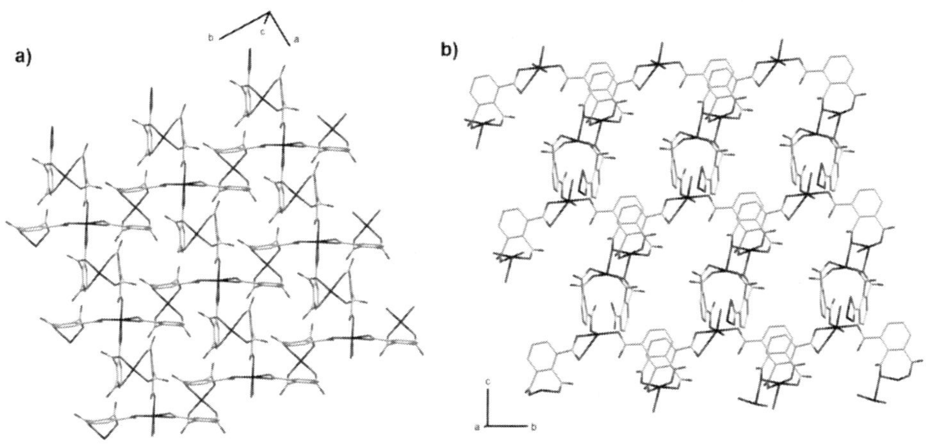

Figure 14. Coordination polymer expanded in crystallographic planes a) *ab* and b) *bc*.

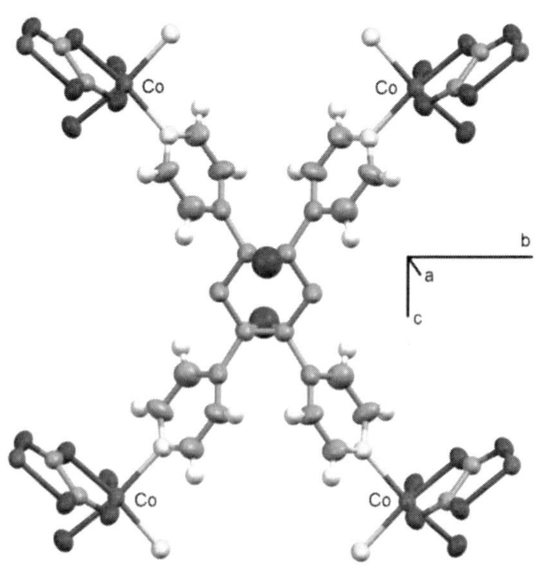

Figure 15. Coordination of ligands with Co^{2+}(Carbon=gray, Oxigen=black, Nitrogen=light gray, Cobalt=black).

In many cases, a 3D coordination polymer can present structures with different applications. For example, a microporous coordination polymer can be applied in gas adsorption [56], drug delivery [57], gas separation [58], and others. A coordination polymer with microporous synthesized by *Hao et al.* can adsorb gas, and this structure is formed by Co^{2+} ion and two types of organic ligands, the 1a,4a-dihydroxy-1e,2e,4e,5e-tetra(4-pyridyl)cyclohexane [dchtpy] and the oxalic acid (Figure 15). This structure belongs to the monoclinic crystalline system, with space group *C2/m* and cell parameters: a = 8.9751(3) Å, b = 24.4278(6) Å, c = 12.0085(3) Å, β = 107.797(3)° and V= 2506.8(1) $Å^3$.

The extending of this structure is based on the coordination of oxalic acid and dchtpy ligand with Co^{2+} ions. Notably, this motif forms 3D channels (A and B) in two directions [59]. The **A** channel is reported as the large one with the size of 10.76 x 10.67 Å; and it can be seen along crystallographic a-axis (Figure 16-a). B channel, with the size of 10.51 x 4.66 Å, disposes along {101} direction (Figure 16-b). These channels together with coordination modes generate a complex topology with simplified Schlafli symbol $(4^2{\cdot}6^2{\cdot}8^2)(4{\cdot}6^4{\cdot}8)_2$, this topology [60] is named as Moganite topology.

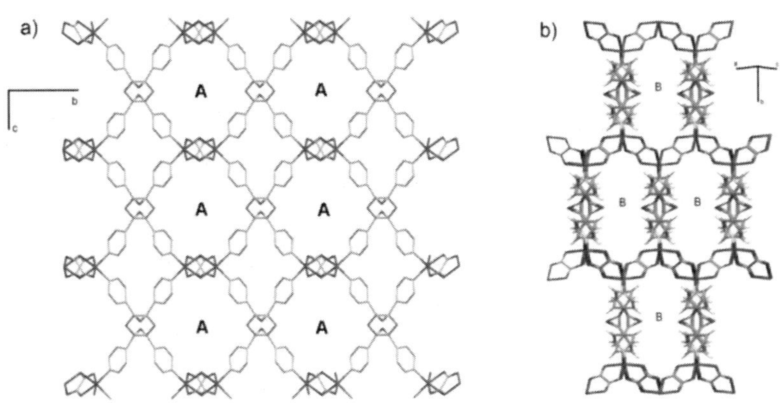

Figure 16. Channels formed by the extended coordination of polymer a) channel A along a axis and b) channel B along {101} direction.

This compound was proven with the capacity of adsorbing gas molecules (H_2, N_2 and CO_2) in channels. It also showed the improvement in CO_2 gas adsorption, when the gas pressure is increased. This property can be justified by the flexibility of this framework.

The flexibility is an interesting property present in this type of structure. The flexibility can promote the adsorption and desorption of guest solvents and the gases in the structure without destabilizing or breaking it. In this compound, this property was proven by powder X-ray diffraction with the channel structure remained unbroken during the heating of guest solvents. The structure modifies itself during this process and changes the size of the channels.

Another important work that involves microporous coordination polymer, is published by An and co-workers in 2009. In their work, a new compound capable of storing drugs can promote slow liberation. This structure was built by two ligands, Adenine and biphenyldicarboxylate, and the Zn^{2+} ions [61]. Pores along crystallographic c-axis can be seen in Figure 17. This topology presents Schlafli symbol $(4^{12}\ 6^3)$, and it can be noted that this complex structure presents a simple topology with close path well defined. This fact can be explained by the high symmetry of the organic ligand and the high symmetry of the unit cell allowing the formation of symmetric pores. The symmetry in unit cell can be confirmed by its parameters, where it belongs to the tetragonal crystalline system and space group $I4_122$, a = 38.237(2) Å, b = 38.237(2) Å, c = 11.1753(12) Å, and V= 16339.1 Å.

This framework is an anionic network and is neutralized by guest ionic solvents (dimethylammonium[$(CH_3)_2NH_2^+$]). This characteristic enables this framework to store and release some cationic drug molecules by cationic exchange. In biological environment, this release can be slow and well-controlled. In this work, this property was tested with procainamide HCl drug, which is an antiarrythmia drug capable of being stored and slowly released. In this way, this coordination polymer could be considered as a good candidate to carrier drugs.

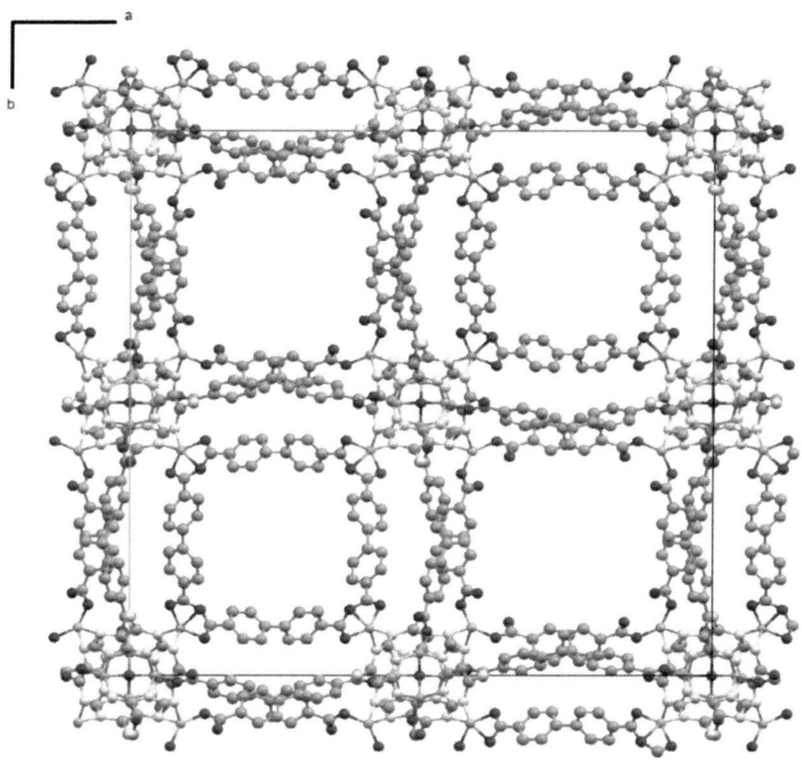

Figure 17. Porous framework formed by the adenine- and biphenyldicarboxylate-coordinated Zn^{2+}.

The knowledge of these structures as well as its applications, allows the analysis on the structural arrangement and its applications. In this way, it is possible to see that the structure arrangement and topology are not main properties to determine it applications. In this case, only the constituent ligands, coordination bond and unit cell symmetry are important. In the other cases, the structure arrangement and topology determine the structure characteristics for the other applications. This study of coordination polymer is an area that should be widely explored by the next years. This is due to the strong potential of obtaining different coordination polymer to develop new applications.

ACKNOWLEDGMENTS

The authors are thankful to the Brazilian Agencies CAPES, FAPEMIG, CNPq for the financial support.

REFERENCES

[1] Eckert, M. *Annalen der Physik*. 2012, 524(5), A83-A85.

[2] Ben-Abraham, S. I. Z. *Kristallogr*. 2007, 222, 310-310.

[3] Chagas, L. H.; Janczak, J.; Gomes, F. S.; Fernandes, N. G.; de Oliveira, L. F. C.; Diniz, R. *J. Mol. Struct*. 2008, 892(1-3), 305-310.

[4] Novak, A. Hydrogen bonding in solids correlation of spectroscopic and crystallographic data; Large Molecules; Springer Berlin Heidelberg. 1974, Vol. 18, 177-216.

[5] Khlobystov, A. N.; Blake, A. J.; Champness, N. R.; Lemenovskii, D. A.; Majouga, A. G.; Zyk, N. V.; Lomonoso, M. V. *Coord. Chem. Rev*. 2001, 222(1), 155-192.

[6] Hosseini, M. W. *Coord. Chem. Rev*. 2003, 240(1-2), 157-166.

[7] Moulton, B.; Zaworotko, M. *J. Chem. Rev*. 2001, 101(6), 1629-1658.

[8] Lehn, J. M. *Science*. 1993, 260(5115), 1762-1763.

[9] Araki, K.; Toma, H. E. Chemistry of supramolecular systems containing porphyrins and metal complexes. *Quim. Nova*. 2002, 25(6A), 962-975.

[10] Connolly, S.; Rao, S. N.; Rizza, R.; Zaccheroni, N. *Coord. Chem. Rev*. 1999, 185-6, 277-295.

[11] Busch, D. H.; Stephenson, N. A. Coord. Chem. Rev.1990, 100, 119-154.

[12] Lehn, J. M. *Pure Appl. Chem*.1978, 50(9-10), 871-892.

[13] Batten, S. R.; Champness, N. R.; Chen, X.-M.; Garcia-Martinez, J.; Kitagawa, S.; Ohrstrom, L.; Reedijk, J. *CrystEngComm*, 2012, 14(9), 3001-3004.

[14] Steed, J. W.; Atwood, J. L. SUPRAMOL CHEM. 2009, England: John Wiley e Sons Ltd.

[15] Dobrawa, R.; Wurthner, F. *J. Polym. Sci., Part A: Polym. Chem*. 2005, 43(21), 4981-4995.

[16] Lohrstrom. 2010, Coordination Polymers and Metal Organic Frameworks: Terminology and Nomenclature Guidelines. IUPAC.

[17] Batten, S. R.; Neville, S. M.; Turner, D. R. Coordination Polymers: Design, Analysis and Application. 2009, Cambridge, England: *The Royal Society of Chemistry*.

[18] Batten, S. R.; Harris, A. R.; Jensen, P.; Murray, K. S.; Ziebell, A. J. Chem. Soc., *Dalton Trans*. 2000, 21: 3829-3837.

[19] Hill, R. J.; Long, D. L.; Turvey, M. S.; Blake, A. J.; Champness, N. R.; Hubberstey, P.; Schroder, M. *Chem. Commun*. 2004, 16: 1792-1793.

[20] Dong, Y. B.; Smith, M. D. *Inorg. Chem*. 2000, 39(21): 4927-4935.

[21] Li, D.; Shi, W. J.; Hou, L. *Inorg. Chem*. 2005, 44(11): 3907-3913.

[22] Abrahams, B. F.; Batten, S. R.; Hamit, H.; Hoskins, B. F.; Robson, A. Angew. *Chem. Int. Ed. Engl*. 1996, 35 (15): 1690-1690.

[23] Carlucci, L.; Ciani, G.; Proserpio, D. M.; Sironi, A. *Chem. Commun*. 1996, 11, 1393-1394.

[24] Abrahams, B. F.; Batten, S. R.; Hoskins, B. F.; Robson, R. *Inorg. Chem.* 2003, 42 (8): 2654-2664.
[25] Wells, A. F. Three dimensional nets and polyhedra. 1977, New York, Wiley.
[26] Carlucci, L.; Ciani, G.; Proserpio, D. M. *Coord. Chem. Rev.* 2003, 246(1), 247-289.
[27] Atwood, J. L.; Stted, J. W. Supramolecular Chemistry. 2009, Chinchester, UK: J. Wiley Sons.
[28] Blatov, V. A.; Shevchenko, A. P.; Serezhkin, V. N. *J. Appl. Crystallogr.* 2000, 33(4), 1193-1193. doi:
[29] Sheldrick, G. M. *Acta Crystallogr., Sect. A: Found. Crystallogr.* 2008, 64, 112-122.
[30] Duchowicz, P. R.; Castro, E. A. Philipp. *J. Sci.* 2004, 33(1), 45-53.
[31] Li, H.; Eddaoudi, M.; O'Keeffe, M.; Yaghi, O. M. *Nature*, 1999, 402(6759), 276-279.
[32] Davies, K.; Bourne, S. A.; Ohrstrom, L.; Oliver, C. L. *Dalton T.* 2010,39(11), 2869-2874.
[33] Du, M.; Zhang, Z.-H.; Zhao, X.-J. *Cryst. Growth Des.* 2005, 5(3), 1247-1254.
[34] Habib, H. A.; Sanchiz, J.; Janiak, C. *Dalton T.* (36), 2008, 4877-4884.
[35] Li, H.; Eddaoudi, M.; Groy, T. L.; Yaghi, O. M. *J. Amer. Chem. Soc.* 1998, 120(33), 8571-8572.
[36] Gutschke, S. O. H.; Price, D. J.; Powell, A. K.; Wood, P. T. Angew. *Chem. Int. Ed.* 2001, 40(10), 1920-1923.
[37] Liu, Y.-H.; Chung, H.-T. Acta *Crystallogr. Sect. E: Struct. Rep.* Online. 2007, 63, M1555-U1101.
[38] Wen, Y.-H.; Zhang, Q.-W.; He, Y.-H.; Feng, Y.-L. *Inorg. Chem. Commun.* 2007, 10(5), 543-546.
[39] Ji, W.-J.; Zhai, Q.-G.; Hu, M.-C.; Li, S.-N.; Jiang, Y.-C.; Wang, Y. *Inorg. Chem. Commun.* 2008, 11(12), 1455-1458.
[40] Zhou, S. X.; Ke, Y. X.; Zou, H. L.; Liu, T. H.; Zhu, F.; Chen, J. M.; Xie, A. J. Z. Naturforsch., B: *Chem. Sci.* 2008, 63(11), 1339-1342.
[41] Chung, H.-T.; Tsai, H.-L.; Yang, E.-C.; Chien, P.-H.; Peng, C.-C.; Huang, Y.-C.; Liu, Y.-H. *Eur. J. Inorg. Chem.* 2009, 3661-3666.
[42] Zhang, Z.; Xiang, S.; Rao, X.; Zheng, Q.; Fronczek, F. R.; Qian, G.; *Chen, B. Chem. Commun.* 2010, 46(38), 7205-7207.
[43] Burrows, A. D.; Frost, C. G.; Kandiah, M.; Keenan, L. L.; Mahon, M. F.; Savarese, T. L.; Warren, J. E. *Inorg. Chim. Acta.* 2011, 366(1), 303-309.
[44] Xie, J.; Han, Z.; Pei, W.; Zou, Y.; Ren, X. *Inorg. Chem. Commun.* 2011, 14(8), 1266-1270.
[45] do Carmo, W. R.; Franco, C. H. J.; Diniz, R. Inorg. *Chim. Acta.* 2012, 385, 119-127.
[46] Huang, G.; Yang, P.; Wang, N.; Wu, J.-Z.; Yu, Y. *Inorg. Chim. Acta.* 2012, 384(0), 333-339.
[47] Luo, F.; Yuan, Z.-Z.; Feng, X.-F.; Batten, S. R.; Li, J.-Q.; Luo, M.-B.; Tian, X.-Z. *Cryst. Growth Des.* 2012, 12(7), 3392-3396.
[48] Liu, M.; Yang, Z.; Sun, W.; Li, X.; Li, J.; Ma, J.; Yang, G. *Inorg. Chim. Acta.* 2009, 362(8), 2884-2889.
[49] Shi, X.; Zhu, G.; Wang, X.; Li, G.; Fang, Q.; Zhao, X.; Qiu, S. P. Cryst. *Growth Des.* 2004, 5(1), 341-346.
[50] Huang, F.-P.; Tian, J.-L.; Gu, W.; Yan, S.-P. *Inorg. Chem. Commun.* 2010,13(1), 90-94.

[51] Ma, L.-F.; Qin, J.-H.; Han, M.-L.; Wang, L.-Y.; Du, M. *Inorg. Chem. Commun.* 2011, 14(10), 1584-1587.

[52] Ren, C.; Hou, L.; Liu, B.; Yang, G. P.; Wang, Y. Y.; Shi, Q. Z. *Dalton T.* 2011, 40(4), 793-804.

[53] Freitas, M. C. R.; Antonio, J. M. S.; Ziolli, R. L.; Yoshida, M. I.; Rey, N. A.; Diniz, R. *Polyhedron*, 2011, 30(11), 1922-1926.

[54] Souza, M. V. N. d.; Vasconcelos, T. R. A. Quim. Nova. 2005, 28(4).

[55] Kaminsky, W.; Responte, D.; Daranciang, D.; Gallegos, J. B.; Tran, B. C. N.; Pham, T. A. *Molecules*, 2010, 15(1), 554-569.

[56] Chun, H.; Dybtsev, D. N.; Kim, H.; Kim, K. *Chem. Eur. J.* 2005, 11(12), 3521-3529.

[57] Sun, C. Y.; Qin, C.; Wang, C. G.; Su, Z. M.; Wang, S.; Wang, X. L.; Wang, E. B. *Adv. Mater.* 2011, 23(47), 5629-5630.

[58] Matsuda, R.; Kitaura, R.; Kitagawa, S.; Kubota, Y.; Belosludov, R. V.; Kobayashi, T. C.; Mita, Y. *Nature*. 2005, 436(7048), 238-241.

[59] Hao, H. Q.; Lin, Z. J.; Hu, S.; Liu, W. T.; Zheng, Y. Z.; Tong, M. L. *CRYSTENGCOMM*. 2010, 2(7), 2225-2231.

[60] Su, C. Y.; Smith, M. D.; Goforth, A. M.; zur Loye, H. C. *Inorg. Chem.* 2004, 43(22), 6881-6883.

[61] An, J. Y.; Geib, S. J.; Rosi, N. L. *J. Am. Chem. Soc.* 2009, 131(24), 8376-8377.

In: X-Ray Diffraction: Structure, Principles and Applications ISBN: 978-1-62808-591-4
Editor: Kaimin Shih

Chapter 3

X-RAY DIFFRACTION ANALYSIS OF MAGNETIC SHAPE MEMORY ALLOYS

Albert Carrillo, Joan Josep Suñol and Maria Lluïsa Escoda*
GRMT, Departament de Física, Universitat de Girona,
Campus Montilivi s/n, Girona

ABSTRACT

Research in the field of magnetic shape memory alloys has increased in the last years for their high interest in technological applications due to the reversible deformations undergone. These applications range from actuators and sensors to magnetic refrigeration. Characterization of the different crystallographic structures of this kind of materials is vital for the understanding of their macroscopic and microscopic behavior. The structures of Heusler magnetic shape memory alloys are high-temperature austenites $L2_1$ and B2 and low-temperature martensites $L1_0$, 5M and 7M. Information of these crystallographic structures has been extracted by analyzing with Rietveld method for the diffractograms of five different samples, obtained through X-ray diffraction, and corresponded each one of them to one of the possible structures exhibited by Heusler magnetic shape memory alloys. CIF files containing theoretical structural information have been generated from the scratch and tested with the experimental diffraction patterns of the five samples. MAUD software has been used to generate CIF files and refine experimental data and was shown to be very useful for austenite phase structures but not ideal for martensites, if modulation is present.

INTRODUCTION

Magnetic shape memory alloys (MSMA), also known as ferromagnetic shape memory alloys (FSMA), have an increasing interest in technological applications, due to the associated reversible deformation and magnetic transformation, which range from actuators and sensors [1] to magnetic refrigeration [2, 3]. The main cause of the deformation of the structure is the

* Correspondence: Tel: +34-972-41-9757;Fax: +34-972-41-318098. Email address: joanjosep.sunyol@udg.edu.

austenite-martensite transition experimented by these materials from a high-temperature ordered cubic austenite phase to a low-temperature tetragonal, orthorhombic or monoclinic martensite phase. The crystallographic structures are usually given using Strukturbericht symbols ($L2_1$, B2, $L1_0$) or specific notation of this alloys (5M, 7M). $L2_1$ consists of four face centered lattice, but B2 is a body centered cubic structure which can be understood as one of the eight cubes that made up $L2_1$ lattice. This change is due to the loss of periodicity of the atoms in the crystal and because of that the alloy cannot be represented with $L2_1$ anymore. $L1_0$ is taken as the central part of $L2_1$. The 5M and 7M can be understood as five and seven $L1_0$ cells making a stack. The interesting fact is that the stack is not making a perfect 'squared' structure but some distortion between the layers is present. The ferromagnetic-paramagnetic transitions are undergone when Curie threshold is over passed. These structural and magnetic transitions can be produced by different causes: mechanical stress, magnetic field application and temperature change.

Most MSMA are those called Heusler alloys, with a chemical composition $X_2YZ$and named after the German mining engineer and chemist Friedrich Heusler, who studied a copper-rich alloy Cu_2MnAl (showing ferromagnetic behavior) at the beginning of the twentieth century [4, 5].

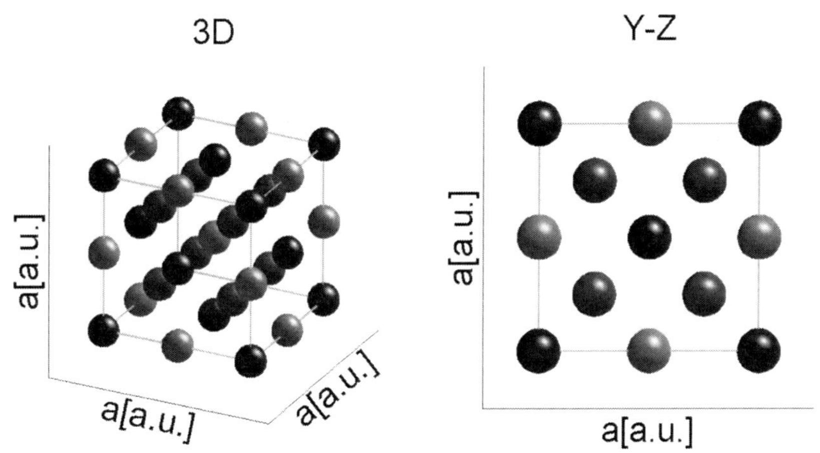

Figure 1. Austenite phase $L2_1$ structure of arbitrary composition viewed in 3D and laterally at left and right respectively.

Cu_2MnAl alloy studied by Heusler exhibited an austenite phase at room temperature whose structure was fully determined three decades later by Bradley and Rodgers [6]. They reported a $L2_1$ cubic structure (Figure 1) with a lattice parameter a = 5.95Å and a unit cell made up by eight atoms of copper and four of manganese and aluminum. The basic cell was constituted by a primitive lattice composed of eight centered cubes with copper atoms in the vertices and alternated atoms of manganese and aluminum in the centers. From crystallography, it is seen as four face centered cubic (fcc) lattices overlapped, two of them consisting in X-position atoms and the other two of Y- and Z-position atoms (Figure 2). It must be noticed that not all materials exhibiting this Heusler $L2_1$ structure undergo martensitic transition.

The loss of periodicity is sometimes present in austenite phase. In these cases, the exhibited structure is B2 rather than $L2_1$. This new structure consists of a body centered cube (bcc) with atoms of X-position in the vertices and atoms of Y- and Z-position in the center with a probability of being Y or Z (Figure 3), which depends on the sample's composition.

Martensite phase is found at lower temperatures and can exhibit three different crystallographic structures. One of them, $L1_0$, is a tetragonal distortion of the austenite $L2_1$ cubic lattice (Figure 4 & Figure 5). Other possible structures are the more complex modulated orthorhombic [7] or monoclinic [8] 5M and 7M (also named 10M and14M). The 6M modulated structure has been also reported [9], but not as widely as the other two.

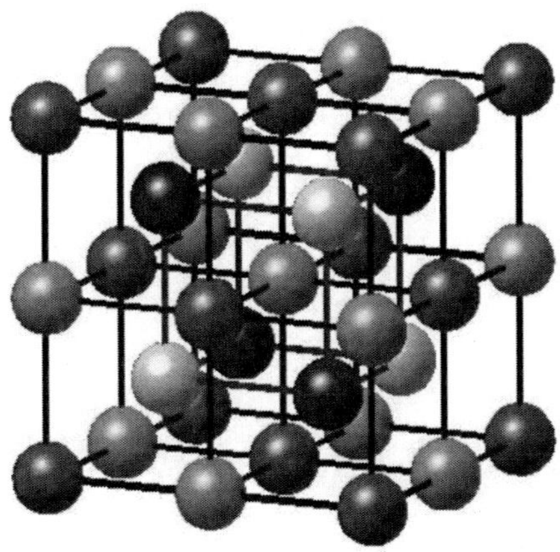

Figure 2. Austenite phase $L2_1$ (Heusler) structure basic cell with four overlapped face centered cubes (fcc). Blue and red atoms correspond to X-position and blue and yellow to Y- and Z-position respectively.

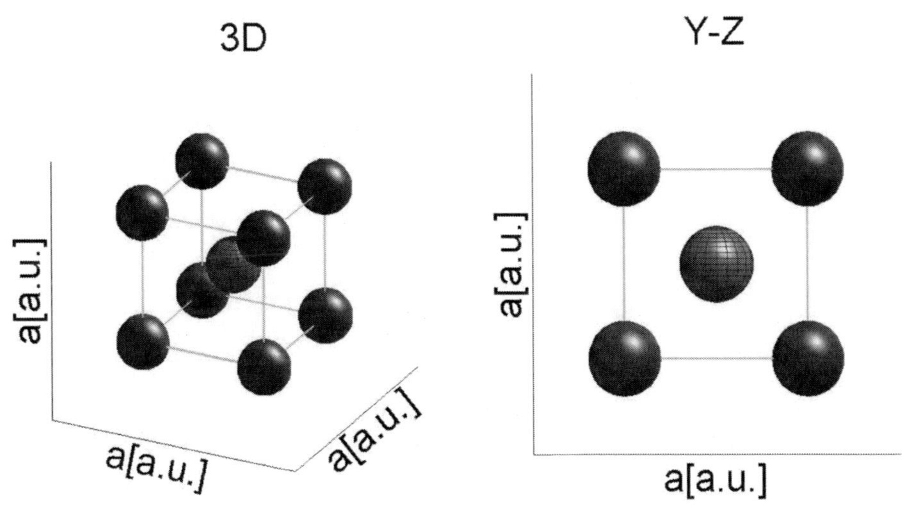

Figure 3. Austenite phase B2 structure of arbitrary composition viewed in 3D and laterally at left and right respectively.

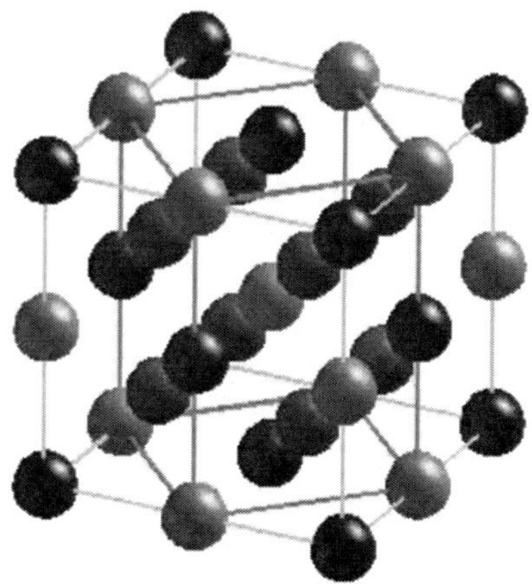

Figure 4. Relation between austenite $L2_1$ and martensite $L1_0$ structures of arbitrary composition in light blue and purple respectively.

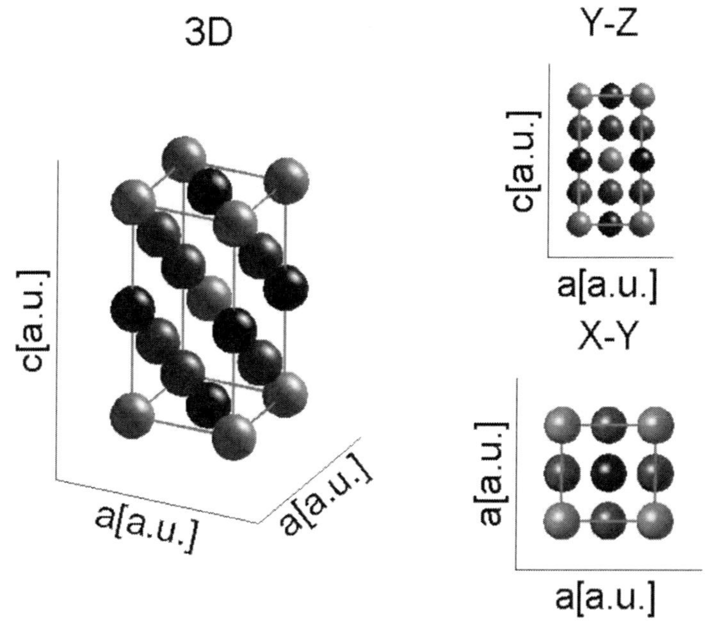

Figure 5. Martensite phase $L1_0$ structure of arbitrary composition viewed in 3D, from above and laterally at left, upper-right and lower-right respectively.

These modulated structures consist of five and seven $L1_0$-style cells for 5M and 7M, respectively. They do not show a perfect 'squared' structure, but some distortion between the different layers is present. It has been reported that modulated 7M structure exhibits a $(5\bar{2})_2$ stacking sequence [10-13] (Figure 6) and 5M structure a $(3\bar{2})_2$ stacking sequence [12-14] (Figure 7) in Zhdanov notation. Some authors simplify the stacking sequence in a real modulation [8, 15-17] in order to make the treatment of the structure easier.

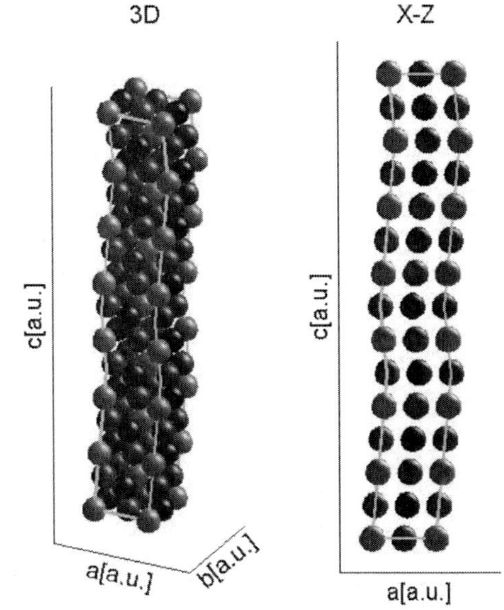

Figure 6. Martensite phase seven-layered modulated structure (7M or 14M) of arbitrary composition with $(5\bar{2})_2$ stacking sequence viewed in 3D and laterally at left and right respectively.

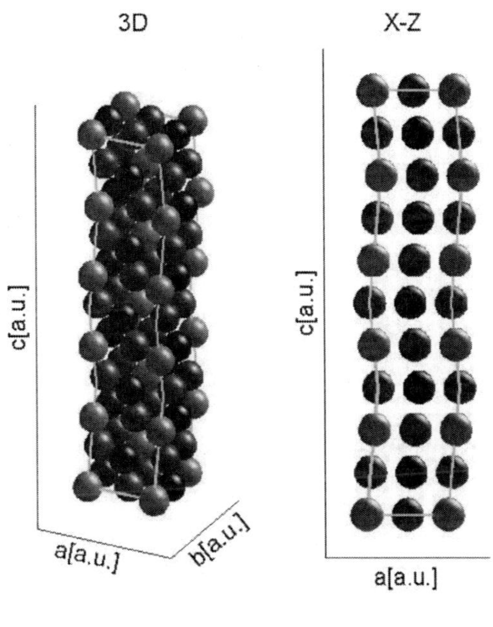

Figure 7. Martensite phase five-layered modulated structure (5M or 10M) of arbitrary composition with $(3\bar{2})_2$ stacking sequence viewed in 3D and laterally at left and right respectively.

Knowing if an alloy is in austenite or martensite phase, irrespectively of the structure, is of great importance. For example, in order to perform a differential scanning calorimetry (DSC) experiment to find the heat associated with the austenite-martensite transition. If the sample is in austenite phase, temperature has to be decreased to find the transition. On the other hand, if it is in martensite phase, temperature must be increased.

Nowadays, Ni_2MnGa is still the most studied Heusler MSMA [18] and shows an interesting fact that it is the only one exhibiting the magnetic properties in its stoichiometric composition. Other Heusler magnetic alloys only show these properties in off-stoichiometry. However, this alloy has some disadvantages. The temperature interval required for austenite-martensite transition is too wide and gallium is too expensive. Because of this, new elements to replace gallium were searched for removing these disadvantages. Krenke et al. studied gallium-free Heusler MSMA by replacing gallium for tin and indium. They reported a full structural study of off-stoichiometry Ni-Mn-Sn [19] and Ni-Mn-In [20] which exhibited good properties. These off-stoichiometry Heusler MSMA have a $X_2Y_{1+x}Z_{1-x}$ (also written as $X_{50}Y_{25+x}Z_{25-x}$) composition, instead of the stoichiometric $X_2YZ(X_{50}Y_{25}Z_{25})$. This means that in off-stoichiometry case some Y-position atoms will be in Z-position. In the Materials and Thermodynamics Research Group (GRMT) at the University of Girona (UdG), non-stoichiometric Heusler MSMA with indium and tin replacing gallium were studied in alloys with ribbon shape [21-24].

The aim of this work is to create the Crystallographic Information File (CIF file), which contains micro-structural information for five different samples. Each of them will be corresponding to one of the five main structures exhibited by Heusler MSMA. Once these files are created, an analysis with the Rietveld method [25, 26] will be made to check if they have been generated correctly and can truly represent the studied samples.

Rietveld method is based on the generation of a theoretical intensity, I_i^C (Equation 1), for every angular position through the information taken from the CIF file, which was put together to generate a theoretical diffraction pattern. The generated pattern must match the experimental diffractogram of the sample, or their difference, M, must be minimized (Equation 2). If the refinement has been done properly, the new theoretical function will give all structural information of the sample from which all physical properties of the material can be derived.

$$I_i^C = S_F \sum_k L_k |F_k|^2 S(2\theta_i - 2\theta_k) P_k A + b_i \quad (1)$$

$$M = \sum_i \frac{1}{I_i^E} \left(I_i^E - I_i^C\right)^2 \quad (2)$$

One must be very careful when working with the Rietveld method, because the results may or may not represent the correct micro-structure of the sample. As the minimization of the difference between theoretical and experimental patterns is done by modifying a large number of parameters of a mathematical method, it may give results with non-physical sense such as lattice parameters smaller than 1Å, negative thermal B factors, negative crystallite sizes, etc. Thus, while refining, the user must be very careful in order not to obtain physical-senseless parameters. Therefore, refining samples by Rietveld method is not straightforward and it requires sufficient experience. Besides not being straightforward, it is impossible to conduct this analysis manually because of the large amount of modifiable parameters. Therefore, there is a need for computational tools, and several software programs are available to perform this analysis. Some of them are with special features and some are for general purposes. Some of them are free of charge (FC) and some are commercial (Com) products. Examples of Rietveld refinement software are WinPLOTR [27] (FC) and FullProf [28] (FC) for general purposes; X'PertHighScore (Com) specialized in phase identification;

Jana98 (FC), Jana2000 [29] (FC) and its new version Jana2006 (FC) specialized in modulated structures; MAUD version 1.993 [30] (FC) and its newest edition MAUD version 2.33 (FC) for general purposes with many texture-related capabilities.

In this chapter, we chose MAUD (Material Analysis Using Diffraction), developed by an Italian engineer Luca Lutterotti [30-32], for the study. Such choice is mainly because of MAUD's user friendly environment, which allows easy access to all features of the program. It allows the real-time observation of the modification for every available parameter, and it also allows the generation of CIF files within the same program.

METHODOLOGY

The procedure of obtaining the ribbon samples has two steps:

- Bulk elemental pieces were cut and introduced in an arc melting device for alloying. It was performed several times in order to obtain homogeneous alloy. The arc melting process was performed under Ar atmosphere.
- Ribbons were produced using the melt spinning technique by rapid solidification. The master alloy is previously melted in an induction system inside the melt spinning device.

Five different samples in the form of ribbon flakes of a medium width of 1.5~2 mm and length of 6~7 mm were produced by melt spinning in argon atmosphere at a wheel linear speed of 48 ms^{-1}. The arc melted alloys were prepared from highly pure elements.

To analyze these samples, the experimental diffractograms, from which their structural information can be extracted, must be obtained. Diffraction patterns were obtained, at room temperature, through an X-ray diffraction experiment using a Bruker AXS D8 Advance diffractometer with CuK_{α} X-ray tube.

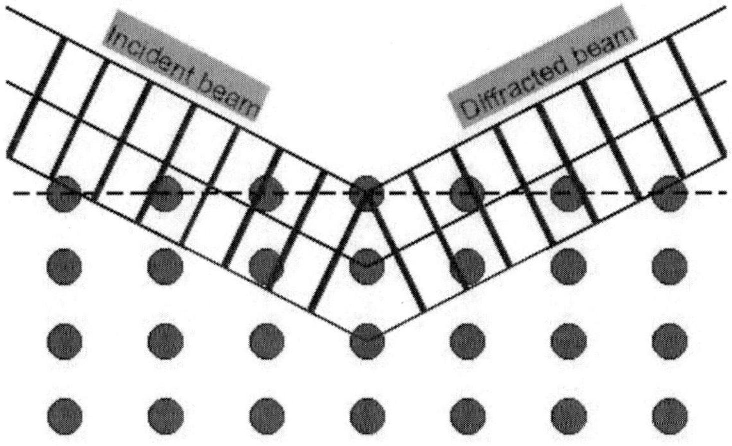

Figure 8. Schematic of the X-ray diffraction technique using Bragg's law shows the incident and diffracted beams and the atomic positions (red circles).

This process consisted of irradiating an X-ray beam to the sample and collecting the diffracted beam in a CCD plate (formerly a photographic plate). The choice of X-ray radiation was not arbitrary. The wavelength of X-rays is of an order of magnitude similar to the inter-atomic distance in the crystallographic structures. The formulation of X-ray diffraction technique (Figure 8) was implemented by William Lawrence Bragg and his father William Henry Bragg in 1913 [33]. The further introduction of this method can be found in several literatures [34, 35].

Once experimental data is obtained, there is a need to generate the CIF file from which the theoretical diffractogram can be derived. A step by step process to generate the CIF file has been developed in the present work for an off-stoichiometry sample. The procedure can be summarized in five steps:

Symmetry: Symmetry selection must be done in order to keep the number of lattice parameters as minimum as possible. It would result in the same cubic symmetry with only one cell parameter instead of an orthorhombic symmetry with three identical lattice parameters. The first choice is the better one because it minimizes the number of modifiable parameters during the refinement.

Space group: A good space group is the one that allows the generation of full basic cell with the minimum number of atomic positions. For example, with a sample of three atomic species, the space group that allows generating the structure with only introducing three atomic positions is preferred.

Initial lattice parameter: A good estimation of the initial lattice parameter needs bibliography. Although it is not compulsory like those two previous parameters, it is very useful to begin the refinement with an acceptable starting point.

Atomic position: If symmetry and space group have been chosen correctly, the introduction of the atoms in their positions (in Cartesian unit coordinates) is straightforward.

Atomic occupancies: Atomic occupancies are very important in the cases of non-stoichiometric samples. If they are not introduced correctly, the theoretical pattern may be completely wrong.

When both experimental and theoretical diffraction patterns are available, the procedure of refinement can begin. Like with the creation of CIF files, the refinement procedure must follow some steps. Although every author has its own protocol [36, 37], some general rules need to be emphasized when performing such analysis. These steps will not be only useful for MAUD, but also for the other Rietveld refinement program. Recommended steps to perform the refinement are as followings:

- Background
- Cell parameters
- Incident radiation intensity
- Crystal size and strain
- Sample displacement
- Pseudo-Voigt parameters

In order to know if the refinement has been done properly, it is not enough to check visually whether the experimental and theoretical diffractograms match. There exist four refinement parameters [26] that give information about the accuracy of the refinement: *R*-structure factor (Equation 3), *R*-Bragg factor (Equation 4), *R*-pattern (Equation 5) and *R*-weighted pattern (Equation 6).

$$R_F = \frac{\sum_k \left| (I_k^E)^{\frac{1}{2}} - (I_k^C)^{\frac{1}{2}} \right|}{\sum_k (I_k^E)^{\frac{1}{2}}} \tag{3}$$

$$R_B = \left| \frac{\sum_k (I_k^E - I_k^C)}{\sum_k I_k^E} \right| \tag{4}$$

$$R_p = \left| \frac{\sum_{i=1}^N (I_i^E - I_i^C)}{\sum_{i=1}^N I_i^E} \right| \tag{5}$$

$$R_{wp} = \left[\frac{\sum_{i=1}^N \left[\frac{1}{(I_i^E)^{\frac{1}{2}}} (I_i^E - I_i^C) \right]^2}{\sum_{i=1}^N \left[\frac{1}{(I_i^E)^{\frac{1}{2}}} I_i^E \right]^2} \right]^{\frac{1}{2}} \tag{6}$$

R parameters of Equations 3 & 4 take into account the intensity of every Bragg reflection k, thus excluding the background. The *R* parameters of Equations 5 & 6 take into account every data point including the background. The most used is the R parameter of equation 6 (R_{wp}) which takes into account all data points and works with weighted intensities. The obtained values for this parameter are typically with R_{wp}~15% for monoclinic and triclinic structures, and with R_{wp}=10% for cubic structures.

The five samples with Rietveld analysis performed were $Mn_{50}Ni_{40}Sn_{10}$, $Mn_{50}Ni_{40}In_{10}$, $Ni_{50}Mn_{45}In_5$, $Ni_{50}Mn_{35.5}In_{14.5}$ and $Mn_{50}Ni_{42.5}Sn_{7.5}$, and their diffraction patterns were obtained using a CuK_α cathode. If another kind of radiation had been used (for example, Fe, Mo, etc.), diffraction peaks would have appeared in different angular positions.

Initial phase identification of the samples, which could not be done with MAUD, was performed using bibliography and comparing their diffractograms to a set of theoretical diffraction patterns. They have been created for arbitrary compositions and using a wavelength equal to that of the copper cathode. If another cathode radiation was used in the experiment, the wavelength in which the theoretical diffractograms were obtained should be changed accordingly to compare them with the experimental ones. Furthermore, depending on rapid solidification conditions, we may obtain different structures. For example, for austenite we can obtain $L2_1$ structure (slower cooling) or B2 structure (faster cooling). It will be interesting to build both structures for the same composition. Theoretical diffractograms have been obtained for austenite structures, $L2_1$ and B2, for a stoichiometric composition of $Mn_{50}Ni_{25}Sn_{25}$, exhibiting $L2_1$ structure pattern (Figure 9) with 16 peaks in the angular interval between 20° and 120°. Their Miller indices have been obtained using the MATLAB code *l21.m*. Lattice parameter has been obtained with the same code using a wavelength that takes into account the double wavelength of copper radiation in a factor of 2:1. The B2 structure, as mentioned in the introduction, corresponds to a loss of periodicity of Y- and Z-position atoms and because of that, it cannot be represented with the typical $L2_1$ cell, and thus a body centered cubic (bcc) structure must be used instead. This new pattern will lose the odd-indexed peaks of $L2_1$ and the remaining peaks' indices will be divided by two. The cell parameter is now the half, since B2 cell is one of the eight cells that built up $L2_1$ structure. Index and lattice parameter have been obtained with *b2.m* MATLAB code, with the same properties than *l21.m*, but specially made to be used in this work. It can be seen graphically in Figure 10.

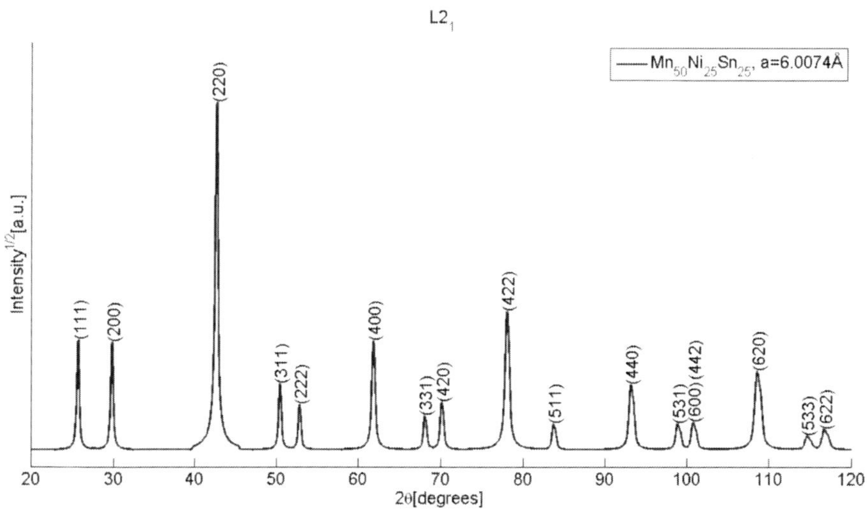

Figure 9. Theoretical diffraction pattern of a sample of austenite phase $L2_1$ structure with stoichiometric composition $Mn_{50}Ni_{25}Sn_{25}$.

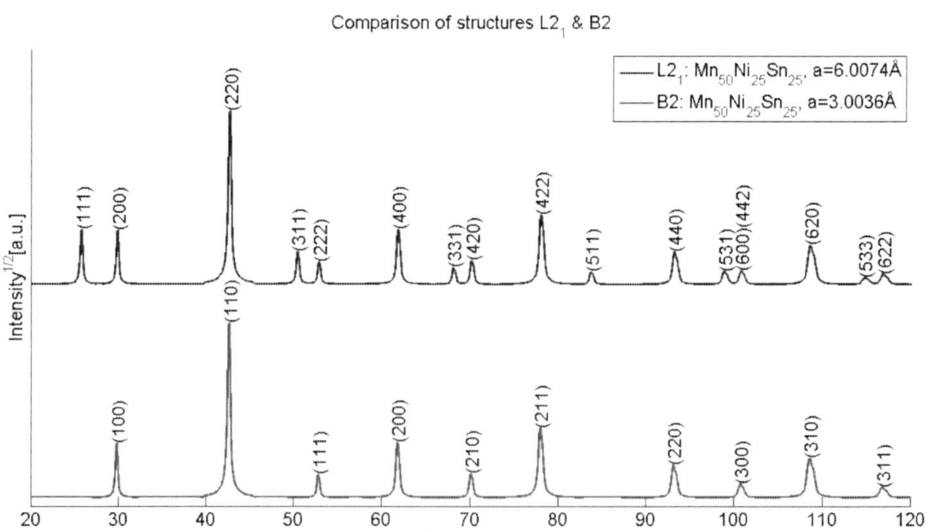

Figure 10. Comparison of theoretical diffraction patterns of two samples of austenite $L2_1$and B2 structures in blue and red respectively with stoichiometric composition $Mn_{50}Ni_{25}Sn_{25}$.

Whilst the austenite phase structures showed a diffractogram that match the theoretical pattern, it is not possible with tetragonal martensite $L1_0$. In austenite phase, only one lattice parameter was available. Decreasing and increasing this parameter made all the peaks move right or left, but the distance between them was fixed all the time. Therefore, the pattern was always the same. In the case of tetragonal $L1_0$, since there are two lattice parameters to be modified, the distance between peaks will not be constant. By changing one of the two lattice parameters, some peaks will move and some others will not. Therefore, it will never be possible to find a 'standard' pattern for this structure, because the peaks will have a different relative position depending on the cell parameters (same problem would be found in all non-cubic structures). Hence, what will be displayed here is a pattern with the typical features of tetragonal $L1_0$ structure of Heusler MSMA. As the aim of this diffraction pattern is to help identify structures, this will be sufficient. A theoretical diffractogram with off-stoichiometry composition $Ni_{50}Mn_{45}Sn_5$ has been created for $L1_0$ (Figure 11). The minor peaks in the diffractogram is due to the fact that the theoretical structure has been assumed to have an atomic thermal displacement factor, B2, equal to zero for simplification (although it never happens in real structures). The appearance of small peaks is due to the fact that the sample composition assigned to the theoretical structure L10 already had these peaks. It must be kept in mind that this is a theoretical diffraction pattern without background noise. If background noise was present, these small peaks would be completely invisible. Because of that, they have been omitted within the theoretical indexation work.

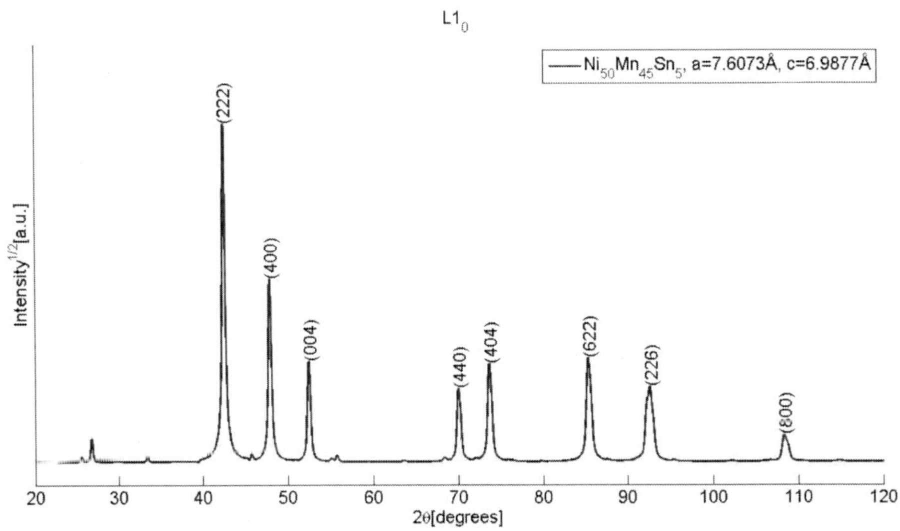

Figure 11. Theoretical diffraction pattern of a sample of martensite phase $L1_0$structure with non-stoichiometric composition $Ni_{50}Mn_{45}Sn_5$.

When B2 factor is increased, smaller peaks tend to be dissimulated because its structure factor, F_k, is reduced (Equation 7). On the other hand, as experimental data have the wide background, these previously dissimulated peaks at the end are practically invisible. As they do not appear in experimental data and cannot be obtained from the refinement software, they have been omitted when performing indexation work with literatures [19, 20]. Cell parameters were obtained with MATLAB *l10.m* code (similar to *l21.m* and *b2.m*) specially designed in this work.

$$|F_k|^2 = m_k \left|\sum_{n=1}^{N} f_n \exp\left(-B_n \frac{\sin^2\theta}{\lambda^2}\right) \exp[2\pi i(hx_n + ky_n + lz_n)]\right|^2 \quad (7)$$

In the case of modulated martensites, it is not possible to show a typical diffraction pattern because of the modulation of crystal structure. Depending on the degree and the form of this modulation, the satellite peaks will appear in different angular positions, and it is not possible to predict where they will be present.

If the structure, orthorhombic or monoclinic, was not modulated, it would be possible to obtain a theoretical diffractogram exhibiting its main features. For example, a typical diffractogram of a non-modulated orthorhombic structure of a Heusler MSMA (Ni_2MnGa) is shown in Figure 12. Its indexation has been performed using the literatures [38], and the diffraction pattern exhibits three peaks instead of one for each of the three main reflections showed by cubic $L2_1$ structure ((220), (400) & (422)). Minor peaks in the diffractogram may significantly be increased with another stoichiometric [39] or non-stoichiometric [38, 40] compositions of the sample. If there was a high thermal B factor, they would completely disappear. In case of being modulated, the additional peaks would appear. With a lower degree of modulation, these additional satellite peaks would appear mainly around the angular positions 40º~45º, adding more peaks to the previous three produced by the true crystal reflections. If the structure was monoclinic rather than orthorhombic, the fourth peak would appear next to the three shown in Figure 12, due to the existence of a-β angle different than 90º. If it was modulated, additional satellite peaks would also appear in random positions, depending on the degree of modulation.

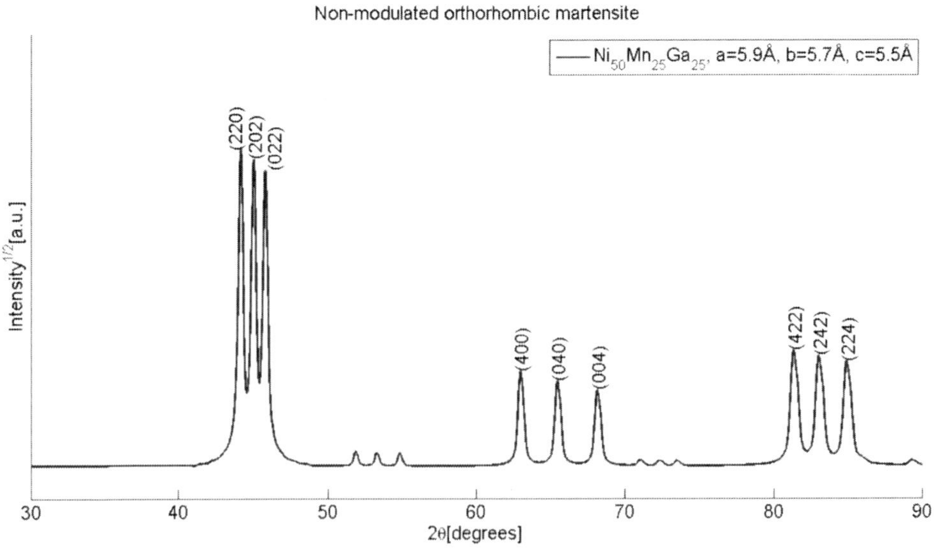

Figure 12. Theoretical diffraction pattern of martensite phase with orthorhombic non-modulated structure and stoichiometric composition of $Ni_{50}Mn_{25}Ga_{25}$.

ANALYSIS AND RESULTS

$Mn_{50}Ni_{40}Sn_{10}$

The first studied sample is an austenite $L2_1$ structure. Its identification can be performed by checking bibliography and also by comparing with the experimental diffractogram of the sample (red crosses in Figure 13). The theoretically generated $L2_1$ structure is displayed in Figure 9. They both share the same peaks in the 27°~110° interval. The fact that odd-indexed peaks of the theoretical diffractogram exhibit higher intensity is due to the fact that the theoretical one is assumed to have a stoichiometric composition. When the degree of stoichiometry is decreased, the intensity of these odd-indexed peaks decreases too. Irrespectively of this fact, there's no doubt that the sample exhibits a $L2_1$ structure. Following the steps given in the previous section, a CIF file for this sample has been generated with its basic parameters given in Table 1.

Table 1. Basic parameters for the CIF file of a sample with non-stoichiometric composition $Mn_{50}Ni_{40}Sn_{10}$

Crystal system: Cubic Space group: $Fm\bar{3}m$ Unit cell parameters: $a=b=c=6$Å,$\alpha=\beta=\gamma=90°$,				
atom	x	y	z	Occup.
Mn1	1/4	1/4	1/4	1.0
Ni1	1/2	1/2	1/2	1.0
Sn1	0	0	0	0.4
Ni2	0	0	0	0.6

After the refinement, it can be seen that the theoretical function fits the experimental data except the height of some peaks (Figure 13). This mismatch in peak height is due to the texture of the sample. The samples under study are not in bulk form, but in ribbon shape. Therefore, these ribbon shape samples exhibit a columnar distribution of the crystallites as can be seen in the images of Figure 14. Preferred orientation can be founded by SEM images included in this work. In this type of materials, columnar grains are usually associated to the texture.

This columnar distribution favors some directions and it causes the peaks to have a different height (intensity) than the one they would have in bulk or powder form. The effects of this texture could be introduced into MAUD and use them in the refinement, if experimental data taken from different incident angles was available. Current diffractometer does not support this feature and, therefore, texture analysis could not be performed for estimating the degree of preferred orientation.

Final refinement parameter obtained for the sample of this work was with R_{wp}= 16.6%, and this is acceptable by taking into account the problem of preferred orientations. Cell parameter obtained after refinement was a = 5.98898(5)Å which is very similar to the initial estimation when creating the CIF file and also to the values found in literatures [19-21]. In addition, an overall thermal factor, which indicates the area where atoms move owing to thermal excitation B = 0.34(2)$Å^2$, was also obtained.

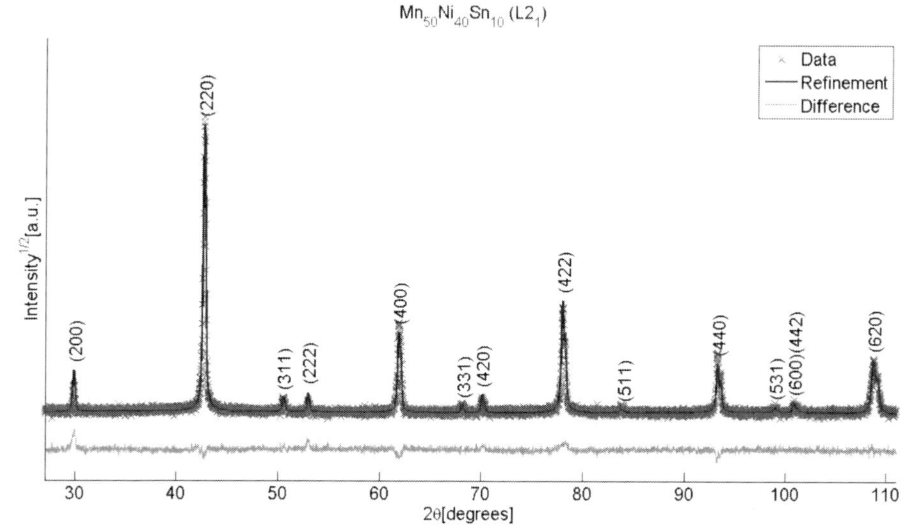

Figure 13. The diffractogram of austenite phase L21 structure with non-stoichiometric composition $Mn_{50}Ni_{40}Sn_{10}$. The experimental data, the refined data and their difference are presented in red crosses, blue and green lines, respectively.

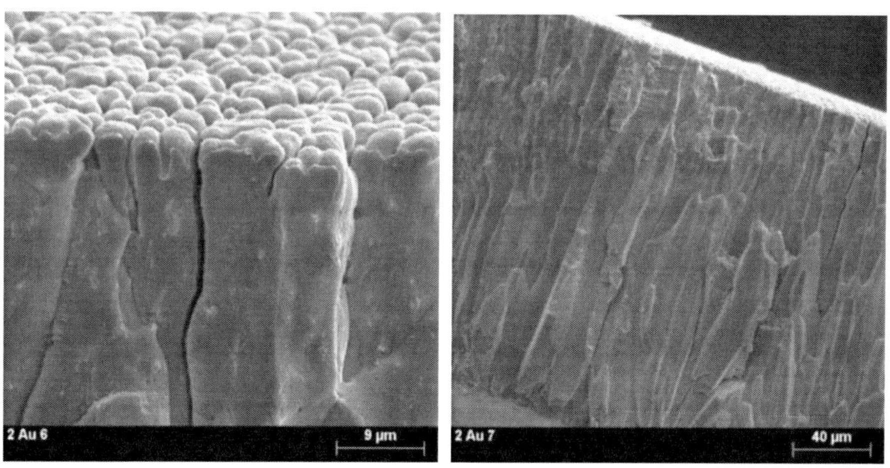

Figure 14. SEM images of a ribbon shape sample from the lateral (left) and the surface-lateral view (right).

$Mn_{50}Ni_{40}In_{10}$

The identification of this second sample has been performed in the same way as the previous one, by comparing its experimental data (Figure 15) with the theoretical patterns. It resulted in a good fit to the characteristics of the B2 structure whose diffractogram is displayed Figure 10 to compare with $L2_1$. As mentioned in the previous section, the only difference between the diffraction patterns of B2 and $L2_1$ is the missing of the odd-indexed peaks in the previous one.

The parameters for generating the CIF file for the sample currently present in this work is shown in Table 2. The refined data is displayed in Figure 15.

Table 2. Basic parameters for the generation of CIF file with non-stoichiometric composition of $Mn_{50}Ni_{40}In_{10}$

Crystal system: Cubic Space group: $Pm\bar{3}m$ Unit cell parameters: $a=b=c=3$Å,$\alpha=\beta=\gamma=90°$,				
atom	x	y	z	Occup.
Mn1	0	0	0	1.0
Ni1	1/2	1/2	1/2	0.8
In1	1/2	1/2	1/2	0.2

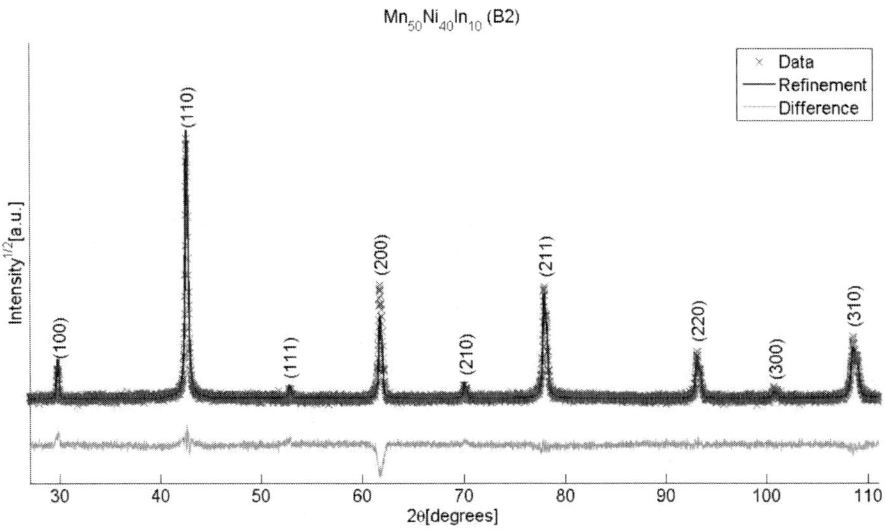

Figure 15. Diffractogram of austenite phase B2 structure with non-stoichiometric composition $Mn_{50}Ni_{40}In_{10}$.The experimental data, the refined data and their difference are presented in red crosses, blue and green lines, respectively.

In general, the fitting is not ideal as it was in the previous sample due to the fact that the present sample exhibits a higher degree of preferred orientation. It can be clearly seen with the lack of intensity of the peak indexed as (200) in the refined data. The angular position of this peak corresponds to the index of (400) in the $L2_1$ structure. Some authors also reported the same influence of this preferred orientation [20].

Final refinement parameter obtained for this sample was with $R_{wp} = 17.5\%$, and this is a little worse than the one obtained with in previous sample. The obtained cell parameter was with a = 2.99826(3)Å and this result agrees with the literatures [19, 20]. Because of the high degree of preferred orientation, the program gave a value for the thermal B factor smaller than zero and this has no physical meaning. Therefore, it was better to fix this value to zero in order to keep it realistic.

If there were no preferred orientation or it was only in a low degree, this problem could have been avoided.

$Ni_{50}Mn_{45}In_5$

As shown in Figure 16, the diffractogram is very different compared to the previous ones. In this case, it is a martensite phase structure, and can be further identified as a martensite $L1_0$, due to the peak positions with those in Figure 11. Structural information for generating the CIF file is provided in Table 3.

Table 3. Basic parameters for the generation of the CIF file with non-stoichiometric composition $Ni_{50}Mn_{45}In_5$ in the form of a $L1_0$ tetragonal cell

Crystal system: Tetragonal Space group: *I4/mmm* Unit cell parameters: $a = b = 3.8$ Å, $c = 7$ Å, $\alpha = \beta = \gamma = 90°$,				
atom	*x*	*y*	*z*	Occup.
Ni1	1/2	0	1/4	1.0
Mn1	0	0	0	1.0
In1	1/2	1/2	0	0.8
Mn2	1/2	1/2	0	0.2

Analysis shown in Figure 16 exhibits the large difference between experimental and theoretical data. It is caused by the background of the experimental data. However, although it is with low quality data, the refinement parameter is obtained with $R_{wp} = 17.7\%$, and it is not as bad as seen in the previous samples.

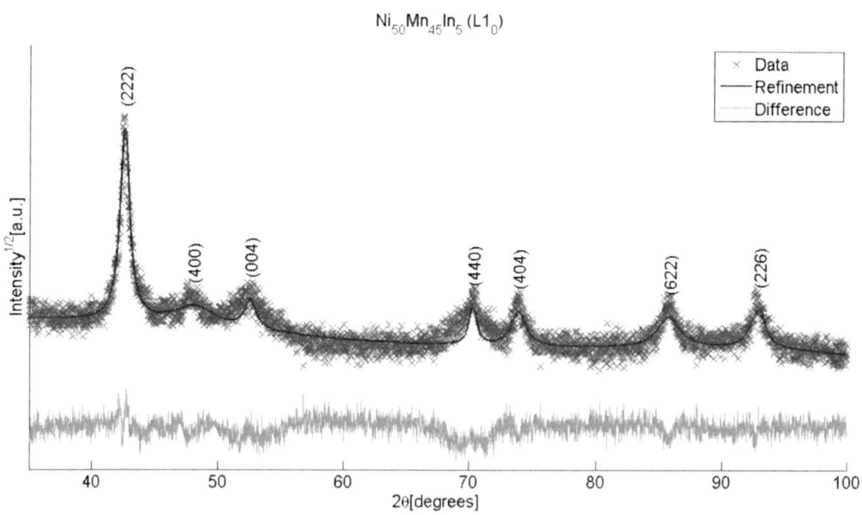

Figure 16. Diffractogram of martensite phase $L1_0$ structure with non-stoichiometric composition $Ni_{50}Mn_{45}In_5$. The experimental data, the refined data and their difference are presented in red crosses, blue and green lines, respectively.

The refined cell parameters for this sample are with a = 3.7826(8)Å and c = 6.963(2)Å and both agree with the literatures. It must be noticed that some authors [19, 20] report a lattice parameter which is the double than the one presented here. It is due to the fact that they use a bigger basic cell, but the resulted diffraction patterns are the same. The same problem with B factor found in the previous analysis was also present in this sample, so it was similarly fixed as zero. The full width at half maximum (FWHM) was much broader than that of other samples. This effect can be caused by lower crystallite size in this alloy and also the higher microstrain usually found in martensite structures.

$Ni_{50}Mn_{35.5}In_{14.5}$

The crystal structure of this sample (Figure 17) corresponds to a structure with a low degree of modulation [41], because there are some extra peaks in the zone around 40° which do not appear in any non-modulated structure. The non-matching peaks are some satellite peaks, and these peaks are caused by the modulation of the structure. The structure is a 5M ($10M$) orthorhombic modulated structure rather than monoclinic. No match was found when comparing the data with the typical five-layered structure composed of orthorhombic $L1_0$-style cells (Table 4).

Table 4. Basic parameters for generating of the CIF file with non-stoichiometric composition $Ni_{50}Mn_{35.5}In_{14.5}$ in the form of a five-layered stack of orthorhombic $L1_0$-style basic cells

Crystal system: Orthorhombic Space group: *Immm* Unit cell parameters: a =4.3Å, b=5.9Å, c=21 Å, $\alpha = \beta = \gamma$ =90°,									
atom	x	y	z	Occup.	atom	x	y	Z	Occup.
Ni1	1/2	1/4	0	1.0	In1	0	1/2	0	0.58
Ni2	1/2	1/4	1/5	1.0	In2	0	1/2	1/5	0.58
Ni3	1/2	1/4	2/5	1.0	In3	1	1/2	2/5	0.58
Mn1	0	0	0	1.0	Mn4	0	1/2	0	0.42
Mn2	0	0	1/5	1.0	Mn5	0	1/2	1/5	0.42
Mn3	0	0	2/5	1.0	Mn6	0	1/2	2/5	0.42

The cause of this unsuccessful matching was due to the incapability of generating a modulation. Therefore, more peaks were in the experimental results and the attempt of only matching those peaks generated by the crystal structure was carried out. To simplify this work, we worked with the orthorhombic $L1_0$ cell with the c parameter being one fifth of that of the five-layered stack, rather than with the five-layered stack itself. It is because there are less parameters to control and the results will be exactly the same. The basic parameters of this structure are provided in Table 5.

When experimental data and the pattern generated by this simplified cell, it was found that the excess of experimental peaks confirmed the existence of modulation. The main problem is due to lack of knowledge in true crystal reflections and the uncertainty of satellite reflections caused by the modulation, especially for the peaks around 40°~45°. A solution to this problem could not be found by the present structure, and another possibility, which has been reported by some authors [7, 38], was also tested. It can use an orthorhombic variation of the $L2_1$ structure, whose basic parameters are in Table 6, instead of the typical $L1_0$-style cell. With this new strategy, it is possible to match the true reflections when realizing that the positions of three peaks between 55°~70° were only modified by one lattice parameter.

Therefore, they were used to lock the rest of the peaks and this can be seen in Figure 17 with acceptable result. All peaks (except three in the central zone (40º~45º) and one around 75º) were generally matched with cell parameters of a = 5.6065(7)Å, b = 5.9935(8)Å and c = 6.3454(9)Å. These are similar to those reported by Wang et al. [40]. The non-matched four peaks were mainly caused by the modulation of the crystal structure.

Table 5. Basic parameters for the generation of the CIF file with non-stoichiometric composition $Ni_{50}Mn_{35.5}In_{14.5}$ in the form of an orthorhombic $L1_0$-style basic cell

Crystal system: Orthorhombic Space group: *Immm* Unit cell parameters: *a* =4.3Å, *b*=5.9Å, *c*=4.2Å, $\alpha = \beta = \gamma$ =90°,				
atom	*x*	*y*	*z*	Occup.
Ni1	1/2	1/4	0	1.0
Mn1	0	0	0	1.0
In1	0	1/2	0	0.58
Mn2	0	1/2	0	0.42

Table 6. Basic parameters for generating the CIF file with non-stoichiometric composition $Ni_{50}Mn_{35.5}In_{14.5}$ in the form of an orthorhombic $L2_1$-style basic cell

Crystal system: Orthorhombic Space group: *Fmmm* Unit cell parameters: *a* =6.3Å, *b*=6.1Å, *c*=5.9Å, $\alpha = \beta = \gamma$=90°,				
atom	*x*	*y*	*z*	Occup.
Ni1	1/4	1/4	1/4	1.0
Mn1	1/2	1/2	1/2	1.0
In1	0	0	0	0.58
Mn2	0	0	0	0.42

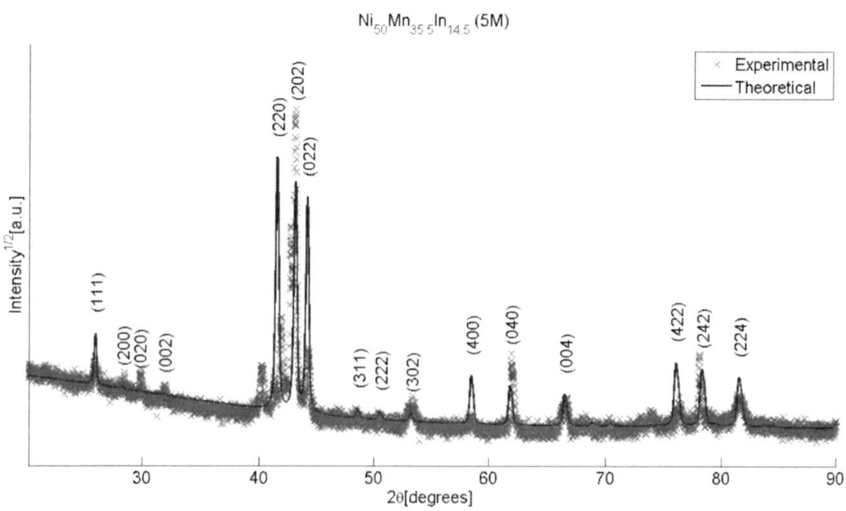

Figure 17. Experimental diffractogram (red) of the five-layered modulated structure (5M or 10M) of martensite phase with non-stoichiometric composition $Ni_{50}Mn_{35.5}In_{14}$, and the overlapped theoretical diffractogram (blue) of an orthorhombic structure of the sample non-stoichiometric composition.

Since there were four peaks not matching the result, it was not possible to carry out the refinement analysis. In these cases, the program may try to match all experimental peaks by interpreting two nearby peaks as one and such operation is not correct.

The reflections of crystal structure have been indexed using literatures [39, 40, 42] and it can be seen that the peaks that within $L2_1$ structure were indexed as (200), (220), (400) and (422) in the diffractogram. Due to the new orthorhombic symmetry, it allows lattice parameters of $a \neq b \neq c$.

$Mn_{50}Ni_{42.5}Sn_{7.5}$

The experimental diffractogram of this last sample is shown Figure 18, and it has been identified by comparing with literatures as a 14M (7M) modulated structure with monoclinic symmetry. This is not the orthorhombic symmetry as in the case of previous samples. In this case, the aim was to match the theoretical peaks generated by the CIF file with the experimental peaks produced by the sample to leave the unmatched peaks for modulation effect. Firstly, the intention was to match a seven-layered structure whose basic properties are displayed in Table 7. However, like in the previous subsections, the software cannot work with modulated structures. A simpler monoclinic $L1_0$-style cell with a c lattice parameter being one seventh of that of the seven-layered structurecould also be obtained. This is with an advantage that there will be a smaller number of parameters to control, and its properties are listed in Table 8.

Table 7. Basic parameters for generating the CIF file for non-stoichiometric composition $Mn_{50}Ni_{42.5}Sn_{7.5}$ in the form of a seven-layered stack of monoclinic $L1_0$-style basic cells

Crystal system: Monoclinic Space group:*C2/m*:b3														
Unit cell parameters: a =4.3Å, b=5.8Å, c=31 Å,α =γ=90°, β =94°,														
atom	*x*	*y*	*z*	Occup.	atom	*x*	*y*	*z*	Occup.	atom	*x*	*y*	Z	Occup.
Mn1	1/2	1/4	0	1	Ni3	0	0	2/7	1	Mn5	0	1/2	0	0.7
Mn2	1/2	1/4	1/7	1	Ni4	0	0	3/7	1	Mn6	0	1/2	1/7	0.7
Mn3	1/2	1/4	2/7	1	Sn1	0	1/2	0	0.3	Mn7	0	1/2	2/7	0.7
Mn4	1/2	1/4	3/7	1	Sn2	0	1/2	1/7	0.3	Mn8	0	1/2	3/7	0.7
Ni1	0	0	0	1	Sn3	0	1/2	2/7	0.3					
Ni2	0	0	1/7	1	Sn4	0	1/2	3/7	0.3					

When trying to match with this monoclinic $L1_0$-style cell, same problem of the orthorhombic 5M was found: there was a lack of knowledge on which peaks were produced by crystallographic structure and which by the modulation effect.

Unlike the previous sample, this problem could not be solved trying an $L2_1$-style basic cell with monoclinic symmetry. The fact of being monoclinic and, therefore, having four independent lattice parameters made it impossible to match the crystal reflections of this sample. Such challenge is the goal of our future work, and similar structures can be also found in the other references [19, 20].

Table 8. Basic parameters for generating the CIF file for non-stoichiometric composition $Mn_{50}Ni_{42.5}Sn_{7.5}$ in the form of a monoclinic $L1_0$-style basic cell

Crystal system: Monoclinic Space group: *C2/m*:b$_3$ Unit cell parameters: a =6.3Å, b=6.1Å, c=5.9Å, $\alpha = \beta = \gamma$=90°,				
atom	*x*	*y*	*z*	Occup.
Mn1	1/2	1/4	0	1.0
Ni1	0	0	0	1.0
Sn1	0	1/2	0	0.3
Ni2	0	1/2	0	0.7

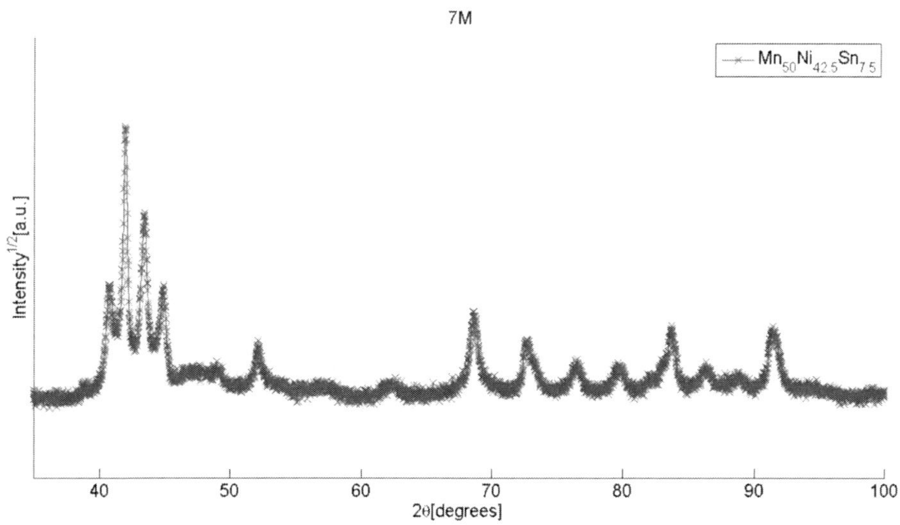

Figure 18. Experimental diffractogram of a seven-layered modulated structure (7M or 14M) of martensite phase with non-stoichiometric composition $Mn_{50}Ni_{42.5}Sn_{7.5}$.

Overall, MAUD has demonstrated its good capability in generating CIF files for both austenite phase structures ($L2_1$ and B2) and the non-modulated martensite phase structure ($L1_0$). For the modulated martensite phase structures (5M and 7M), it was found that MAUD does not allow the generation of non-symmetric structures. As the program works with symmetry systems, when one atom is located in the cell by the user, the other atoms of the same kind are automatically located using the chosen symmetry. This means that their positions cannot be controlled, and only the positions of atoms generating these symmetric positions can be located. Such situation makes it very difficult to generate a modulation effect, because only certain atoms positions can be shifted, while other atoms (the majority) will change their position accordingly. Because of this major disadvantage, it was not possible to generate the $(5\bar{2})_2$ and $(3\bar{2})_2$ stacking sequences exhibited by 7M and 5M structures, and neither the 'perfect' modulation. This problem could be resolved if all atomic positions were introduced individually and not by symmetry. Although it was not possible to generate the modulated structures, MAUD has been very useful to distinguish the true reflections from the satellite peaks for orthorhombic modulated structure 5M. However, up to this moment, it still is not for the monoclinic 7M structure shown in this chapter.

In the refinement analysis, although once can only perform with samples whose CIF files generated a pattern that matched the experimental peaks, MAUD still demonstrated a very powerful application. Good results were obtained when refining both structures of the austenite phase in subsections A and B, although better results could have been obtained, if the experimental data collected from different angular positions was available. To observe the degree of preferred orientation, the crystallographic direction (200) found in the B2 structure can be an example as shown in Figure 15. Excluding the effects of the preferred orientations, which cannot be introduced in the CIF file, the refinement demonstrated that CIF files for $L2_1$ and B2 structures were correctly implemented. In martensite phase, only the $L1_0$ structure of subsection C could be refined because it was the only one with all peaks matched. Although this refinement could not be performed as successfully as the previous two cases, due to the lack of quality data, acceptable results were still obtained. This demonstrated that, just like for austenites, the CIF file could be correctly generated for $L1_0$.

Refinement of modulated structures in subsections D and E could not be performed, because the patterns created through their CIF files did not generate the satellite peaks caused by the modulation. If they could be generated by introducing all atomic positions individually, the present modulation can be commensurate and it means that it fulfills a complete cycle within the cell. By refining the spatial positions of all the atoms of the 10 or 14 stacks individually, the matching of all the peaks could be achieved and refinement could be performed. However, this is not a straightforward task to perform. On the other hand, if the modulation was incommensurate (the modulation cycle exceeds the basic cell), this method would not be useful. It is a major disadvantage of X-ray diffraction for MSMA modulated structures, because the major part of them exhibits the incommensurate modulation [9]. There exists an alternative method to deal with this kind of modulations, and that can be also applied to commensurate ones. This is the super-space approach [43, 44]. In this approach, the added fourth dimension to the structure allows taking into account the modulation, regardless to be commensurate or incommensurate. In super-space approach, Miller indices are four: (hklm) instead of the typical three: (hkl). Here, m index stands for the satellite peaks. For example, the satellite peaks related to the true reflection (210) will be named "(210m)", where m is the number assigned to each one of its satellite peaks. Recommended software that can work with the super-space approach is Jana (any version), which is a program specialized in the refinement of modulated structures.

CONCLUSION

Rietveld method is a good approach to characterize austenite and martensite phases of magnetic shape memory alloys. The software used in this chapter is MAUD, which has demonstrated its very powerful capability for both structure refinement and CIF file generation. The feature that allows changing every parameter manually and seeing its real-time operation status is very useful. In the preliminary steps of the refinement, it is better to be done manually rather than automatically by thc program. The option of the program that allows users to refine multiple parameters with only one click is very useful and indicates the future design trend. The way MAUD allows to create CIF files from the scratch makes it easier and faster to generate CIF files, comparing to check the crystal databases for similar samples and to modify them for the needs.

The disadvantage of MAUD is mainly on its lack of crystal cell and phase determination capability. However, it is compensated by more advantages in its refinement procedure. The use of the programs FullProf and X'PertHighScore is recommended for crystal cell and phase determination, respectively. Taking the generated CIF files by MAUD and executing the refinement program, one can often find the good fitting results. The only main drawback for MAUD is that it does not allow working with the modulated structures. Under such condition, the Jana program is often employed instead.

REFERENCES

[1] Marioni, M.A.; O'Handley, R.C.; Allen, S.M.; Hall, S.R.; Paul, D.I.; Richard, M.L.; Feuchtwanger, J.; Peterson B.W.; Chambers, J.M.; Techapiesancharoenkij, *R. J. Magn. Magn. Mater.* 2005, 290, 35.

[2] Brown, G.V. *J. Appl. Phys.*1976, 47, 3673.

[3] Pecharsky V.K.; Gschneider, K.A. *J. Int. J. Refrig.*2006, 29, 1239.

[4] Heusler, F.; Stark, W.; Haupt, E. V. Deut. *Phys. Ges.*1903, 5, 229.

[5] Heusler, F.; Richarz, F. Z. *Anorg. Chem.*1909, 61, 265.

[6] Bradley, A.J.; Rodgers, J.W. Proc. R. soc. Lond. Ser. A-Contain. *Pap. Math. Phys. Character,* 1934, 144, 0340.

[7] Ranjan, R.; Banik, S.; Barman, S.R.; Kumar, U.; Mukhopadhyay, P.K.; Pandey, D. *Phys. Rev. B*, 2006, 74, 224433.

[8] Righi, L.; Albertini, F.; Villa, E.; Paoluzi, A.; Calestani, G.; Chernenko, V.; Besseghini, S.; Ritter, C.; Passaretti, F. *Acta Mater.*2008, 56, 4529.

[9] Righi, L.; Lazpita, P.; Gutierrez, J.; Barandiaran, J.M.; Chernenko V.A.; Calestani, G. *Scr. Mater.*2010, 62, 383.

[10] Kaufmann, S.; Roessler, U.K.; Heczko, O.; Wuttig, M.; Buschbeck, J.; Schultz, L.; Faehler, S.; *Phys. Rev. Lett.*2010, 104, 145702.

[11] Pons, J.; Santamarta, R.; Chemenko V.A.; Cesari, E. *Mater. Sci. Eng. A-Struct. Mater. Prop. Microstruct. Process.*2006, 438, 931.

[12] Hamilton, R.F.; Sehitoglu, C.; Efstathiou, H.; Maier, H.J. *Acta Mater.*2007, 55, 4867.

[13] Acet, M.; Mañosa, L.; Planes, A. Chapter Four - *Magnetic-Field-Induced Effects in Martensitic Heusler-Based Magnetic Shape Memory Alloys*, Ed. Elsevier, UK,2011, Vol. 19, p. 231.

[14] Planes, A. *Physics* 2010, 3, 36.

[15] Righi, L.; Albertini, F.; Calestani, G.; Pareti, L.; Paoluzi, A.; Ritter, C.; Algarabel, P.A.; Morellon, L.; Ibarra, M.R. J. *Solid State Chem.*2006, 179, 3525.

[16] Righi, L.; Albertini, F.; Pareti, L.; Paoluzi, A.; Calestani, G. Acta Mater.2007, 55, 5237.

[17] Mogylnyy, G.; Glavatskyy, I.; Glavatska, N.; Soderberg, O.; Ge, Y.; Lindroos, V.K. *Scr. Mater.*2003, 48, 1427.

[18] Ullakko, K.; Huang, J.K.; Kantner, C.; O Handley, R.C.; Kokorin, V.V.; *Appl. Phys. Lett.*1996, 69, 1966.

[19] Krenke, T.; Acet, M.; Wassermann, E.F.; Moya, X.; Manosa, L.; Planes, *A. Phys. Rev.* B2005, 72, 014412.
[20] Krenke, T.; Acet, M.; Wassermann, E.F.; Moya, X.; Manosa, L.; Planes, *A. Phys. Rev.* B2006, 73, 174413.
[21] Coll, R.; Escoda, L.; Saurina, J.; Sanchez-Llamazares, J.L.; Hernando, B.; Suñol, J.J. *J. Therm. Anal. Calorim.*2010, 99, 905.
[22] Sanchez-Llamazares, J.L.; Hernando, B.; Suñol, J.J.; Garcia, C.; Ross, C.A.; *J. Appl. Phys.*2010, 107, 09A956.
[23] Hernando, B.; Sanchez-Llamazares, J.L.; Santos, J.D.; Escoda, L.; Suñol, J.J.; Varga, R.; Baldomir, D.; Serantes, D. *Appl. Phys. Lett.*2008, 92, 042504.
[24] Sanchez-Llamazares, J.L.; Sanchez, T.; Santos, J:D.; Perez, M.J.; Sanchez, M.L.; Hernando, B.; Escoda, L.; Suñol, J.J.; Varga, R.; Appl. Phys. Lett.2008, 92, 012513.
[25] Rietveld, H.M. *J. Appl. Crystallogr.*1969, 2, 65.
[26] Young, R.A. The Rietveld method, Oxford University Press, New York, NY, 1993, p. 312.
[27] Roisnel, T.; Rodriguez-Carvajal, J. Epdic 7: European Powder Diffraction, Pts 1 and 2, 2001, 378-3, 118.
[28] Rodriguez-Carvajal, *J. Physica* B1993, 192, 55.
[29] Dusek, M.; Petricek, V.; Wunschel, M.; Dinnebier R.E., van Smaalen, S. *J. Appl. Crystallogr.*2001, 34, 398.
[30] Lutterotti, L.; Matthies, S.; Wenk, H.R.; Schultz, A.S.; Richardson, J.W. *J. Appl. Phys.*1997, 81, 594.
[31] Lutterotti, L.; Chateigner, D.; Ferrari, S.; Ricote, *J. Thin Solid Films*2004, 450, 34.
[32] Lutterotti, L. *Nucl. Instrum. Methods Phys. Res. Sect. B-Beam Interact. Mater. Atoms* 2010, 268, 334.
[33] Bragg, W.L.; Thomson, J.*J. Proc. Canbridge Philos. Soc.* 1914, 17, 43.
[34] Kittel, C, *Introduction to solid state physics,* Wiley, NY, 2005, p. 680.
[35] Ashcroft N.W.; Mermin, N.D. *Solid state physics*, Saunders College, Philadelphia PA. 1988, p. 826.
[36] McCusker, L.B.; Von Dreele, R.B.; Cox, D.E.; Louer, D.; Scardi, P. *J. Appl. Crystallogr.*1999, 32, 36.
[37] Giacovazzo, C.; Monaco, H.; Artioli, G.; Viterbo, D.; Ferraris, G.; Gilli, G.; Zanotti, G.; Catti, M. Fundamentals of crystallography, Oxford University Press, New York, NY, 1992, p. 674.
[38] Santos, J.D.; Sanchez, T.; Alvarez, P.; Sanchez, M.L.; Sanchez-Llamazares, J.L.; Hernando, B.; Escoda, L.; Suñol, J.J.; Varga, R. *J. Appl. Phys.*2008, 103, 07B326.
[39] Liu, Z.H.; Zhang, M.; Cui, Y:T.; Zhou, Y.Q.; Wang, W.H.; Wu, G.H.; Zhang, X.X.; Xiao, G. *Appl. Phys. Lett.*2003, 82, 424.
[40] Wang, W.H.; Ren, X.; Wu, G.H.; *Phys. Rev.* B2006, 73, 092101.
[41] Khan, M.; Pathak, A.K.; Paudel, M.R.; Dubenko, I.; Stadler, S.; Ali, N. *J. Magn. Magn. Mater.* 2008, 320, 21.
[42] Sun, X.; Song, J.; Zhang, X.; Xie, C. *J. Alloys Compounds* 2011, 509, 4868.
[43] van Smaalen, S. Z. *Kristall.* 2004, 219, 681.
[44] van Smaalen, S. Incommensurate crystallography, Oxford University Press, New York, NY, 2007, p. 288.

In: X-Ray Diffraction: Structure, Principles and Applications ISBN: 978-1-62808-591-4
Editor: Kaimin Shih

Chapter 4

CRYSTAL STRUCTURES OF NEW RARE EARTH INTERMETALLIC COMPOUNDS

Wei He*

Key Laboratory of Nonferrous Metal Materials and New Processing Technology, Ministry of Education (Guangxi University), Nanning, China
College of Materials Science and Engineering, Guangxi University, Nanning, China

ABSTRACT

The chapter provides the crystal structures of some novel intermetallic compounds determined by X-ray powder diffraction techniques and reports the crystallographic data of these new intermetallic compounds obtained from the Rietveld refinement. The chapter also illustrates how to identify the existence of a new rare earth compound during the research of alloy phase diagrams, and the sample preparations for the crystal structure determination involving new compounds. It is very important to obtain accurate lattice parameters for the determination of a crystal structure. In order to illustrate how to obtain accurate lattice parameters of the new compounds, this chapter includes detailed experimental methods, and also describes the determination of the lattice thermal expansion properties of intermetallic compounds from high temperature X-ray diffraction (HTXRD). The chapter will also include the lattice thermal expansion properties of some novel compounds in the systems of rare earth-transition metals.

It is well known that rare earth (RE) alloys have excellent properties and there exist hundred thousand intermetallic compounds in the systems of rare earth-transition metals. There are compounds with excellent properties among them, such as the famous magnetocaloric materials $Gd_5Ge_2Si_2$, $La(FeCo)_xSi_{13-x}$ and hard magnetic garnet $Nd_2Fe_{14}B$, etc. [1, 2]. Properties of rare earth intermetallic compounds are closely related to their crystal structure. X-ray powder diffraction technique is one of the important methods for determining crystal structure, especially when the single crystal is difficult to grow. In recent decays, numerous new compounds in the systems of rare earth-transition metals are discovered and their crystal structures are determined from X-ray powder

* Corresponding author: Tel: +86-771-3275918. E-mail address: wei_he@gxu.edu.cn.

diffraction. This chapter introduces structures of about 20 new ternary rare earth compounds found in the systems of rare earth-transition metals.

1. Discovery of New Compounds

Novel intermetallic compounds are often found during the investigations of alloy phase diagrams. In order to illustrate the method for discovering new compound, here we take a new ternary intermetallic compound discovered in a rare earth-transition metal ternary system during the study on the phase diagram of the system as examples. There exists a new ternary intermetalllic compound in a ternary alloy system at a certain temperature (in a certain isothermal section of a ternary phase diagram). When one analyzes the X-ray diffraction patterns of the alloys with compositions near the new compound, he/she would observe one extra set of X-ray diffraction pattern in the alloy samples, except those patterns belonging to the known compounds with compositions around the new compounds. The most importance is that this extra set of X-ray diffraction pattern existing in all alloys with compositions close to the new compound. This set of X-ray diffraction pattern cannot be explained by using the patterns of the known binary or ternary compounds in the system after taking off the influence of metal oxide and/or the possibility of introducing other impurity phases. In other words, the observed new set of X-ray diffraction pattern neither can be explained by the patterns of known binary or ternary compounds nor metal oxides or impurity phases introduced during the samples preparation. Normally, the patterns of metal oxides are very easy to be excluded from those of the alloy samples. For the case of introducing impurity phases in the samples, the new set pattern are only observed in some alloy samples but not in all ternary alloy samples with the compositions near the new compound, since the introducing of the impurity phase is occurred by accident. According to the Gibbs phase rule, only the X-ray diffraction patterns of two compounds are observed in the patterns of the equilibrium sample after taking off the new pattern set, and at the same time the sample's composition is located in a three-phase region of the phase diagram at a certain temperature (i.e. a three-phase region in an isothermal section). Then, the new set pattern should be considered to assign to a new compound. Furthermore, if the new pattern set can be index successfully, then there is no doubt that this new pattern set stands for a new compound in the investigated alloy system. There are some other methods to discover novel intermetallic compounds. For example, there are two structure types of compound DyAlSi, i.e., the orthorhombic YAlGe-type structure with space group of Cmcm (No. 63) and the tetragonal α-$ThSi_2$-type structure with space group of $I4_1/amd$ (No. 141). The tetragonal phase is the high temperature phase and was found in an annealing DyAlSi sample [3].

After confirming the existence of a new compound by X-ray diffraction method, one should try to find out the exact composition of the new compound. Usually, the exact composition of the new compound can be found out by using the scanning electronic microscopy with the aid of emission of X-ray spectrum (SEM/EDS) or the microprobe material analysis (EPMA). According to the obtained composition, new sample with the single phase of the new compound can be prepared. In general case, it is very difficult to grow single crystal of rare earth intermetallic compound, and the crystal structures of new rare earth intermetallic compounds can often be determined by X-ray powder diffraction method.

2. Experimental Methods

The alloy samples of new compounds can be prepared by arc-melting the initial materials of at least of 99.9 wt.% purity under high-purity argon atmosphere. Each sample should be turned over and remelted three times to ensure good homogeneity, and the weight loss after melting should be less than 0.5%. In order to obtain samples with homogeneous microstructure and composition as well as well crystallization, homogenizing annealing at appropriate temperature is necessary. The alloy samples are wrapped with a Ta foil in an evacuated quartz tube and the homogenizing annealing is carried out. Then, the sample was quenched into liquid nitrogen. Each sample is ground in an agate mortar and pestle to a particle size of less than 10μm and annealed in an evacuated small quartz tube at appropriate temperature for 5 hours to eliminate stress, then quenched into liquid nitrogen.

The composition and the phases of each sample are examined by means of scanning electron microscopy with energy dispersive X-ray spectroscopy using ZAF correction before X-ray diffraction. Only the samples with single phase can be used for structure determination. Occasionally, sample with tracing impurity phase is used when it is difficult for obtaining samples with single phase. If the rough structure is known and the impurity phase is very limit, then it might be possible to determine the structure by using the sample with impurity phase [4].

The crystal structures and the lattice thermal expansion properties of new compounds were determined from powder X-ray diffraction patterns. Powder diffraction data of new compounds for structure determination were collected by using a Rigaku D/max 2500V diffractometer with a CuK_{α} radiation. The scan range was 10° to 110° (2θ) with a step size of 0.02° and a count time of 2 s per step. The index of the sample and the initial lattice parameters determination was carried out by using a Jade 5.0 program [5]. The crystal structural refinement was performed by using Rietveld method [6, 7] from X-ray powder diffraction data. The study on lattice thermal expansion of new compounds was investigated by using high temperature X-ray diffraction (HTXRD) technique. The diffraction data at different temperatures were collected for the investigations of thermal expansion properties.

3. Structure Determination

The formula unit (Z) in each unit cell can be calculated from the density, formula and unit cell parameters of the new compound. Structure determination can be performed by using isomorphous replacement method, trial and error method and ab-initio structure determination method, etc.

It is the easiest and rapidest way to determine the crystal structure of a new compound if a known isostructure compound is available.

Therefore, one should check whether the new compound is isostructural with a known compound. Once the known isostructure compound is found, the rough structure of the new compound is obtained, and then the structure of the new compound can be refined from this rough structure. In fact, many structures of new compounds are determined by the isomorphous replacement method. Ab-initio structure determination [8, 9] is a method determining structure from powder diffraction data. There are three steps for carried out the method.

- Collecting X-ray diffraction pattern of a compound and indexing accurately to give the unit cell parameters and the space group of the compound.
- Using Patterson method/direct method and Fourier/difference Fourier to solve the structure and obtain the initial structure model.
- Refine the structure by the Rietveld's powder diffraction profile-fitting technique.

An important step is necessary to perform after the determination of crystal structure. That is to calculate the bond lengths and check whether these bond lengths fall in a reasonable range.

The Rietveld method [10] is a least squares refinement procedure where the experimental step-scanned values are adapted to the calculated ones. The profiles are considered to be known, and a model for a crystal structure is available. In the Rietveld method the least-squares refinements are carried out until the best fit is obtained between the entire observed powder diffraction patterns and the calculated pattern, based on the simultaneously refined models for the crystal structure(s), diffraction optics effects, instrumental factors and other specimen characteristics (e.g. lattice parameters). A key feature is the feedback during the refinement, which can improve the knowledge of the structure and the allocation of observed intensity. The quantity minimized in the least-squares refinement is the residual, S_y:

$$S_y = \sum_i w_i \, (y_i - y_{ci})^2 \quad (4\text{-}1)$$

where $w_i = 1/y_i$, y_i =observed (gross) intensity at the ith step, y_{ci} =calculated intensity at the ith step, and the sum is over all data points.

The calculated counts y_{ci} are determined by summing the contributions from neighboring Bragg reflections, K, for all phases, p, plus the background, b_i:

$$y_{ci} = s \sum_K L_K \, |F_K|^2 \phi(2\theta_i - 2\theta_k) P_K A + y_{bi} \quad (4\text{-}2)$$

where s is the scale factor, K represents the Miller indices, $h\ k\ l$, for a Bragg reflection, L_K contains the Lorentz, polarization and multiplicity factors, ϕ is the reflection profile function, P_K is a preferred orientation function, A is an absorption factor, F_K is the structure factor for the Kth Bragg reflection, and b_i is the background contribution.

When describing a diffraction pattern, the width of the diffraction peaks is another important parameter and variable. The peak width, described as "full width at half maximum", *FWHM*, is in general a function of the diffraction angle 2θ.

$$(FWHM)_k = U\tan^2\theta_k + V\tan\theta_k + W \quad (4\text{-}3)$$

with U, V and W adjustable parameters. The best fit sought is the best least squares fit to all of the thousands of y_i simultaneously. The quantity minimized therefore is in general terms (4-1). By using the Rietveld refinement, the crystal structure details can be determined. The Rietveld refinement process will adjust the refinable parameters until the residual equation (4-

1) is minimized in some sense. That is a "best fit" of the entire calculated pattern to the entire observed pattern. Several R-values listed below are now commonly used as the criteria of fit:

$$\text{R-structure factor: } R_p = \frac{\sum |Y_i(obs) - Y_i(calc)|}{\sum Y_i(obs)} \tag{4-4}$$

$$\text{R-weighted pattern: } R_{wp} = \left\{ \frac{\sum \omega_i [Y_i(obs) - Y_i(calc)]^2}{\sum \omega_i [Y_i(obs)]^2} \right\}^{1/2} \tag{4-5}$$

$$\text{R-expected factor: } R_{\exp} = \left[(N-P) / \sum_i W_i Y^2{}_{iobs} \right]^{1/2} \tag{4-6}$$

$$\text{R-Bragg factor: } R_{Bragg} = \frac{\sum |I_k(obs) - I_k(calc)|}{\sum I_k(obs)} \tag{4-7}$$

$$\text{The Goodness-of-fit indicator: } S = [S_y/(N-P)]^2 = R_{wp} / R_{\exp} \tag{4-8}$$

R_{wp} is the most meaningful one among these R-factors, and it is also the one that best reflects the progress of refinement. In principle, the parameters that can be adjusted in the least squares refinement for each phase include:

- x_j, y_j, z_j, B_j, N_j, (x_j, y_j, z_j are atomic positions, B_j is an isotropic thermal parameter, and N_j is the site-occupancy multiplier. All for the jth atom in the unit cell.)
- Overall scale factor (making quantitative phase analysis feasible)
- Specimen-profile breadth parameters
- Lattice parameters (a, b, c, α, β, γ)
- Temperature factor (thermal parameters)
- Individual anisotropic thermal parameters
- Preferred orientation
- Crystallite size and microstrain (through profile parameters)
- Extinction

The global refinable parameters are:

- 2θ-zero
- Instrumental profile
- Profile asymmetry
- Background (parameters in analytical function, e.g. 5th order polynomial, sum of exponentials, RDF for amorphous component, etc.)
- Wavelength

- Specimen displacement
- Specimen transparency
- Absorption

The input information for a Rietveld refinement program requires:

- Initial values of all variable parameters (listed above)
- Step-scan data in the same increments of 2θ
- 2θ limits, starting and ending values of 2θ, and the regions which shall be excluded in the data
- Wavelength data

With a complete and good structural model (together with the good starting values), Rietveld refinement can begin. The good starting values are needed for the unit cell, background, profile parameters and the approximate atomic positions. One must have the correct space group, fairly good lattice parameters, atomic coordinates, and estimated *N*'s and *B*'s. Stoichiometric N_j values and zero B_j are usually used to get it started. Young suggested a possible good turn-on sequence, which is listed in Table 1.1.

Table 1.1. A suggested parameter turn-on sequence in a single-phase, single-pattern Rietveld refinement using constant wavelength x-ray or neutron data [10]

Parameters	Linear	Stable	Comment	Sequence
Scale	Yes	Yes	Note 1	1
Specimen displacement	No	Yes	Note 1	1
Flat background	Yes	Yes		2
Lattice parameters	No	Yes	Note 3	2
More background	No	Yes	Note 4	2 or 3
W	No	Poorly	Note 5	3 or 4
Coordinates of atoms *x, y, z*	No	Fairly	Note 6	3
N's and *B*'s	No	Unknown	Correlated	4
U, V, etc	No	No	Note 5	Last
Anisotropic thermal	No	No, or unknown		Last
Zero point	No	Yes	Note 2	1, 4 or not

Note 1: If the scale factor is very far off, or the crystal structure model is very bad, the scale factor may get worse, such as too small, during the refinement. It is because the difference between a pattern and nothing is less than the difference between two badly mismatched patterns.

Note 2: For properly aligned and mechanically stable diffractometers, the zero point error should be and remain inconsequential. In any event, it cannot change from sample to sample whereas the effective specimen displacement can be changed. The displacement parameter will also take up some of the specimen transparency effect, which occurs in non-infinitely absorbing specimen.

Note 3: One or more incorrect lattice parameters may cause one or more calculated peaks to "lock on" to the wrong observed peaks. The result can be a false minimum. Artificially broadening the calculated profiles (temporarily) may help.

Note 4: If more background parameters are turned on than those needed to model the angular dependence of the background, the result will be of high correlations for large shifts. They are mostly mutually compensated but may lead to erratic behavior and the failure of refinement. The high order ones should be turned off sequentially until the problem is corrected.

Note 5: *U, V,* W tend to be highly correlated. Various combinations of different values can lead to essentially the same peak widths.

Note 6: Graphics and the reflection indices should be used now to check whether preferred orientation should be modeled at this point.

4. Structures of Some New Rare Earth Compounds

4.1. $RE_5Al_{14}Si$ (RE = Nd, Gd, Dy)

The new ternary $Al_{14}Nd_5Si$ compound was synthesized by arc-melting and the crystal structure was refined by means of Rietveld method from X-ray powder diffraction data. All X-ray diffraction peaks can be indexed successfully on the basis of hexagonal structure for $Al_{14}Nd_5Si$ with the lattice parameters a = 0.64416 (2) nm and c = 0.45894 (2) nm. The Smith and Snyder figure of merit for the index, F_N, is F_{30} = 97.8(30) [11]. The reflection conditions yield three possible space groups: $P6_3/mmc$ (No. 194), $P\bar{6}2c$ (No. 190) and $P6_3mc$ (No. 186) for the structure of $Al_{14}Nd_5Si$. The isostructural compound $Al_{14}Y_5Si$ [12] (with hexagonal Ni_3Sn-type structure, space group $P6_3/mmc$, No. 194) was found for the new compound $Al_{14}Nd_5Si$. Then, the space group $P6_3/mmc$, No. 194 was assigned to the structure of $Al_{14}Nd_5Si$. The atomic position parameters of Ni_3Sn were chosen as the initial atomic positions for the structural refinement.

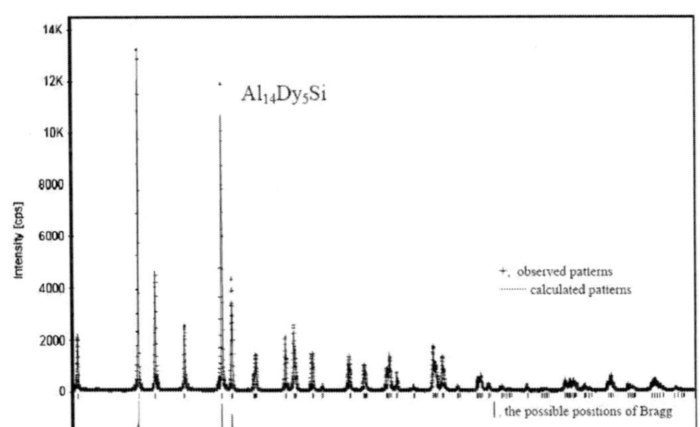

Figure 1. The observed, calculated data and differences of the X-ray powder diffraction pattern of $Al_{14}Nd_5Si$ [13].

The Rietveld structural refinement was carried out. The refinement reliability factors, the pattern R-factor (R_p) and the weighted pattern R-factor (R_{wp}) are 0.088 and 0.120, respectively. The lattice parameters after refinement are a = 0. 64470(2) nm, c = 0.45926(1) nm. The observed, calculated data and differences of the powder diffraction patterns of $Al_{14}Nd_5Si$ are shown in Figure 1. In the hexagonal structure of $Al_{14}Nd_5Si$, Nd atoms are in the 2c positions, while Al and Si atoms occupy the 6h positions. Figure 2 shows the crystal structure of $Al_{14}Nd_5Si$. The environments of atoms Nd are at 2c positions; and the environments of atoms M (M =Al, Si) are at 6h positions. Table 1 gives the atomic parameters of $Al_{14}Nd_5Si$ [13].

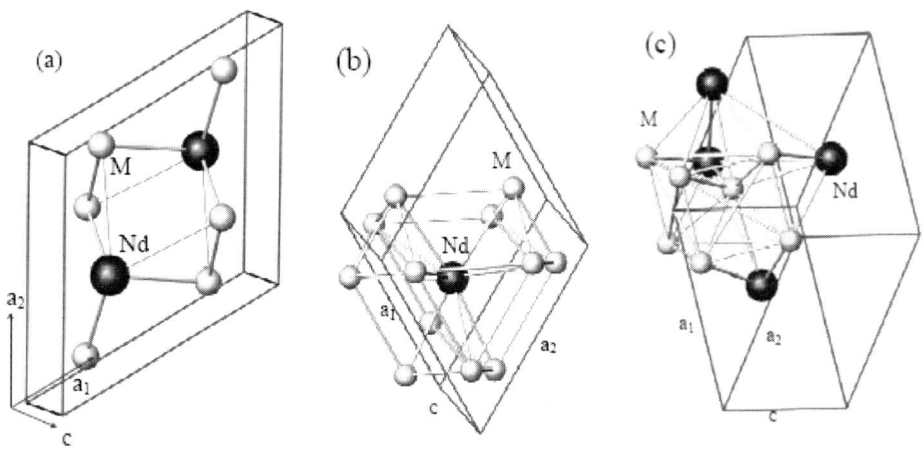

Figure 2. Crystal structure of the crystal structure of $Al_{14}Nd_5Si$ [13]. This figure shows the (a) crystal structure of $Al_{14}Nd_5Si$, (b) environments of atoms Nd at 2c positions, and (c) environments of M atoms (M =Al, Si) at 6h positions.

Table 1. Atomic parameters of $Al_{14}Nd_5Si$ [13]

$Al_{14}Nd_5Si$; $P6_3/mmc$ (No. 194); a = 0. 64470(2) nm, c = 0.45926(1) nm ; D_{calc}= 4.526g/cm^3 R-factors(%): R_p=0.088, R_{wp}=0.12					
Atoms	Positions	Fractional coordinates			occupancy
		x	y	z	
Gd	2c	1/3	2/3	1/4	1
M = Al, Si	6h	0. 8564(4)	0. 7128(4)	1/4	(0.933Al+0.067Si)

The crystal structure of the high temperature phase of $Al_{14}Gd_5Si$ is isostructural with $Al_{14}Nd_5Si$ and the lattice parameters are a = 0.63022 (1)nm, c = 0.45856 (1)nm. Table 2 lists the atomic parameters of $Al_{14}Gd_5Si$ and Table 3 gives the bond lengths and the number of the nearest neighbors of each atom for the structure of $Al_{14}Gd_5Si$[14].

The indexation of the X-ray diffraction pattern $Al_{14}Dy_5Si$ was carried out successfully on the basis of hexagonal structure with the lattice parameters a = 0.61505(9) nm and c = 2.09798(1) nm. The Smith and Snyder figure of merit [11] for the index, F_N, is F_{30} = 152.0(45). The calculated density is 5.301g/cm^3. The reflection conditions yield three possible space groups: $R\bar{3}m$ (No. 166), $R3m$ (No. 160) and $R32$ (No. 155) for the structure of $Al_{14}Dy_5Si$.

Although the composition of $Al_{14}Dy_5Si$ is similar to $Al_{14}Gd_5Si$, its structure is different from $Al_{14}Gd_5Si$. It is isostructural with Al_3Y and adopts the trigonal $BaPb_3$-type structure with space group $R\bar{3}m$ (No. 166) [12].

The Rietveld refinement was carried out from the initial atomic parameters of Al_3Y for $Al_{14}Dy_5Si$. The reliability R-factors are R_p=0.091 and R_{wp}=0.120. Table 4 gives the atomic parameters of $Al_{14}Dy_5Si$ and Figure 3 presents the observed, calculated data and differences of the X-ray powder diffraction pattern of $Al_{14}Dy_5Si$. In the structure of $Al_{14}Dy_5Si$, Dy atoms are in the 3a and 6c positions, some Al atoms are in 9e positions, while other Al atoms and all of Si atoms occupy the 18h positions. The crystal structure of $Al_{14}Dy_5Si$ is given in Figure 4 [15].

Table 2. The atomic parameters for $Al_{14}Gd_5Si$ [14]

Atoms	Positions	x	y	z	Occupancy
Gd	2c	1/3	2/3	1/4	1
Al, Si	6h	0.8552	0.7104	1/4	(0.933Al+0.067Si)

Table 3. The bond lengths and the number of the nearest neighbors of each atom for the trigonal structure of $Al_{14}Gd_5Si$ [14]

Atoms	Neighbor atoms	Bond lengths (nm)	Number of the nearest atoms
Gd	-M=(Al,Si)	0.3160(1)	6
	-M	0. 3080(2)	6
M	-Gd	0.3160(1)	2
	-Gd	0.3080(2)	2
	-M	0. 2785(4)	4
	-M	0.3565(3)	2
	-M	0.2738(4)	2

Table 4. The atomic parameters for $Al_{14}Dy_5Si$ [15]

$Al_{14}Dy_5Si$; $R\bar{3}m$ (No. 166); a = 0.61482(1) nm, c = 2.09780(2) nm ; D_{calc}= 5.301 g/cm^3 R-factors(%): R_p=0.091 R_{wp}=0.12					
Atoms	Positions	x	y	z	Occupancy
Dy1	3a	0	0	0	1
Dy2	6c	0	0	0.2172(2)	1
Al	9e	0.5	0	0	1
M=(Al, Si)	18h	0. 4791(3)	0. 5209 (3)	0.2225(1)	(0.9Al+0.1Si)

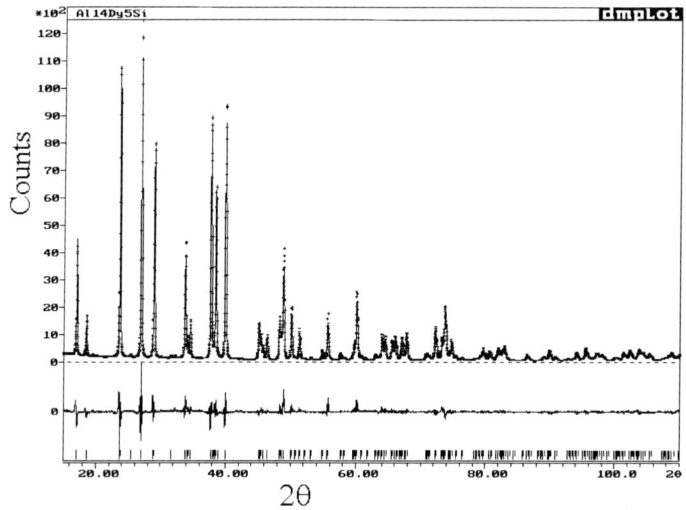

+, observed patterns; Solid line, calculated patterns; |, possible positions of Bragg reflections; The bottom curve, the difference between the observed and calculated patterns.

Figure 3. The X-ray powder diffraction pattern of $Al_{14}Dy_5Si$ [15].

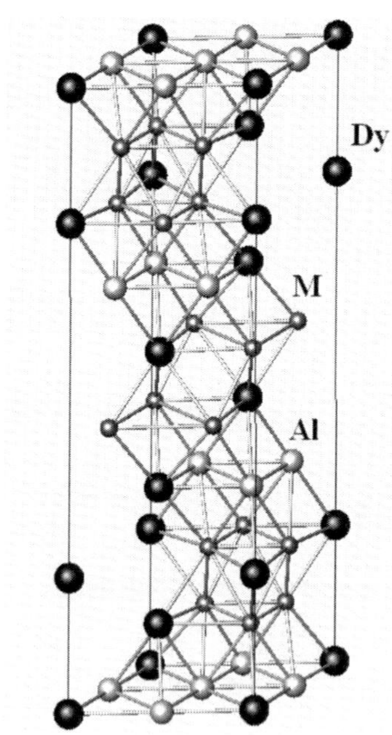

Figure 4. The crystal structure of compound $Al_{14}Dy_5Si$ [15].

4.2. $La_8Al_{13}Sn_3$

New ternary compound $La_8Al_{13}Sn_3$ was found to exist at 540°C and its crystal structure was studied by means of X-ray powder diffraction. The alloy sample annealed at 540°C with single phase of $La_8Al_{13}Sn_3$ was broken into two parts and one part was ground into powder and analyzed by X-ray diffraction technique. All of the X-ray diffraction peaks of this new compound were indexed successfully on the basis of hexagonal structure with lattice parameters *a* = 0.44892(2) nm, *c* = 0.43810(2) nm by using Jade 5.0 program. The Smith and Snyder figure of merit [11] for the index, F_N, is F_{30} = 197.3(30), and the calculated density is 4.925 g/cm^3. The reflection conditions yield five possible space groups: *P6/mmm* (No. 191), $P\bar{6}2m$ (No. 189), $P\bar{6}m2$ (No. 187), *P6mm* (No. 183) and *P622* (No. 177). The sample composition, reflection intensities, and the calculated lattice parameters proved that this compound is isostructural with the compound $Ce_8Al_{13}Sn_3$ (space group *P6/mmm* (No. 191)) [12]. So the space group *P6/mmm* (No. 191) for the hexagonal structure of $La_8Al_{13}Sn_3$ was chosen.

The Rietveld refinement was performed and the structural refinement supports that the atomic positions of $La_8Al_{13}Sn_3$ are similar to those of $Ce_8Al_{13}Sn_3$. The pattern R-factor (R_p) and the weighted pattern R-factor (R_{wp}) are 0.114 and 0.148, respectively. The lattice parameters after refinement are *a* = 0.44919(1) nm and *c* = 0.43835(1) nm. Figure 5 shows the X-ray powder diffraction pattern of $La_8Al_{13}Sn_3$. In the hexagonal structure, the La atoms are in the 1a positions, while the Al and Sn atoms occupy the 2d positions, as given in Table 5. Figure 6 presents the crystal structure of the compound and the environment of atoms at 1a and 2d positions [16].

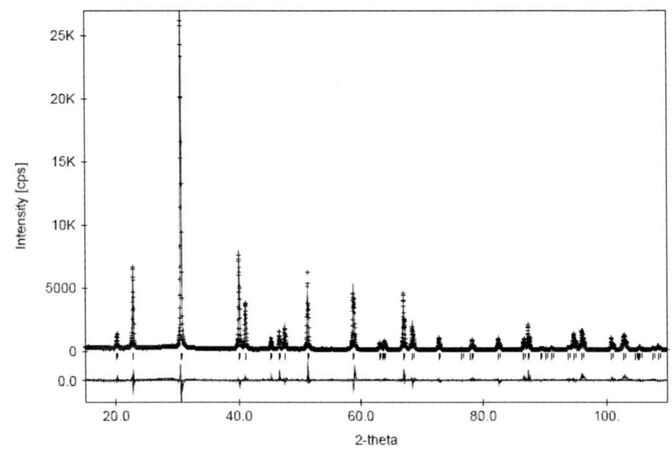

Figure 5. X-ray powder diffraction pattern of $La_8Al_{13}Sn_3$ compound. +, observed patterns; Solid line, calculated patterns; |, possible positions of Bragg reflections; The bottom curve, the difference between the observed and calculated patterns [16].

Table 5. Atomic parameters of $La_8Al_{13}Sn_3$ [16]

$La_8Al_{13}Sn_3$; P6/mmm (No. 191); a = 0.44919(1) nm, c = 0.43835(1) ; D_{calc}= 4.925g/cm^3 R-factors(%): R_p= 0.114 and R_{wp}= 0.148						
Atoms	Positions	Fractional coordinates			Occupancy	B_{eq}
		x	y	z		
La	1a	0	0	0	1	0.0059
M = Al, Sn	2d	0.3333	0.6667	0.5	(0.8125Al+0.1875Sn)	0.0268

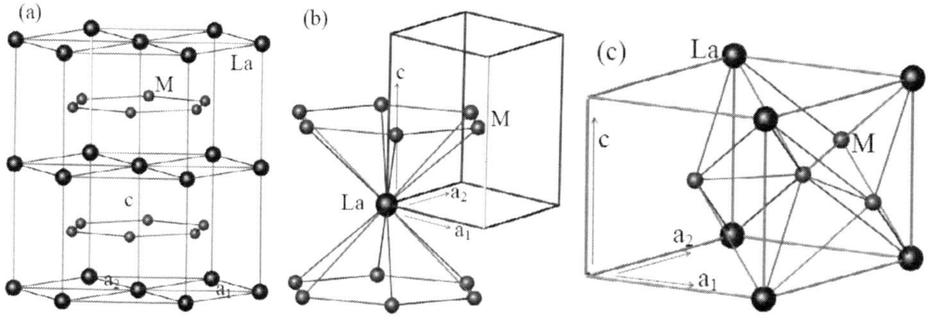

Figure 6. Crystal structure of compound $La_8Al_{13}Sn_3$ [16]. The figure shows (a) crystal structure; (b) environment of La atoms at 1a positions; and (c) environment of M (= Al, Sn) atoms at 2d positions.

The other part of the alloy sample was re-annealed in an evacuated quartz tube at 720°C for 20 days, quenched into liquid nitrogen, and then ground into powder for XRD analysis. The phase identification result pointed out that there were three phases in the sample, as shown in Figure 7. It was found that one set of the X-ray diffraction pattern in the sample is similar to those of the compound $LaAl_2$ with a cubic structure, space group $Fd\overline{3}m$ (227) and a = 0.79373 (2) nm, while the other set of the X-ray pattern is similar to those of the compound $LaAl_3$ with a hexagonal structure, space group of $P6_3/mmc$ (194) and a = 0.63475(9) nm and c = 0.46280(6) nm.

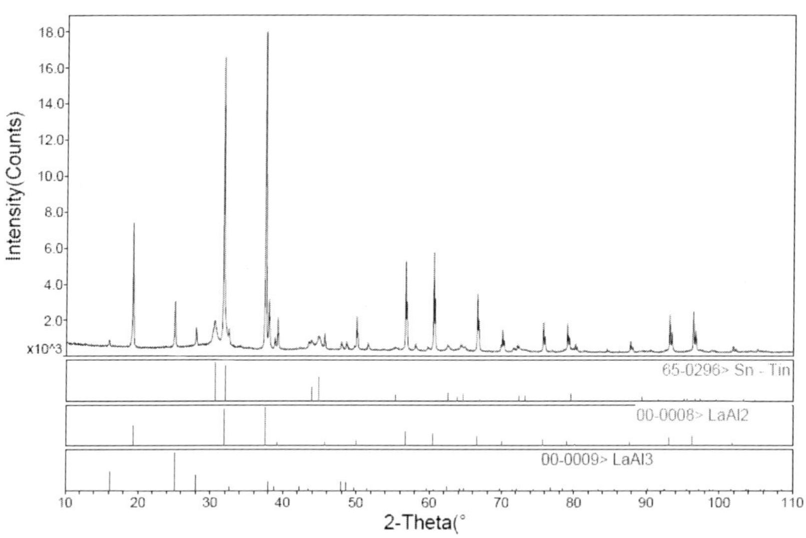

Figure 7. X-ray powder diffraction pattern of the sample re-annealed at 720°C [16].

There is a small amount of Sn in the sample. Compared the unit cell volumes of these two phases obtained by Jade 5.0 program with those of $LaAl_2$ and $LaAl_3$ available in Powder Diffractions Files (PDF) [17], it is found that the unit cell volumes of $LaAl_2$ and $LaAl_3$ in the sample are smaller than those of $LaAl_2$ (a = 0.8125nm) and $LaAl_3$ (a = 0.6667 nm and c = 0.4616 nm) available in PDF files. The smaller unit cells of the $LaAl_2$ and $LaAl_3$ (in fact, the unit cell of $LaAl_3$ shrinks at the [100] direction and almost keeps unchanged in [001] direction) in the sample might be caused by the inefficiency of Al in the lattices or some solid solubility of Sn in these two phases (the radii of R_{Sn} = 0.1405 nm is smaller that of R_{Al} = 0.1431 nm). Hence, it can be inferred that the compound $La_8Al_{13}Sn_3$ decomposed into three phases, i.e., Sn, $LaAl_2$ and $LaAl_3$ at 720°C. The intensity of the phases $LaAl_2$ and $LaAl_3$ in Figure 7 were calculated by using Powder Cell program [18] with the lattice parameters obtained by Jade 5.0 program. The Rietveld analysis results showed that the percentage of the three phases in the sample at 720°C were 67.1% of $LaAl_2$, 15.6% of $LaAl_3$, and 17.3% of Sn. The R factors of the Rietveld refinement are R_p= 0.088 and R_{wp}= 0.112, respectively. The Rietveld analysis also pointed out that the occupancy of Al in the 6h site in the structure of $LaAl_3$ is only about 0.97 and the Al and Sn atoms (0.891Al +0.109Sn) occupy the 16d site in the structure of $LaAl_2$. This is the reason causing the smaller unit cells of the $LaAl_2$ and $LaAl_3$ in the sample at 720°C. This result also confirmed that the compound $La_8Al_{13}Sn_3$ decomposed into three phases of Sn, $LaAl_2$ and $LaAl_3$ at 720°C [16].

4.3. Al_2REGe_2 (RE = Dy, Er)

New ternary compounds Al_2REGe_2 (RE = Dy, Er) were found and determined by using the Rietveld method from X-ray powder diffraction data [19]. All reflections of Al_2ErGe_2 were well indexed on the basis of a hexagonal unit cell with the lattice parameters a = 0.41801(9) nm and c = 0.66704(2) nm. Each unit cell contains one formula unit with 5 atoms. The compound is isostructural with the Al_2REGe_2 (RE = Sc, Y, La, Ce, Sm, Eu, Gd, Tb, Yb, Lu) composition and in Al_2CaSi_2-type structure [20, 21]. The Rietveld refinement for the compound Al_2ErGe_2 was performed by using the DBWS9807 program. The refined parameters are a = 0.41781(9) and c = 0.66698(1), while R-factors for the refinement are R_p = 9.97%, R_{wp} = 13.74%, $R_{expected}$ = 5.52%. The atomic coordinates, occupancy and thermal factors are given in Table 6. Figure 8 is the X-ray powder diffraction pattern of Al_2ErGe_2 and Figure 9 is a projection of the Al_2ErGe_2 structure [19].

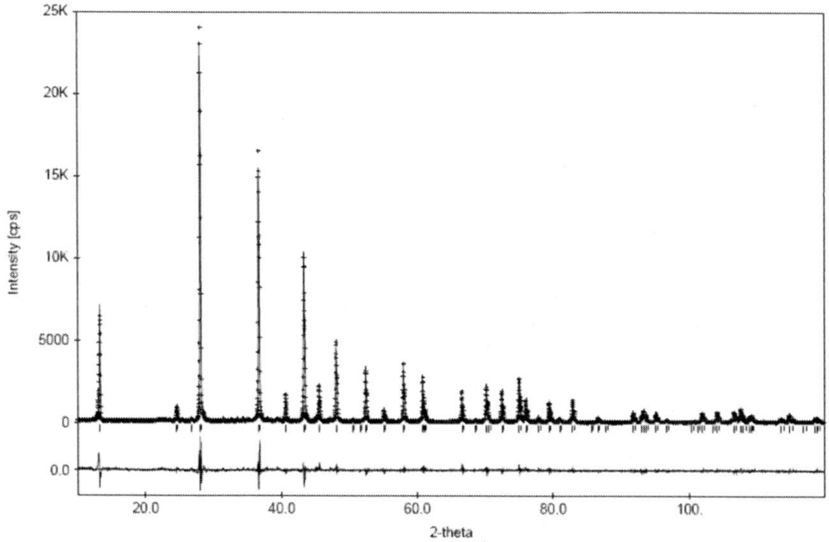

+, observed patterns; Solid line, calculated patterns; |, Possible positions of Bragg reflections; The bottom curve, the difference between the observed and calculated patterns.

Figure 8. X-ray powder diffraction pattern of Al_2ErGe_2 [19].

It was reported that the compound Al_2DyGe_2 crystallizes in the tetragonal Mo_2FeB_2 type structure with space group of P4/mbm (No. 127) and lattice parameters of a = 0.7019 nm and c = 0.4291 nm[22]. Later, the trigonal La_2O_3 structure type (with space group of $P\bar{3}m1$, No. 164) of the compound was found [23]. The crystal structure of Al_2DyGe_2 is isostructural with Al_2ErGe_2 and the lattice parameters of Al_2DyGe_2 are a = 0.4204(3) nm and c = 0.6681(2) nm.

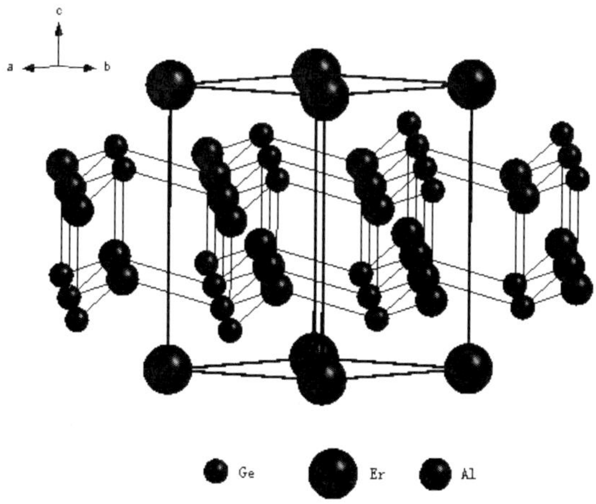

Figure 9. A projection of the Al_2ErGe_2 structure [19].

Table 6. Atomic and thermal factors for the compound Al_2ErGe_2 [19]

Al_2ErGe_2; $P\bar{3}m1$ (No. 164); a = 0.41781(9) nm and c = 0.66698(1) nm. and D_{calc}= 6.032g/cm^3 R-factors(%): R_p=0.0997 R_{wp}=0.1374, R_{exp}=0.0552						
Atom	Position	x	y	z	Occupancy	B_{eq} (nm^2)
Er	1a	0	0	0	1	0.29(8)
Al	2d	1/3	2/3	0.6411(8)	1	0.3(3)
Ge	2d	1/3	2/3	0.2547(3)	1	0.66(4)

4.4. RE_6FeSb_2 (RE = Tb, Dy, Ho)

The crystal structure of novel compounds RE_6FeSb_2 (RE = Tb, Dy, Ho) was investigated and determined by X-ray powder diffraction method [24-26]. All reflections of $FeHo_6Sb_2$ were well indexed in a hexagonal unit cell with the lattice parameters a = 0.8139(1) nm, c = 0.41345(8) nm. Each unit cell contains nine atoms, based on the lattice parameters and the measured density (D_m = 8.928 g/cm^3). Reflection conditions (hh2−hl: no conditions; −hh0l: no conditions; 000l: no conditions) pointed to the five possible space groups *P6/mmm* (No. 191), $P\bar{6}2m$ (No. 189), $P\bar{6}m2$ (No. 187), *P6mm* (No. 183), *P622* (No. 177). The further calculated results indicate that $FeHo_6Sb_2$ and Fe_2P have the same structure type. The structure refinement was performed from the initial structure of Fe_2P [27]. A total of 25 parameters were refined and the refined lattice parameters are a = 0.81403(3) nm, c = 0.41346(2) nm. The atomic coordinates, occupancy and thermal parameters are summarized in Table 7. Figure 10 shows the crystal structure of the $FeHo_6Sb_2$.

Table 7. Atomic and thermal parameters for the $FeHo_6Sb_2$ compound [25]

$FeHo_6Sb_2$; $P\bar{6}m2$ (No. 187); a = 0.81403(3) nm and c = 0.41346(2) nm. D_{calc}= 9.019g/cm^3 and D_{meas}= 8.928g/cm^3; R-factors(%): R_p=0.0791, R_{wp}=0.1001, R_{exp}=0.0351						
Atom	Position	*x*	*y*	*z*	Occupancy	B_{eq}(nm^2)
Fe	1b	0.0	0.0	1/2	1	0.006(4)
Sb	2c	1/3	2/3	0.0	1	0.002(4)
Ho1	3f	0.2380(2)	0.0	0.0	1	0.007(4)
Ho2	3g	0.6005(2)	0.0	1/2	1	0.006(4)

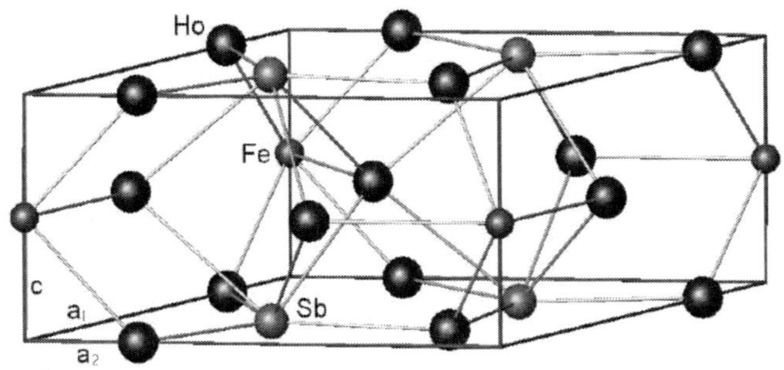

Figure 10. The crystal structure of the $FeHo_6Sb_2$.

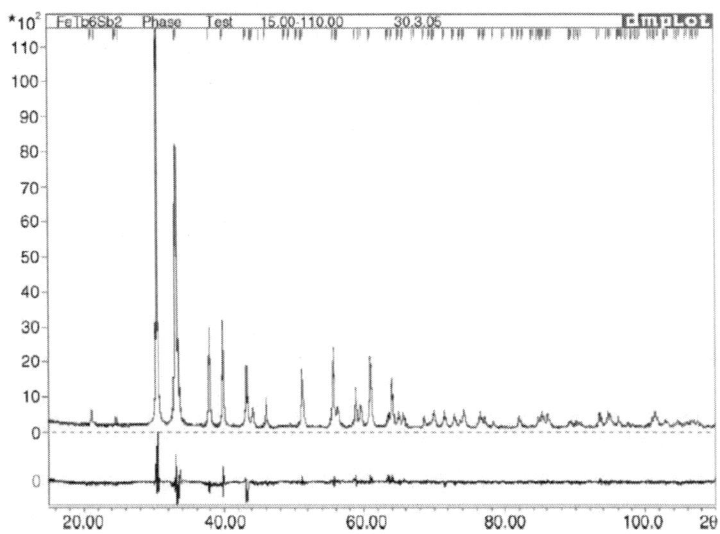

Figure 11. The observed and calculated data and residuals of the powder diffraction pattern of Tb_6FeSb_2 [26].

Analysis on all the X-ray powder diffraction reflections for Tb_6FeSb_2 pointed out that the compound was crystallized in a hexagonal unit cell with isostructural Ho_6FeSb_2. It has the lattice parameters of a = 0.8193(8) nm and c = 0.4175(5) nm. The X-ray diffraction pattern of the new compound Tb_6FeSb_2 is given in Figure 11. Table 8 gives the atomic parameters of the novel compound. The bond lengths and the numbers of the nearest neighbors of each atom in the hexagonal structure of Tb_6FeSb_2 are listed in Table 9. The novel compound Dy_6FeSb_2 with similar composition is also isostructural with Ho_6FeSb_2, which has the lattice parameters of a=0.81459(5) nm and c=0.41647(8) nm [24]. The atomic and thermal parameters of Dy_6FeSb_2 are given in Table 10.

Table 8. Atomic and thermal shake parameters for the Tb_6FeSb_2 compound [26]

Atom	Position	*x*	*y*	*z*	Occupancy	B_{eq} (nm^2)
Fe	1b	0	0	1/2	1	0.008(4)
Sb	2c	1/3	2/3	0	1	0.006(4)
Tb1	3f	0.2376(2)	0	0	1	0.006(3)
Tb2	3g	0.5978(2)	0	1/2	1	0.008(4)

Table 9. The bond lengths (<0.35 nm) and the number of the nearest neighbors of each atom for $FeTb_6Sb_2$ [26]

Atoms	Neighbor atoms	Bond lengths (nm)	Atoms	Neighbor atoms	Bond lengths (nm)
Tb1	2Tb1	0.3349	Sb	3Tb1	0.3206
	4Tb2	0.3548		6Tb2	0.3255
	2Tb2	0.3626		2Fe	0.2846
	1Fe	0.3296		2Sb	0.3206
Tb2	4Tb1	0.3548	Fe	6Tb1	0.2846
	2Tb1	0.3626		3Tb2	0.3296
	4Sb	0.3255			

Table 10. Atomic and thermal parameters for the Dy_6FeSb_2 compound [24]

Atom	Position	x	y	z	Occupancy	B_{eq}(nm2)
Fe	1b	0.0	0.0	1/2	1	0.28(9)
Sb	2c	1/3	2/3	0.0	1	0.26(9)
Dy1	3f	0.2379(2)	0.0	0.0	1	0.26(9)
Dy2	3g	0.5992(2)	0.0	1/2	1	0.25(9)

4.5. $CuSb_2Ti_5$

The sample of new compound $CuSb_2Ti_5$ was prepared and the X-ray diffraction pattern of the compound was indexed successfully based on tetragonal symmetry. The indexed lattice parameters were used as the starting lattice parameters for the Rietveld refinement. According to the similarity of the powder diffraction pattern of $CuSb_2Ti_5$ with those of $FeSb_2Ti_5$ [28], the space group I4/mcm (No. 140) was chosen. The goodness of profile refinement was observed from the calculated R values (R_p=8.96%, R_{wp}=12.05% and S=1.97). The lattice parameters after refinement are a = 1.04804(2) nm and c = 0.52424(2) nm. Formula units in each unit cell is Z = 4. Refinement results for the atomic coordinates, occupancy and thermal parameters are given in Table 11. The X-ray powder diffraction pattern of $CuSb_2Ti_5$ is shown in Figure 12 [29].

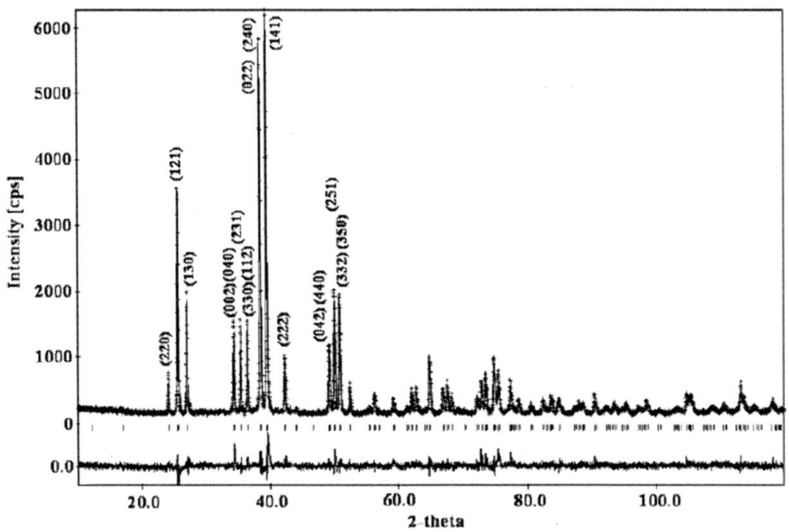

Figure 12. The X-ray powder diffraction pattern of $CuSb_2Ti_5$ [29].

Table 11. Atomic and thermal parameters for the $CuSb_2Ti_5$ compound [29]

$CuSb_2Ti_5$; I4/mcm (No. 140); a = 1.04804(2) nm and c = 0.52424(2) nm Reliability-factors (%): R_p=8.96%, R_{wp}=12.05% and S=1.97						
Atoms	Positions	x	y	z	Occupancy	B_{eq} (nm^2)
Cu	4a	0	0	0.25	1	0.0010
Ti(1)	4b	0	0.5	0.25	1	0.0049
Sb	8h	0.1634(1)	0.6634(1)	0	1	0.0035
Ti(2)	16k	0.0774(3)	0.2224(3)	0	1	0.0032

4.6. $SmCo_9Si_4$

The new compound was found during the investigation on the Sm-Co-Si system [30, 31]. Figure 13 presents the X-ray diffraction pattern of novel compound $SmCo_9Si_4$ [31], which was crystallized in the $Ce_2Ni_{17}Si_9$ tetragonal structure type with the space group *I4/mcm* (No. 140) and lattice parameters a = 0.77815(2) nm and c = 1.14993(5) nm. According to the indexation of the X-ray diffraction pattern of the compound, the reflection conditions (*hkl*: $h + k + l = 2n$, *hk*0: $h + k = 2n$, 0*kl*: $l, k = 2n$, *hhl*: $l = 2n$, 00*l*: $l = 2n$, *h*00: $h = 2n$) yield three possible space groups: *I4/mcm*, *I4c*2 and *I4cm*. Based on the similarity of the X-ray powder diffraction pattern of $SmCo_9Si_4$ with those of $PrCo_9Si_4$ and $NdCo_9Si_4$ [32, 33], the space group *I4/mcm* was chosen. The lattice parameters after Rietveld refinement are a = 0.77858(1) nm and c = 1.15054(2) nm. Table 12 gives the atomic parameters of the compound and Figure 14 shows the projection of $SmCo_9Si_4$ structure on the *yz* plane [31].

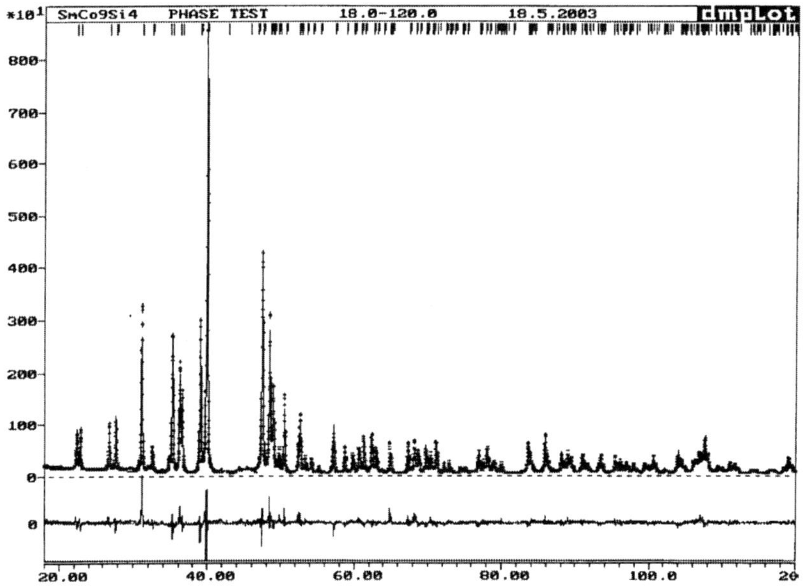

Figure 13. The observed, calculated data and residuals of the powder diffraction pattern of $SmCo_9Si_4$ [31].

Table 12. Atomic and thermal parameters for the $SmCo_9Si_4$ compound [31]

$SmCo_9Si_4$; *I4/mcm* (No. 140); a = 0.77858(1) nm and c = 1.15054(2) nm, and D_{calc}= 7.553 g/cm^3 ; R-factors(%): R_p = 0.1034, R_{wp} = 0.1311, R_{exp} = 0.0658						
Atom	Position	*x*	*y*	*z*	Occupancy	B_{eq} (nm^2)
Sm	4a	0.0	0.0	0.25	1	0.0083(4)
Co1	4d	0.0	0.5	0.0	1	0.0081(7)
Co2	16l	0.1287(2)	0.6287(2)	0.1826(1)	1	0.0069(4)
Co3	16k	0.1993(2)	0.0710(2)	0.0	1	0.0076(4)
Si	16l	0.3296(3)	0.8296(3)	0.1225(3)	1	0.0093(8)

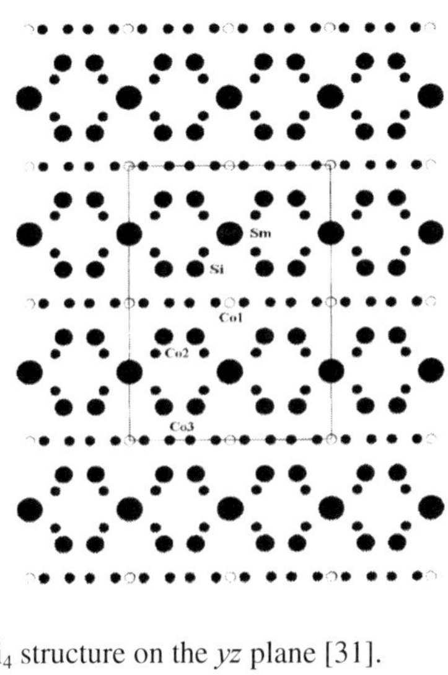

Figure 14. Projection of $SmCo_9Si_4$ structure on the *yz* plane [31].

4.7. REAlSi (RE = Dy, Er)

Most of the equiatomic REAlSi compounds can be divided into two groups. Compounds of group one (RE = La-Nd, Gd and Tb) crystallizes with a tetragonal, α-$ThSi_2$ type structure with space group $I4_1/amd$ (No. 141); while compounds in the other group (RE = Y, Dy-Lu) have an orthorhombic, YAlGe type structure with space group Cmcm (No. 63). It was reported that the ternary compound REAlSi (RE = Dy, Er) has an orthorhombic, YAlGe type structure with space group Cmcm (No. 63) at 873K [34, 35]. Later, the new crystal structure was found to be a high-temperature phase of REAlSi, which is different from that of low temperature form [35], as reported in the literatures [3, 4]. The arc-melting alloy samples were enclosed in an evacuated quartz tube and annealed at 993K for 20 days, then quenched into liquid nitrogen. The high temperature phase of REAlSi (RE = Dy, Er) was then obtained.

Figure 15 presents the X-ray diffraction pattern of the new structure as a high temperature phase of DyAlSi. All reflections for this high temperature phase of DyAlSi were well-indexed in a tetragonal unit cell with the lattice parameters a = 0.41243(3) nm and c = 1.44036(1)nm. The reflection conditions yield the only possible space group: $I4_1/amd$ (No. 141). The Rietveld's profile-fitting technique was used to analyze the X-ray powder diffraction pattern of the new tetragonal structure of DyAlSi. The structural refinement was performed by using the DBWS9411 program [36]. The indexed lattice parameters and the atomic position parameters of $CeAl_{1.2}Si_{0.8}$ (with α-$ThSi_2$ type structure) [34] were taken as the starting values to refine the structural parameters of DyAlSi. The refinement indicates that the high temperature phase of DyAlSi was crystallized in a tetragonal α-$ThSi_2$ type structure with space group $I4_1/amd$ (No. 141), Z = 4. Table 13 gives the Rietveld refinement results. In the tetragonal structure of DyAlSi, Dy located in 4a position and Al and Si occupied the 8e position, as seen in Table 14. Table 15 lists the bond lengths and the number of the nearest neighbors of each atom for the tetragonal structure of DyAlSi [3]. One can noted that these bond lengths are reasonable compared with the radius of Dy, Al and Si atoms (R_{Dy} = 0.177nm, R_{Al} = 0.1431 nm and R_{Si} = 0.134nm).

Table 13. Rietveld refinement data of the tetragonal structure of DyAlSi and ErAlSi [3, 4]

Formula	DyAlSi	ErAlSi
Space group	$I4_1/amd$ (No. 141)	$I4_1/amd$ (No. 141)
Cell Parameters (nm)	a= 0.41254(5), c=1.44079(3)	a=0.42244 (1), c=1.45257(2)
Volume of unit cell (nm^3)	0.24521(4)	0.25922
Formula units per unit cell	4	4
Calculated density (g/cm^3)	5.891	5.689
Number of refined parameters	22	34
R_p	0.106	0.089
R_{wp}	0.135	0.116

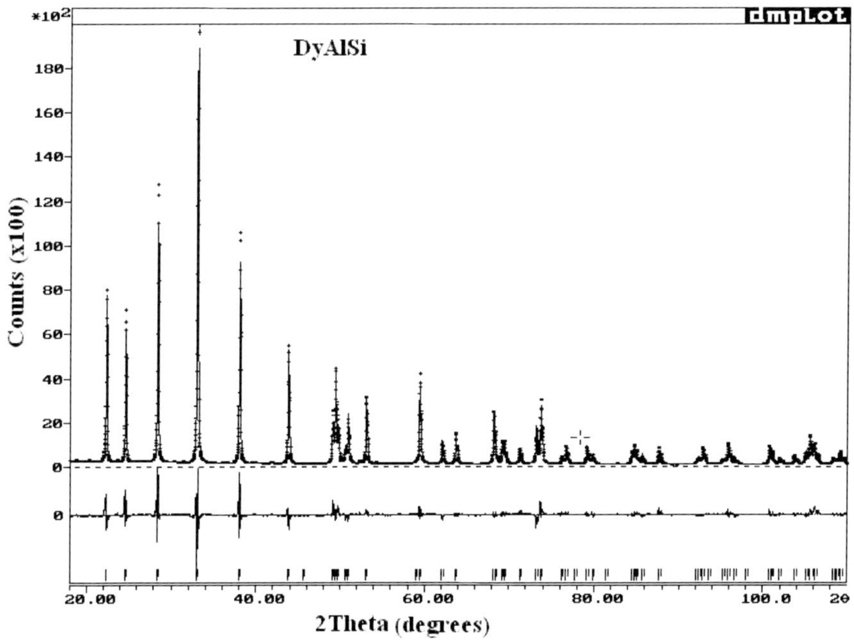

Figure 15. The X-ray diffraction pattern of DyAlSi. •, observed patterns; Solid line, calculated patterns; |, Possible positions of Bragg reflections [3].

Table 14. Atomic parameters for the tetragonal structure of DyAlSi [3]

Atoms	Positions	x	y	z	Occupancy
Dy	4a	0	0.75	0.125	1
M(=Al,Si)	8e	0	0.25	0.2910(2)	(0.5Al+0.5Si)

Table 15. The bond lengths and the number of the nearest neighbors of each atom for the tetragonal structure of DyAlSi [3]

Atoms	Neighbor atoms	Bond lengths (nm)	Number of the nearest atoms
Dy	-M	0.3158(1)	12
M=(Al,Si)	-Dy	0.3158(1)	6
	-M	0.2376(1)	2
	-M	0.2420(1)	1

Pukas et al. have reported that the ternary compound ErAlSi [35] was crystallized in an orthorhombic, YAlGe-type structure with space group Cmcm (No. 63) and lattice parameters a = 0.39740(3) nm, b = 1.0158(1) nm, and c = 0.6554(4). The relation between the YAlGe-type orthorhombic and α-$ThSi_2$-type tetragonal structures of equiatomic alumosilicides and alumogermanides was also discussed in their work.

Figure 18 shows projections of the crystal structure of ErAlSi with the orthorhombic and tetragonal structures on the bc plane.

The unit cell elongates in the [010] direction and shrinks in the [001] direction with different annealing conditions, and the crystal structure changes from the orthorhombic structure to the tetragonal structure.

It was found that the unit cell volume of ErAlSi compound crystallized in the tetragonal structure is 0.25922(nm^3), larger than that of crystallizing in the orthorhombic-type structure 0.2283(nm^3) [35].

The interatomic distances $d_{Er\text{-}M}$ = 0.32154 (nm) (M = Al, Si) in the α-$ThSi_2$-type tetragonal structure are larger than those ($d_{Er\text{-}M}$ = 0.31378 (nm)) in the YAlGe-type orthorhombic structure calculated from the data in Ref. [35]. In the other hand, the interatomic distances $d_{Al\text{-}Si}$ = 0.24379(2) (nm) in the α-$ThSi_2$-type tetragonal structure become shorter than those in the YAlGe-type orthorhombic structure ($d_{Al\text{-}Si}$ = 0.26400 nm) [4].

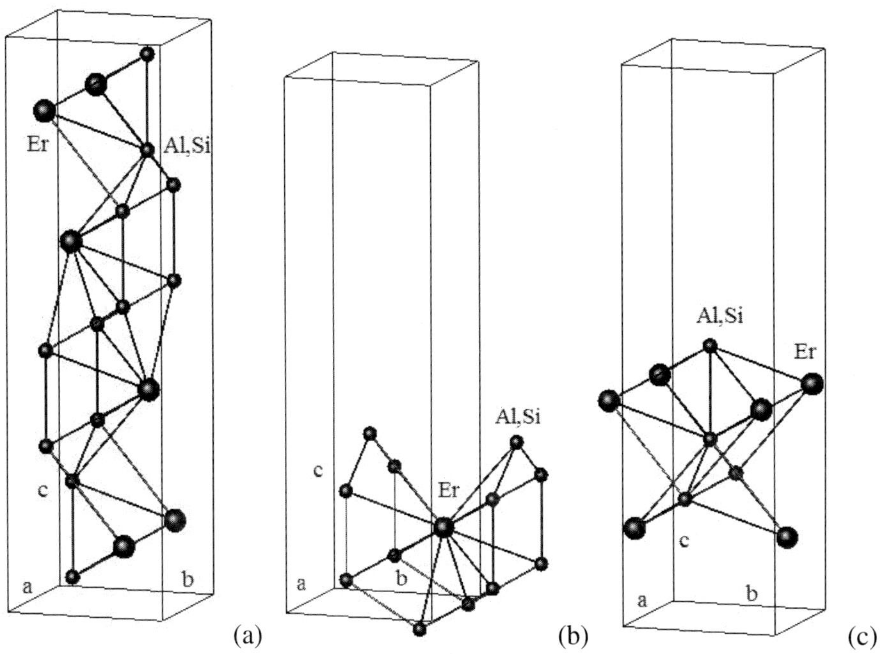

Figure 16. The (a) crystal structure of the tetragonal structure of ErAlSi; (b) environment of atom Er at 4a postions; and (c) environment of atom Al, Si at 8e positions [4].

Table 16. The atomic position parameters for the tetragonal structure of ErAlSi [4]

Atoms	Positions	x	y	z	Occupancy	$B_{eq}(nm^2)$
Er	4a	0	0.75	0.125	1	0.0127(1)
Al, Si	8e	0	0.25	0.2930(1)	(0.5Al+0.5Si)	0.0170(9)

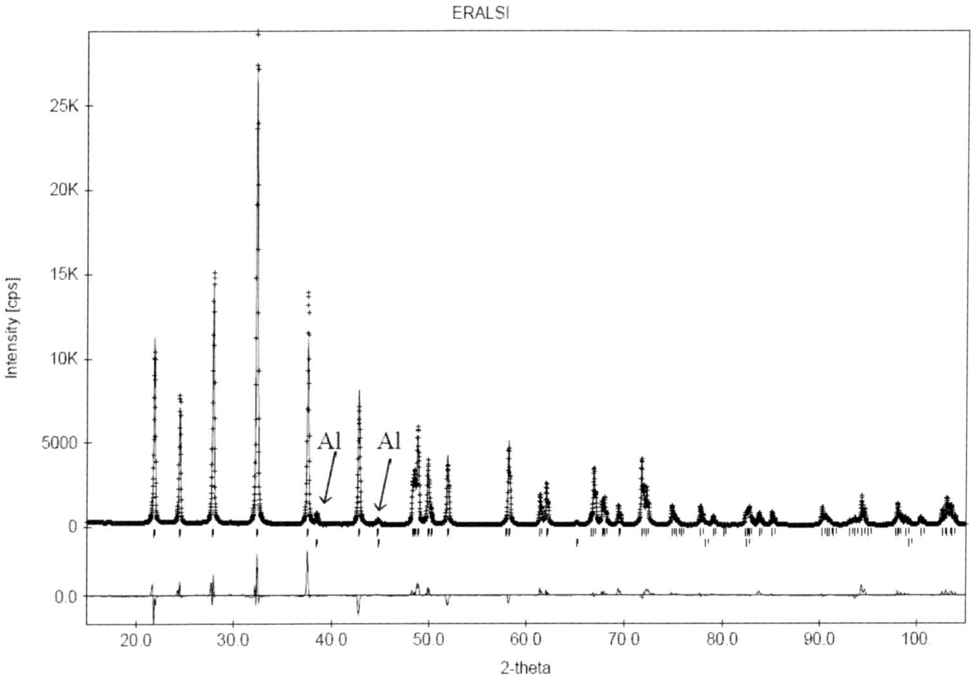

Figure 17. The X-ray powder diffraction pattern of the tetragonal structure of ErAlSi [4].

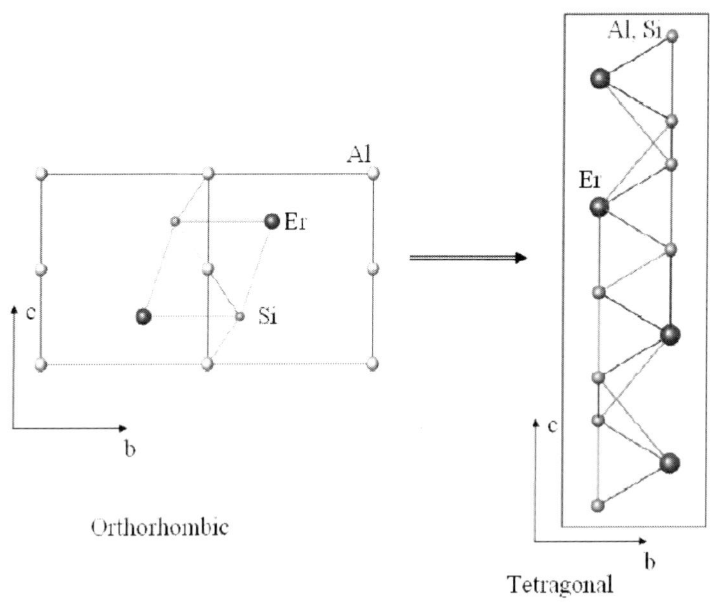

Figure 18. Projections of the crystal structure of ErAlSi with the orthorhombic and tetragonal structures on the bc plane [4].

4.8. $Dy_5Co_6Sn_{18}$

New ternary compound $Dy_5Co_6Sn_{18}$ was found and the sample of the compound was prepared. The indexation of the powder X-ray diffraction pattern indicated that the compound was crystallized in a tetragonal unit cell with the lattice parameters a = 1.35598(3) nm and c = 2.71470(5) nm.

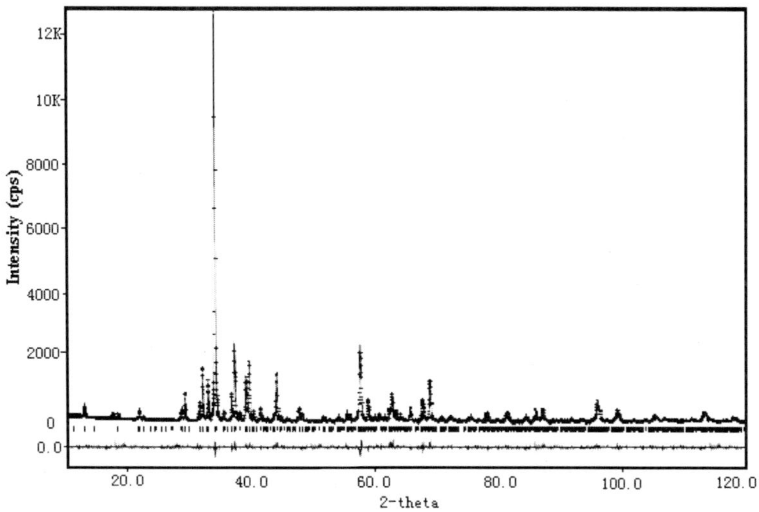

Figure 19a. The X-ray powder diffraction pattern of $Dy_5Co_6Sn_{18}$ [38].

The figure of merit is F_{30} = 41.36 (0.0078, 93). Reflection conditions (*hkl*: *h+k+l=2n*; *hk0: h, k=2n; 0kl: k, l=2n; hhl: 2h+l=4n; 00l: l=4n; h00: h=2n; hh0: h=2n*) pointed to the only space group $I4_1$/acd (No. 142). The compound was found to be isostructural with the $Tb_5Rh_6Sn_{18}$ [37] structure type. The Rietveld refinement results are listed in Table 17 and some selected interatomic distances for the new compound $Dy_5Co_6Sn_{18}$ are given in Table 18 [38].

Table 17. Refined atomic parameters for $Dy_5Co_6Sn_{18}$ [38]

$Dy_5Co_6Sn_{18}$; $I4_1$/acd (No.142); a=1.35598(3) nm, c=2.71470(5) nm, and D_{calc}=8.789 g/cm^3 R-factors(%): R_P=5.42, R_F=4.39, R_{Bragg}=5.67, R_{WP}=8.07,						
Atom	Position	x	y	z	Occupancy	B_{eq}(nm^2)
Dy1	8b	0	0	1/4	1	0.0196(1)
Co1	16d	0	0	0.1231(2)	1	0.0108(6)
Sn1	16e	1/4	0.2081(3)	1/8	1	0.0145(2)
Sn2	16f	0.1721(1)	0.1721(1)	1/4	1	0.0089(4)
Sn3	16f	0.3289(4)	0.3289(4)	1/4	1	0.0027(8)
Co2	32g	0.2451(3)	0.0013(2)	0	1	0.0108(7)
Sn4	32g	0.0899(2)	0.0908(3)	0.0447(2)	1	0.0029(6)
Sn5	32g	0.3290(2)	0.0089(4)	0.0854(3)	1	0.0084(1)
Sn6	32g	0.9987(2)	0.3249(1)	0.0861(2)	1	0.0035(2)
Dy2	32g	0.1288(3)	0.1354(2)	0.9318(1)	1	0.0161(7)

Table 18. Selected the interatomic distances (Å) in the crystal structure of $Dy_5Co_6Sn_{18}$ [38]

Atoms	Neighbor atoms	Bond lengths (nm)	Atoms	Neighbor atoms	Bond lengths (nm)
Dy1	2Sn2	3.384(2)	Sn6	Dy1	3.324(3)
	2Sn3	3.292(3)		Dy2	3.272(2)
	4Sn5	3.297(1)			3.253(2)
	4Sn6	3.344(3)		Co1	2.611(2
Co1	2Dy2	2.954(2)		Sn2	3.281(1)
	2Sn4	2.743(2)		Sn3	3.298(2)
	2Sn5	2.572(1)			3.293(2)
	2Sn6	2.612(2)		Sn5	3.209(3)
Sn1	2Dy2	3.109(3)	Dy2	Co1	2.944(2)
	2Sn2	3.550(1)		Co2	3.054(2)
	2Sn4	3.463(2)			2.967(2)
	2Sn5	3.102(2)		Sn1	3.109(1)
	2Sn6	3.560(2)		Sn2	3.254(2)
Sn2	Dy1	3.394(1)		Sn3	3.271(2)
	Dy2	3.352(3)		Sn4	3.114(3)
		3.264(2)			3.175(2)
	Sn3	2.951(1)			3.181(2)
	Sn5	3.212(2)		Sn5	3.233(2)
		3.266(2)			3.261(1)
	Sn6	3.281(3)		Sn6	3.272(2)
		3.337(1)			3.253(3)
	Co2	2.544(2)	Sn4	Dy2	3.181(2)
		2.650(3)			3.175(2)
Sn3	2Dy2	3.271(2)			3.114(1)
	Dy1	3.282(2)		Co1	2.742(1)
	Sn2	2.951(1)		Co2	2.712(3)
	Sn5	3.358(2)			2.719(2)
		3.362(2)	Sn5	Dy1	3.287(3)
	Sn6	3.294(3)		Dy2	3.261(2)
		3.298(2)			3.233(2)
Co2	Dy2	3.054(2)		Co1	3.571(1)
		2.967(1)		Co2	2.585(2)
	Sn2	2.544(2)		Sn1	3.101(2)
	Sn3	2.597(2)		Sn2	3.212(1)
	Sn4	2.712(3)		Sn3	3.358(1)
		2.719(2)		Sn6	3.209(2)
	Sn5	2.585(2)			
	Sn6	2.566(1)			

4.9. $Al_{0.35}GdGe_2$, $Al_{0.33}DyGe_2$, $HoAl_{0.34}Ge_2$ and $Al_{0.32}ErGe_2$

The RE-Al-Ge (RE = rare earths) alloys exhibit a variety of interesting and outstanding physical properties. To discover new ternary compounds with excellent properties, many efforts have been done by a large number of researchers. New ternary compounds Al_xREGe_2 (RE = Gd, Dy, Ho, Er) with slightly different x ($0.32 < x < 0.35$) were found [39-42].

All reflections for $HoAl_{0.34}Ge_2$ were well indexed using the Jade 5.0 program in an orthorhombic unit cell with the lattice parameters a = 0.40914(2) nm, b = 1.61624(5) nm, and c = 0.39927(1) nm. Reflection conditions (*h k l*: *h* + *k*=2*n*; 0*k l*: *k*=2*n*; *h* 0 *l*: *h*=2*n*, *l*=2*n*; *hk*0: *h* + *k*=2*n*; *h*00: *h*=2*n*; 0*k* 0: *k*=2*n*; 00*l*: *l*=2*n*) pointed to two possible space groups *Cmcm* (No. 63) and *Cmc*21 (No. 36). Formula units per unit cell is Z = 4. It was found that $HoAl_{0.34}Ge_2$ is isostructural with $CeNiSi_2$ with space group *Cmcm* (No. 63) [43]. Thus the space group *Cmcm* (No. 63) was chosen for $HoAl_{0.34}Ge_2$. The Rietveld refinement of $HoAl_{0.34}Ge_2$ was performed and the indexing results were used as the starting lattice parameters. The atomic positions of the unit cell were taken from the structure of $CeNiSi_2$. Figure 19b shows the X-ray diffraction pattern of $HoAl_{0.34}Ge_2$ [41].

The refinement reliability factors are R_p=10.51%, R_{wp} = 6.45%, R_F = 6.57% and the lattice parameters after refinement are a = 0.40905(1) (nm), b = 1.61575(5) (nm) and c =0.39395(1) (nm). The occupancy of Al atom is 0.34 for the new compound. Calculation shows that the shortest Al–Ge distance is 0.2082(2) nm, which indicates a contraction of about 16% relative to the sums of the radii. Refs. [44, 45] also observed a similar contraction results in the RE–M–Ge system (M=Fe, Co, Ni). This result probably relates to the partial occupancy of the Al atom position, as given in Table 19. Table 20 lists the selected interatomic distances (nm) in the crystal structure of $HoAl_{0.34}Ge_2$ [41].

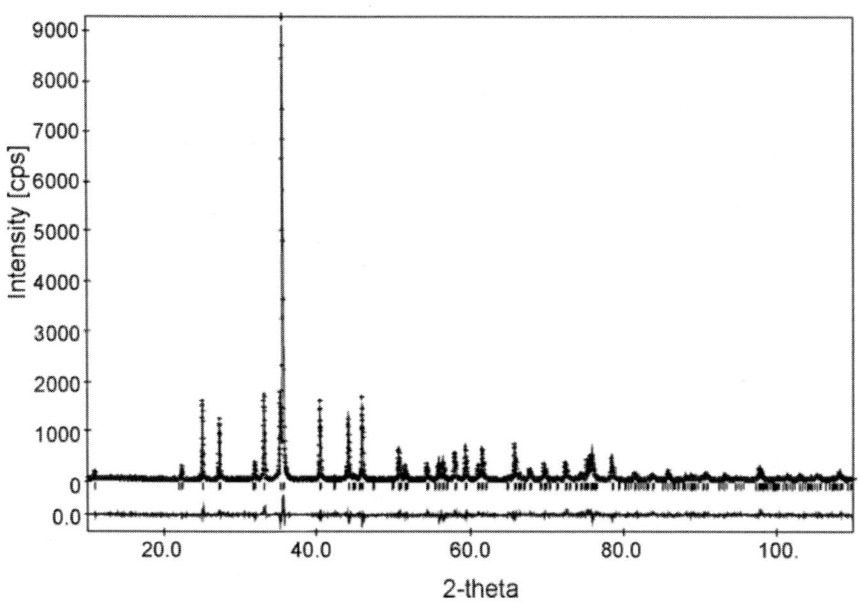

Figure 19b. The X-ray diffraction pattern of $HoAl_{0.34}Ge_2$ [41].

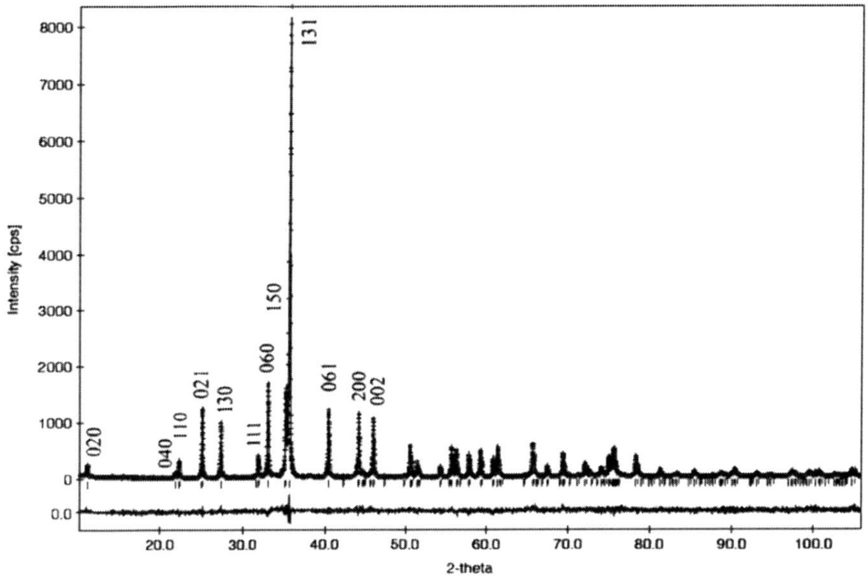

Figure 20. The X-ray diffraction pattern of $DyAl_{0.33}Ge_2$ [40].

The compound $ErAl_{0.32}Ge_2$ with similar composition is also isostructural with $HoAl_{0.34}Ge_2$, but the occupancy of Al atoms at 4c position is 0.32. The lattice parameters of $ErAl_{0.32}Ge_2$ are a = 0.40701(2) nm, b = 1.60401(9) nm, and c = 0.39240(2) nm. Table 21 gives the atomic parameters of the compound [42]. The isostructural compound $DyAl_{0.33}Ge_2$ has lattice parameters of a = 0.41018(2) nm, b = 1.62323(6) nm, c = 0.39463(1) nm and the occupancy of Al atoms at 4c position is 0.33, which is very close to that of $HoAl_{0.34}Ge_2$. [40]. Figure 20 presents the X-ray diffraction pattern of $DyAl_{0.33}Ge_2$. Similarly, the new compound $Al_{0.35}GdGe_2$ was crystallized in the orthorhombic with the $CeNiSi_2$ structure type.

It is isostructural with $HoAl_{0.34}Ge_2$, except the slightly different occupancy of Al atoms. The lattice parameters of $Al_{0.35}GdGe_2$ are a = 0.40874(2) nm, b = 1.61499(5) nm, and c = 0.39372(1) nm [39].

Table 19. Atomic coordinates and thermal parameters for the compound $HoAl_{0.34}Ge_2$ [41]

$HoAl_{0.34}Ge_2$; *Cmcm* (No. 63); a = 0.40905(1) nm, b = 1.61575(5) nm,c =0.39395(1) nm; D_{calc}=8.789 g/cm^3 R-factors(%): R_p=0.1051, R_{wp} = 0.0645, R_F = 0.0657							
Atom	Position	*x*	*y*	*z*	Occupancy	B_{eq} (nm^2)	
Ho	4c	0	0.1035(1)	1/4	1	0.0050(4)	
Ge1	4c	0	0.4512(1)	1/4	1	0.0093(5)	
Ge2	4c	0	0.7479(1)	1/4	1	0.0109(6)	
Al	4c	0	0.2959(8)	1/4	0.34(1)	0.0047(8)	

Table 20. Selected interatomic distances (nm) in the crystal structure of $HoAl_{0.34}Ge_2$ [41]

Atoms	Bond lengths (nm)	Number of neighbor atoms	Atoms	Bond lengths (nm)	Number of neighbor atoms
Ho–Ho	0.3866(2)	2×	Ho–Al	0.3285(7)	4×
Ho–Ge(1)	0.2971(1)	4×	Ge(1)–Ge(1)	0.2535(4)	2×
Ho–Ge(1)	0.3209(2)	2×	Ge(1)–Al	0.2529(6)	1×
Ho–Ge(2)	0.3102(1)	2×	Ge(2)–Ge(2)	0.2840(1)	4×
Ho–Ge(2)	0.3099(3)	2×	Ge(2)–Al	0.2176(2)	2×
Ho–Al	0.3074(5)	1×	Ge(2)–Al	0.2082(2)	2×

Table 21. The atomic parameters of the compound $ErAl_{0.32}Ge_2$. [42]

Atom	Position	*x*	*y*	*z*	Occupancy	$B_{eq}(nm^2)$
Er	4c	0	0.1039(1)	1/4	1	0.0133(6)
Ge1	4c	0	0.4478(2)	1/4	1	0.0156(6)
Ge2	4c	0	0.7468(2)	1/4	1	0.0261(6)
Al	4c	0	0.303(3)	1/4	0.32	0.0613(9)

4.10. $NdCo_2Al_8$

New ternary compound $NdCo_2Al_8$ has been found and investigated by X-ray diffraction technique [46]. All of the X-ray diffraction peaks for the compound $NdCo_2Al_8$ were indexed successfully on the basis of orthorhombic structure. The Smith and Snyder figure of merit for the index, F_N, is F_{30} = 117.3(47). There are four formulas in each unit cell. The reflection conditions yield two possible space groups: Pbam (No. 55) and Pba2 (No. 32) for the structure of $NdCo_2Al_8$. The compound $NdCo_2Al_8$ is isostructural with $PrCo_2Al_8$ compound (with an orthorhombic $CaCo_2Al_8$-type structure, space group Pbam (No. 55)) [47]). So the space group Pbam (No. 55) was assigned for the orthorhombic structure of $NdCo_2Al_8$. The X-ray powder diffraction pattern of $NdCo_2Al_8$ is shown in Figure 21.

The reliability R factors for the Rietveld refinement are R_p = 0.093, R_{wp} = 0.126, and R_{exp} = 0.064. The refined lattice parameters for $NdCo_2Al_8$ are a = 1.24517(2) nm, b = 1.43621(3) nm, and c= 0.40055(1) nm. The atomic parameters are listed in Table 22 and the bond lengths (<0.33nm) and the nearest neighbors of each atom for $NdCo_2Al_8$ are presented in Table 23. Projections on different planes of the crystal structure for $NdCo_2Al_8$ are given in Figure 22 [46].

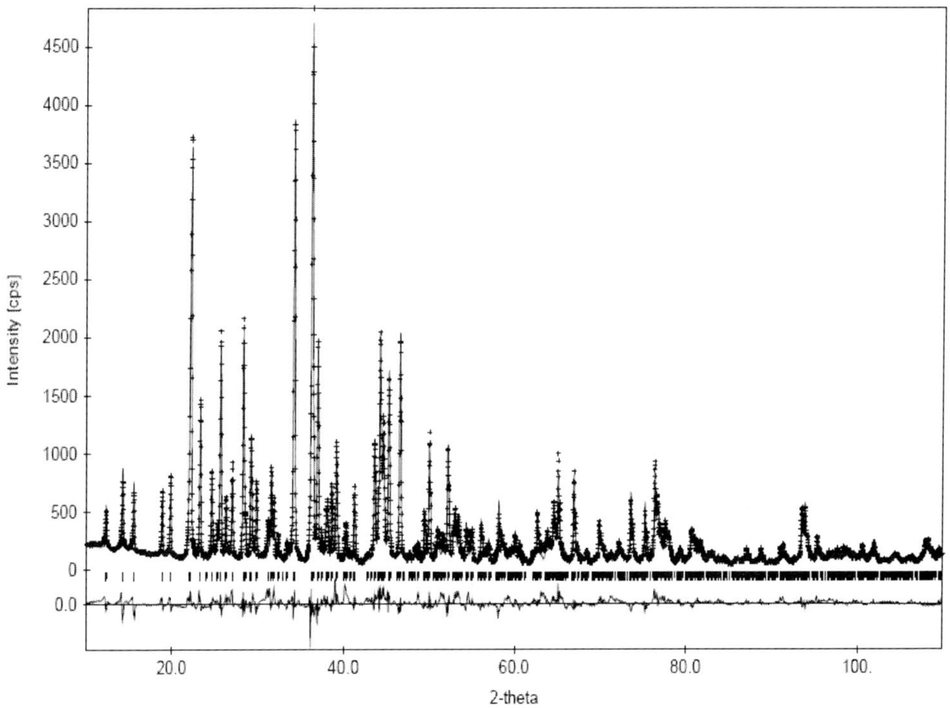

Figure 21. The X-ray powder diffraction pattern of $NdCo_2Al_8$ [46].

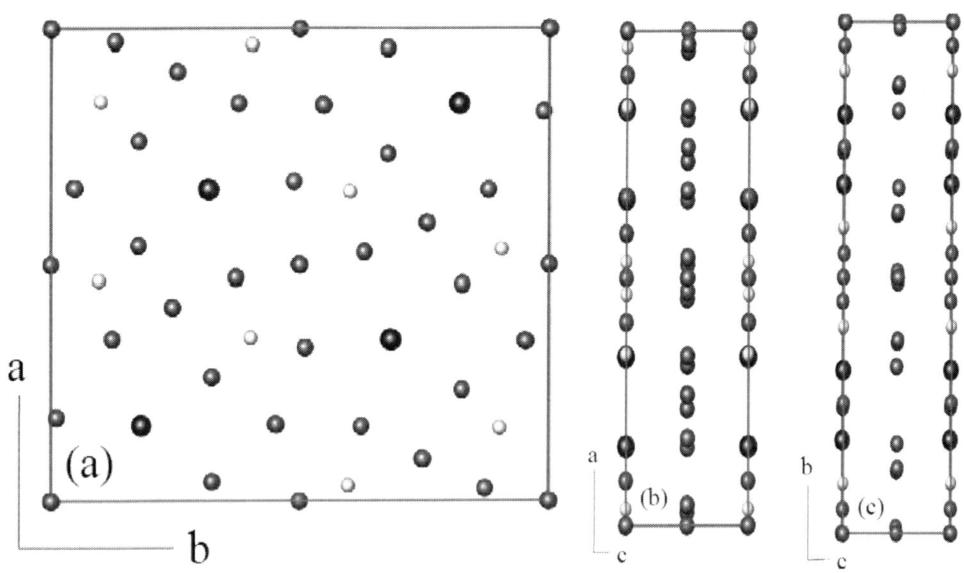

Figure 22. Projection of $NdCo_2Al_8$ structure on ab, ac, and bc planes [46]. The figure shows (a) projection on ab plane; (b) projection on ac plane; and (c) projection on bc plane.

Table 22. The atomic parameters for the $NdCo_2Al_8$ compound [46]

$NdCo_2Al_8$; Pbam (No. 55); a = 1.24517(2) nm, b = 1.43621(3) nm,c =0.40055(1) nm; D_{calc}=4.430 g/cm^3 R-factors(%): R_p=0.093, R_{wp} = 0.126, R_{exp}= 0.064						
Atoms	Positions	x	y	z	Occupancy	B_{eq}(nm^2)
Nd	4g	0.3407(2)	0.3183(1)	0	1	0.00941
Co(1)	4g	0.0334(3)	0.4045(3)	0	1	0.00062
Co(2)	4g	0.1545(6)	0.0985(3)	0	1	0.01142
Al(1)	4h	0.0271(7)	0.1277(5)	0.5	1	0.00009
Al(2)	4h	0.158(1)	0.3779(5)	0.5	1	0.00009
Al(3)	4h	0.2361(6)	0.1753(7)	0.5	1	0.00010
Al(4)	4h	0.3239(7)	0.4887(5)	0.5	1	0.00010
Al(5)	4h	0.4586(6)	0.1748(7)	0.5	1	0.00010
Al(6)	4g	0.0900(9)	0.2542(6)	0	1	0.00010
Al(7)	4g	0.3380(9)	0.0470(6)	0	1	0.00011
Al(8)	2d	0	0.5	0.5	1	0.00009
Al(9)	2a	0	0	0	1	0.00010

4.11. $Al_2RE_{11}Ge_8$ (RE = Y, Gd, Tb, Dy, Ho, Er)

A new series of new ternary rare earth compounds $Al_2RE_{11}Ge_8$ (RE = Y, Gd, Tb, Dy, Ho, Er) were found and the crystal structure of these new compounds was studied using X-ray powder diffraction data. The indexation of the powder X-ray diffraction pattern of $Al_2Er_{11}Ge_8$ indicated that the compound was crystallized in a tetragonal unit cell with the lattice parameters a = 1.0917(3) nm and c = 1.6207(1) nm. Each unit cell contains 84 atoms based on the lattice parameters and the density (D_m = 8.69 g/cm3) measured by a specific gravity bottle method.

Table 23. The bond lengths (<3.3Å) and the nearest neighbors of each atom for $NdCo_2Al_8$ [46]

Atoms	Neighbor atoms and numbers	Bond lengths (Å)	Atoms	Neighbor atoms and numbers	Bond lengths (Å)	Atoms	Neighbor atoms and numbers	Bond lengths (Å)
Nd-	Al(1)×2	0.31622	Co(1)-	Co(1)	0.28665	Co(2)-	Al(1)×2	0.25891
	Al(2)×2	0.31494		Al(2)×2	0.25620		Al(3)×2	0.25020
	Al(3)×2	0.31505		Al(5)×2	0.24851		Al(4)×2	0.25632
	Al(4)×2	0.31692		Al(6)	0.22708		Al(6)	0.23760
	Al(5)×2	0.32270		Al(7)	0.25308		Al(7)	0.24016
	Al(6)	0.32546		Al(7)	0.25986		Al(9)	0.23879
	Al(6)	0.32742		Al(8)×2	0.24628			
	Al(9)	0.32779						
Al(1)-	Nd×2	0.31622	Al(2)-	Nd×2	0.31494	Al(3)-	Nd×2	0.31505
	Co(2)×2	0.25891		Co(1)×2	0.25620		Co(2)×2	0.25020

Table 23. (Continued)

Atoms	Neighbor atoms and numbers	Bond lengths (Å)	Atoms	Neighbor atoms and numbers	Bond lengths (Å)	Atoms	Neighbor atoms and numbers	Bond lengths (Å)
	Al(3)	0.26907		Al(3)	0.30680		Al(1)	0.26907
	Al(4)	0.30326		Al(4)	0.26076		Al(2)	0.30680
	Al(4)	0.27253		Al(5)	0.25957		Al(4)	0.27822
	Al(5)	0.29620		Al(6)×2	0.28079		Al(5)	0.27705
	Al(6)×2	0.28152		Al(7)×2	0.31483		Al(6)×2	0.29334
	Al(9)×2	0.27365		Al(8)	0.26355		Al(7)×2	0.30027
Al(4)-	Nd×2	0.31692	Al(5)-	Nd×2	0.32270	Al(6)-	Nd	0.32546
	Co(2)×2	0.25632		Co(1)×2	0.24851		Nd	0.32742
	Al(1)	0.30326		Al(1)	0.29620		Co(1)	0.22708
	Al(1)	0.27253		Al(2)	0.25957		Co(2)	0.23760
	Al(2)	0.26076		Al(3)	0.27705		Al(1)×2	0.28152
	Al(3)	0.27822		Al(6)×2	0.27799		Al(2)×2	0.28079
	Al(7)×2	0.29624		Al(7)×2	0.31040		Al(3)×2	0.29334
	Al(9)×2	0.29741		Al(8)	0.25629		Al(5)×2	0.27799
Al(7)-	Co(1)	0.25308	Al(8)-	Co(1)×4	0.24628	Al(9)-	Nd×2	0.32779
	Co(1)	0.25986		Al(2)×2	0.26355		Co(2)×2	0.23879
	Co(2)	0.24016		Al(5)×2	0.25629		Al(1)×4	0.27365
	Al(2)×2	0.31483		Al(7)×4	0.29216		Al(4)×4	0.29741
	Al(3)×2	0.30027						
	Al(4)×2	0.29624						
	Al(5)×2	0.31040						
	Al(8)×2	0.29216						

Reflection conditions (hkl: h + k + l = 2n; hk0: h + k = 2n; 0kl: k + l = 2n; hhl: l = 2n; 00l: l = 2n; h00: h = 2n) pointed to five possible space groups *I4/mmm* (No. 139), *I*422 (No. 97), $I\bar{4}m2$ (No. 119), $I\bar{4}2m$ (No. 121), and *I4mm* (No. 107). By comparing crystallographic characteristics of the $Al_2Er_{11}Ge_8$ compound with those presented in the structure type database, it was found that $Al_2Er_{11}Ge_8$ and $Ho_{11}Ge_{10}$ [48] have the same structure type. Therefore, the space group *I4/mmm* (No. 139) and the atomic position parameters of $Ho_{11}Ge_{10}$ were taken as the starting values to refine the structural parameters of $Al_2Er_{11}Ge_8$. Figure 23 shows the X-ray powder diffraction pattern of $Al_2Er_{11}Ge_8$. The Er atoms were assigned to the Ho sites in $Ho_{11}Ge_{10}$, and the Ge atoms together with the Al atoms were assigned to the Ge sites in $Ho_{11}Ge_{10}$.

Trial and test method was applied to find out the mode of Al and Ge at Ge sites in the structure of $Ho_{11}Ge_{10}$. Several compound modes for Al and Ge at Ge sites were tested. When eight Al atoms occupy the Ge (8h) sites in $Ho_{11}Ge_{10}$, the refined results are optimized, as given in Table 24. Figure 24 presents the projection of the $Al_2Er_{11}Ge_8$ structure on the *xy*-plane. The selected interatomic distances for the new compound $Al_2Er_{11}Ge_8$ are listed in Table 25 [49].

Table 26 lists the lattice parameters and unit cell volumes of $Al_2RE_{11}Ge_8$ (RE = Y, Gd, Tb, Dy, Ho, Er) compounds as obtained from XRD patterns corrected by adding standard reference materials SRM 640c. Figure 25 shows the cell parameters and unit cell volume for these compounds as a function of the rare earth radius. The unit cell volume of these compounds indicates that the data are in good agreement with the behavior of the atomic radii of the respective rare earth elements [49].

Table 24. Refined atomic parameters for $Al_2Er_{11}Ge_8$ [49]

$Al_2Er_{11}Ge_8$; *I4/mmm*; a = 1.08242(5) nm, c = 1.60001(5) nm, D_{calc} = 8.765 g/cm^3						
R-factors (%): R_P = 7.13, R_F = 3.87, R_B = 6.10, R_{WP} = 10.68						
Atom	Position	x	y	z	Occupancy	$B_{eq}(nm^2)$
Ge(1)	4d	0	1/2	1/4	1	0.0416(2)
Ge(2)	4e	0	0	0.1201(5)	1	0.0595(2)
Al	8h	0.3806(3)	0.3806(3)	0	1	0.0221(3)
Ge(3)	8i	0.3596(4)	0	0	1	0.0385(5)
Ge(4)	16m	0.2070(2)	0.2070(2)	0.3240(2)	1	0.0413(3)
Er(1)	4e	0	0	0.3351(2)	1	0.0469(4)
Er(2)	8h	0.1756(2)	0.1756(2)	0	1	0.0351(2)
Er(3)	16n	0	0.2509(2)	0.1888(2)	1	0.0299(1)
Er(4)	16n	0	0.3293(2)	0.3980(1)	1	0.0334(1)

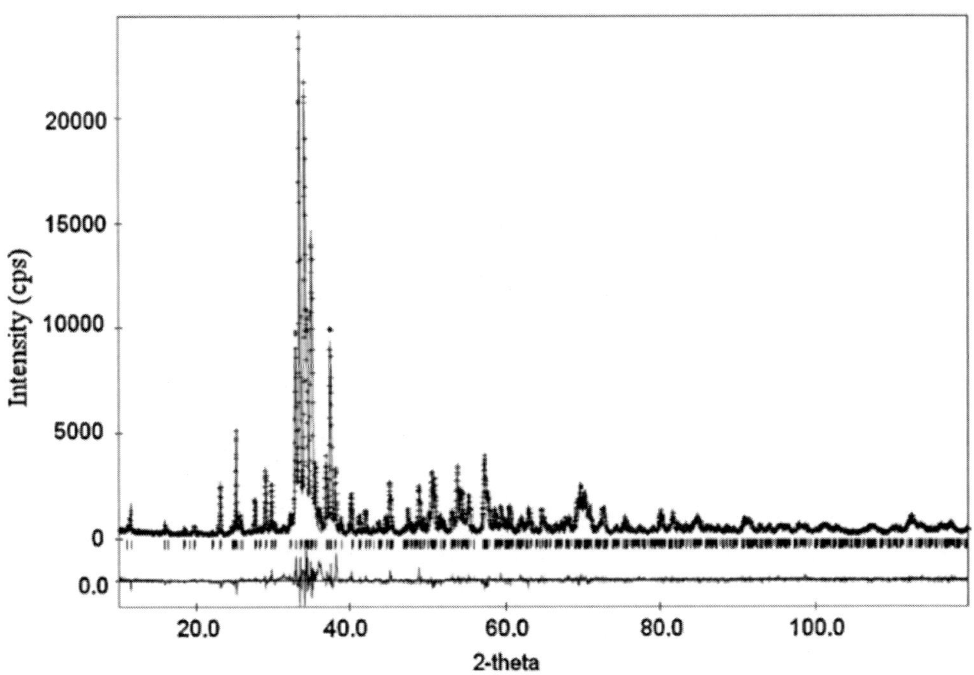

Figure 23. The X-ray powder diffraction pattern of $Al_2Er_{11}Ge_8$ [49].

Table 25. Selected distances in the crystal structure of $Al_2Er_{11}Ge_8$ [49]

Atom-atom		d (nm)	Atom–atom		d (nm)
Ge(2)	4Er(3)	0.2933(5)	Er(1)	4Ge(4)	0.3176(2)
	4Er(2)	0.3307(1)		4Al	0.3211(4)
	Er(1)	0.3442(3)		Ge(2)	0.3442(3)
Ge(3)	2Er(2)	0.2756(4)	Er(2)	2Ge(3)	0.2756(4)
	4Er(4)	0.2898(3)		Al	0.3141(2)
	Ge(3)	0.3042(5)		2Ge(2)	0.3307(1)
	2Er(3)	0.3242(4)	Er(4)	2Ge(4)	0.3342(3)
Er(3)	Ge(1)	0.2871(5)		2Ge(4)	0.2861(5)
	Ge(2)	0.2933(5)		2Ge(3)	0.2898(3)
	2Ge(4)	0.3153(4)		Ge(1)	0.3004(2)
	2Ge(4)	0.3214(2)			
	Ge(3)	0.3242(4)	Al	2Al	0.3085(6)
	Er(4)	0.3454(3)		Er(4)	0.3266(5)
				Er(3)	0.3454(3)
Ge(4)	Ge(4)	0.2712(4)		2Al	0.2587(2)
	Er(4)	0.2861(5)		4Er(4)	0.3085(6)
	Al	0.3120(2)		2Ge(4)	0.3120(2)
	2Er(3)	0.3153(4)		Er(2)	0.3141(2)
	Er(1)	0.3176(2)		2Er(1)	0.3211(4)
	2Er(3)	0.3214(2)			
	Er(2)	0.3342(3)	Ge(1)	4Er(3)	0.2871(5)
				4Er(4)	0.3004(2)

Table 26. Lattice constants and unit cell volumes of $Al_2RE_{11}Ge_8$ (RE = Er, Ho, Dy, Tb, Gd, Y) [49]

Compound	a(nm)	c(nm)	V(nm^3)	c/a
$Al_2Er_{11}Ge_8$	1.08242(4)	1.60001(6)	1.87473	1.478
$Al_2Ho_{11}Ge_8$	1.08536(4)	1.61136(6)	1.89819	1.485
$Al_2Dy_{11}Ge_8$	1.09240(5)	1.62039(7)	1.93368	1.483
$Al_2Tb_{11}Ge_8$	1.09371(3)	1.62855(3)	1.94808	1.489
$Al_2Y_{11}Ge_8$	1.09580(4)	1.62653(2)	1.95311	1.484
$Al_2Gd_{11}Ge_8$	1.09924(5)	1.63026(5)	1.96990	1.483

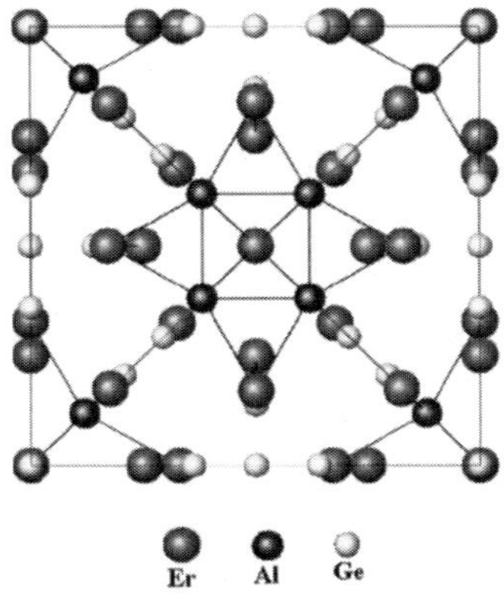

Figure 24. Projection of the $Al_2Er_{11}Ge_8$ structure on the *xy*-plane [49].

4.12. Other New Compounds in Al-Dy-Ge System

During the investigation on the isothermal section of the phase diagram of the Al-Dy-Ge system at 673K, five novel compounds, i.e., $Al_{0.33}DyGe_2$, $AlDy_2Ge_6$, Al_2DyGe_2, $AlDy_3Ge_3$, $Al_{3-x}Dy_{11}Ge_{7+x}$ (x = 0 - 0.7), and one pseudo-binary compound $Al_{3-x}DyGe_x$ in the system were found. The compounds $Al_{0.33}DyGe_2$ and Al_2DyGe_2 are discussed in the previous sections. Table 27 lists the crystallographic data for these novel compounds [23].

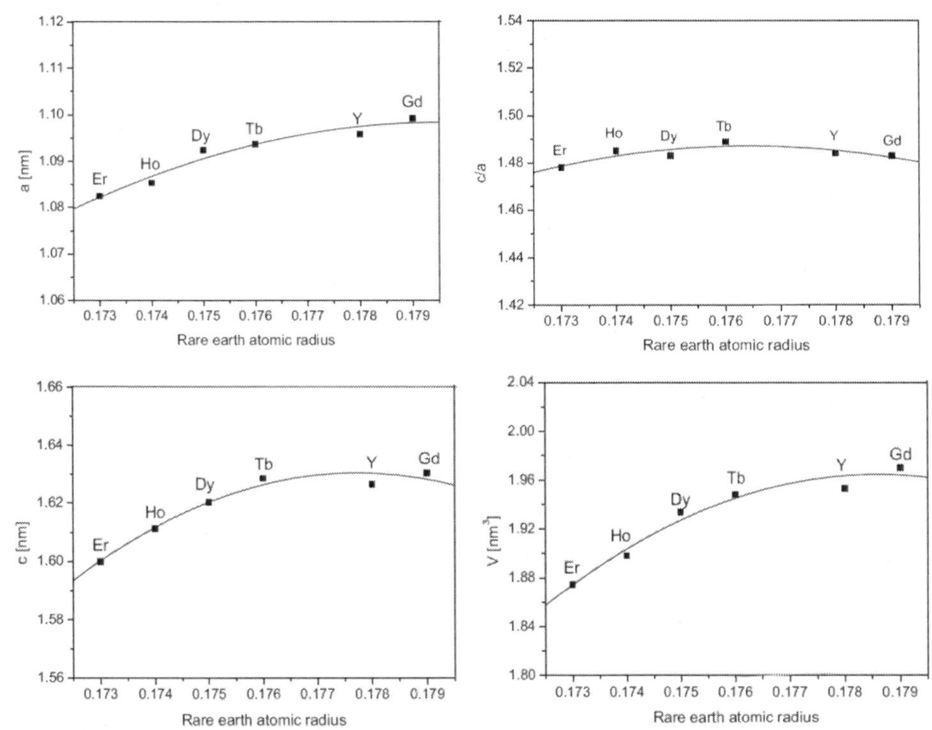

Figure 25. Cell parameters of $Al_2RE_{11}Ge_8$ (RE = Y, Gd, Tb, Dy, Ho, Er) versus rare earth atomic radius [49].

Table 27. Crystallographic data for the novel compounds in the Al-Dy-Ge system [23]

Compounds	Space group	Structure type	Lattice parameters (nm)		
			a	b	c
Al_2DyGe_2	P-3ml (164)	La_2O_3	0.4204(3)	–	0.6681(2)
$Al_{0.33}DyGe_2$	Cmcm (63)	$CeNiSi_2$	0.41018(2)	1.62323(6)	0.39463(1)
$AlDy_2Ge_6$	Amm2 (38)	Ce_2CuGe_6	0.4144(4)	0.3992(3)	2.0576(3)
$AlDy_3Ge_3$	Pmcn (62)	unknown	1.7266(8)	0.8175(2)	0.5999(3)
$Al_{3-x}Dy_{11}Ge_{7+x}$ (x=0)	I4/mmm(139)	unknown	1.09816(4)	–	1.61037(8)
$Al_{3-x}DyGe_x$ (x=0)	$P6_3/mmc(194)$	Ni_3Ti	0.6095(1)	–	0.9531(1)

4.13. $Dy_{1.2}Fe_4Si_{9.8}$

The indexation of the X-ray diffraction pattern of the novel compound $Dy_{1.2}Fe_4Si_{9.8}$ shows that the compound was crystallized in hexagonal structure [50]. The reflection conditions yield three possible space groups: $P6_3/mmc$ (No. 194), $P\bar{6}2c$ (No. 190) and $P6_3mc$ (No. 186) for the structure of compound $Dy_{1.2}Fe_4Si_{9.8}$. It was found that this compound is isostructural with the $Er_{1.2}Fe_4Si_{9.8}$ and $Gd_{1.2}Fe_4Si_{9.8}$ (space group $P6_3/mmc$ (No. 194) [51, 52]. So the space group $P6_3/mmc$ (No. 194) was chosen for the hexagonal structure of $Dy_{1.2}Fe_4Si_{9.8}$. A total of 23 parameters were refined during the Rietveld refinement. Figure 26 presents the X-ray powder diffraction pattern of $Dy_{1.2}Fe_4Si_{9.8}$. The atomic parameters are listed in Table 28.

The Rietveld refinement results pointed out that Dy atoms occupy 60% of the 2c site positions, and Fe, Si(1) and Si(2) atoms are in the 4f, 4f and 4e positions, respectively, as given in Table 28. The Si(3) atoms occupy 30% of the 6h positions. The crystal structure of $Dy_{1.2}Fe_4Si_{9.8}$ is given in Figure 27. In this figure, all sites are fully occupied. Table 29 gives the bond lengths and the number of the nearest neighbors of each atom for the hexagonal structure of $Dy_{1.2}Fe_4Si_{9.8}$.

However, it can be found from Table 29 that the shortest distances of Dy-Si(3) = 0.14505nm and Si(3)-Si(3) = 0.14292nm are not reasonable compared to the radii of Dy and Si atoms (R_{Dy} = 0.177nm and R_{Si} = 0.117nm). From Figure 27(a), it can be seen that the Si(3) atoms (or 6h site positions) can be divided into two groups, each with three Si(3) atoms in the unit cell. These three positions of each group in the 6h site are very close to each other and among three Dy atoms in 2c sites as shown in Figure 27(b). Compared to the radii of Dy and Si atoms, it is not possible to locate these three Si(3) atoms in their positions at the same time. From the Rietveld refinement result, it also can be noted that only 30% of the 6h positions were occupied by the Si(3) atoms. Hence, it is more reasonable to consider that these three positions of 6h site are occupied by only one Si(3) atom. The only Si(3) atom is located in the space of one group of the close three positions [50].

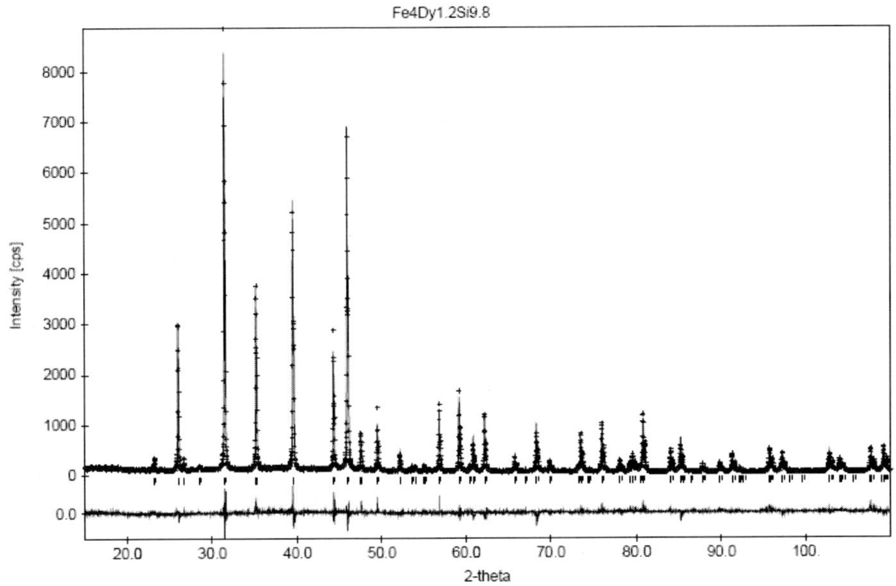

+, observed patterns; Solid line, calculated patterns; |, Possible positions of Bragg reflections; The bottom curve, the difference between the observed and calculated patterns.

Figure 26. The X-Ray powder diffraction pattern of $Dy_{1.2}Fe_4Si_{9.8}$ [50].

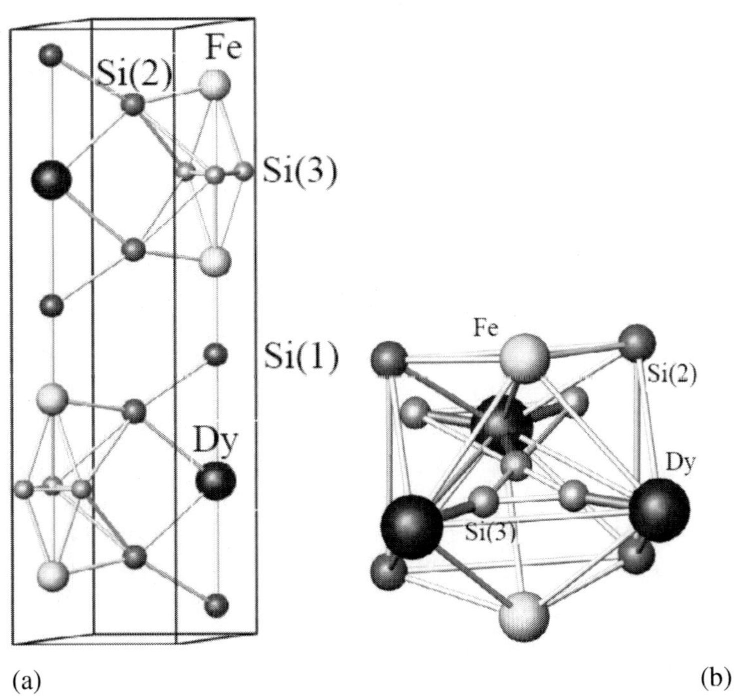

Figure 27. The drawing of crystal structure of compound $Dy_{1.2}Fe_4Si_{9.8}$[50]. The figure shows the full occupancy of all sites. The occupancy factor is 0.6 for Dy atoms in the 2c site positions (a), and 0.3 for the Si(3) atoms in the 6h site positions (b). The environment of atom Si(3) is in 6h site.

Table 28. The atomic parameters for $Dy_{1.2}Fe_4Si_{9.8}$ compound[50]

$Dy_{1.2}Fe_4Si_{9.8}$; $P6_3/mmc$ (No. 194); a = 0.39415(1) nm and c = 1.52771(3) nm, and D_{calc}= 5.614 g/cm^3 R-factors(%): R_p=0.104, R_{wp}=0.134, R_{exp}=0.075						
Atoms	Positions	x	y	z	Occupancy	$B_{eq}(nm^2)$
Dy	2c	1/3	2/3	1/4	0.60 (1)	0.0080
Fe	4f	1/3	2/3	0.8933(1)	1	0.0042
Si(1)	4f	1/3	2/3	0.0465(2)	1	0.0076
Si(2)	4e	0	0	0.1326(2)	1	0.0062
Si(3)	6h	0.5458(9)	-0.5458(9)	1/4	0.30 (5)	0.0026

4.14. $Pr_{117}Co_{57}Sn_{112}$

A new compound $Pr_{117}Co_{57}Sn_{112}$ with a gigantic unit cell in the Pr-Co-Sn system was synthesized and studied by X-ray diffraction technique [53]. The compound was crystallized in a cubic cell with lattice parameter a = 3.08283(3) nm. The reflection conditions (*hkl: h+k, h+l, k+l =2n; 0kl: k,l = 2n; hhl: h+l =2n; h00:h=2n*) yielded three possible space groups: $Fm\bar{3}m$ (No. 225), $F\bar{4}3m$ (No. 216) and $F432$ (No. 209) for $Pr_{117}Co_{57}Sn_{112}$. The space group $Fm\bar{3}m$ (No. 225) has the highest value of the figure of merit, F_N, for the index (F_{30} = 80.4 (30)) among the three space groups. Moreover, it is noted that the composition of the sample, the intensities of reflections, and the calculated lattice parameters of the $Pr_{117}Co_{57}Sn_{112}$ are similar to those of $Dy_{117}Co_{57}Sn_{112}$ compound (with a cubic structure and space group $Fm\bar{3}m$ (No. 225)) [54]. Hence, the space group $Fm\bar{3}m$ (No. 225) was chosen. There are four formula units in each unit cell (Z = 4) with a large number of 1114 atoms in a unit cell. Values of reliability factors for the successful structural refinement are R_p = 0.077, R_{wp} = 0.101, and R_{exp} = 0.032. The Rietveld refinement results indicate that $Pr_{117}Co_{57}Sn_{112}$ is isostructural with $Dy_{117}Co_{57}Sn_{112}$.

The refined lattice parameter is a = 3.08202(5) nm. The refined atomic parameters of $Pr_{117}Co_{57}Sn_{112}$ are listed in Table 30. The X-ray powder diffraction pattern of $Pr_{117}Co_{57}Sn_{112}$ is shown in Figure 28. It can be seen in Table 30, the occupancies of Co2 atoms at the 96k site and Co6 at 32f sites are 50%. Further calculation of the bond lengths for the compound found that the shortest distances of Co2-Co2 = 0.10171 nm and Co2-Co6 = 0.14494 nm are not reasonable, compared to the radius of Co atoms (R_{Co} = 0.125 nm). Figure 29 shows the environment of the atoms Co2 at 96k site and Co6 at 32f site with full occupancies of the 96k site and 32f site. It can be seen in Figure 29(a) that a pair position of the Co2 atoms at 96k site are very close to each other and located between two Co6 atoms at 32f site. It was found that three pairs of such Co2 atoms at 96k site surround one Co6 atom at 32f site, as seen in Figure 29(a) and (b). Compared to the radius of Co atoms, it is not possible to locate a pair of Co atoms in these positions (96k site) at the same time. From the Rietveld refinement result shown in Table 30, it also can be noted that only 50% of the 96k sites was occupied by the Co2 atoms. Hence, it is more reasonable to consider that one pair position of the Co2 atoms at the 96k sites are occupied by only one Co2 atom. The only Co2 atom is located in the space of one pair position of the closest two positions in the 96k site [53].

Table 29. The bond lengths (< 3.0nm) and the nearest neighbors of each atom for $Dy_{1.2}Fe_4Si_{9.8}$ (The Dy atoms randomly occupy 60% of the 2c sites) [50]

Atoms	Neighbor atoms	Number of the nearest atoms	Bond lengths (nm)	Atoms	Neighbor atoms	Number of the nearest atoms	Bond lengths (nm)
Dy-	Dy	6	0.39415	Fe-	Dy	3	0.31577
	Fe	6	0.31577		Fe	6	0.39415
	Si(1)	2	0.31089		Fe	3	0.39758
	Si(2)	6	0.28975		Si(1)	1	0.23405
	Si(3)	6	0.27815		Si(1)	3	0.24545
	Si(3)	3	0.14505		Si(2)	3	0.23098
					Si(3)	3	0.23396
Si(1)-	Dy	1	0.31089		Si(3)	6	0.39212
	Fe	1	0.23405				
	Fe	3	0.24545	Si(3)-	Dy	2	0.27815
	Si(1)	6	0.39415		Dy	1	0.14505
	Si(1)	3	0.26827		Fe	4	0.39212
	Si(2)	3	0.26284		Fe	2	0.23396
	Si(2)	3	0.35588		Si(1)	2	0.34306
	Si(3)	3	0.34306		Si(2)	4	0.26830
					Si(2)	2	0.35821
Si(2)-	Dy	3	0.28975		Si(3)	6	0.39415
	Fe	3	0.23098		Si(3)	4	0.34561
	Si(1)	3	0.26284		Si(3)	2	0.14292
	Si(1)	3	0.35588		Si(3)	2	0.25123
	Si(2)	6	0.39415				
	Si(2)	1	0.35871				
	Si(3)	6	0.26830				
	Si(3)	3	0.35821				

The compound with giant unit cell $Tb_{117}Fe_{52}Ge_{112}$ [55] was crystallized in cubic structure with space group $Fm\overline{3}m$ (No. 225) and its reflection conditions are very similar to those of $Dy_{117}Co_{57}Sn_{112}$. The crystal structures of these two compounds with different Wyckoff positions have been reported [11]. For comparison, a second Rietveld refinement of $Pr_{117}Co_{57}Sn_{112}$ was also carried out using the atomic position parameters of $Tb_{117}Fe_{52}Ge_{112}$ as initial atomic positions. Large values of liability factors for R_p and R_{wp} were obtained. This suggests that the structure of $Pr_{117}Co_{57}Sn_{112}$ is different from that of $Tb_{117}Fe_{52}Ge_{112}$. In these two structures, the Wyckoff positions of rare earth elements and the third metal (Ge or Sn) are the same but those of transition metals (Fe or Co) are different. In the structure of $Tb_{117}Fe_{52}Ge_{112}$, Fe occupied four Wyckoff positions (32f, 32f, 96k and 96k), while Co located at six Wyckoff positions in the structure of $Pr_{117}Co_{57}Sn_{112}$ as shown in Table 30. Detailed difference between the two structures was discussed in Ref. [11].

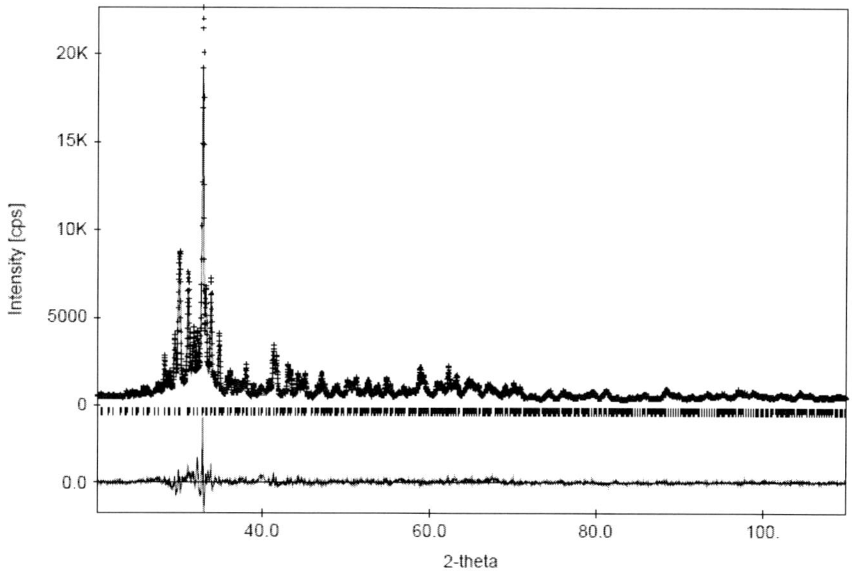

Figure 28. The X-ray powder diffraction pattern of $Pr_{117}Co_{57}Sn_{112}$ [53].

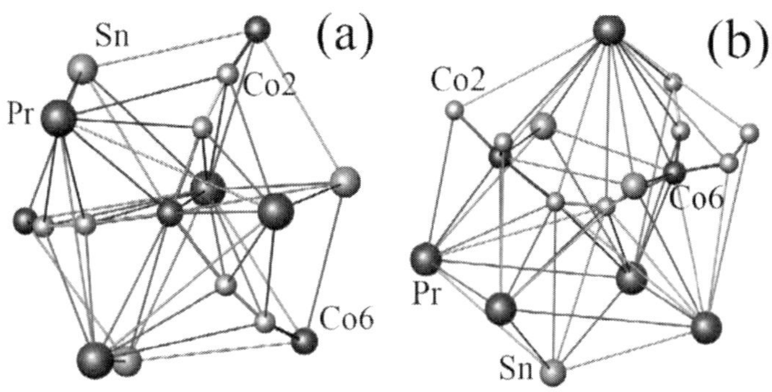

Figure 29. The environments of the atoms Co6 at 32f sites (a) and Co2 at 96k sites (b) in the structure of $Pr_{117}Co_{57}Sn_{112}$ compound [53].

Table 30. Atomic parameters for $Pr_{117}Co_{57}Sn_{112}$ (space group $Fm\overline{3}m$ (No. 225) [53]

$Pr_{117}Co_{57}Sn_{112}$; $Fm\overline{3}m$ (No. 225); a = 3.08202(5) nm and D_{calc}= 7.514 g/cm^3 R-factors(%): R_p = 0.077, R_{wp} = 0.101, R_{exp} = 0.032						
Atoms	Positions	x	y	z	Occupancy	$B_{eq}(nm^2)$
Pr1	96k	0.0674(2)	0.0674(2)	0.1559(2)	1	0.011459
Pr2	96k	0.1793(2)	0.1793(2)	0.4033(2)	1	0.004444
Pr3	96k	0.2005(2)	0.2005(2)	0.0669(2)	1	0.008451
Pr4	96j	0.2525(2)	0.1047(2)	0	1	0.009858
Pr5	48i	0.1212(3)	0.1212(3)	1/2	1	0.008885
Pr6	24e	0.3398(5)	0	0	1	0.015295
Pr7	8c	1/4	1/4	1/4	1	0.001074

$Pr_{117}Co_{57}Sn_{112}$; $Fm\bar{3}m$ (No. 225); a = 3.08202(5) nm and D_{calc}= 7.514 g/cm^3 R-factors(%): R_p = 0.077, R_{wp} = 0.101, R_{exp} = 0.032						
Atoms	Positions	x	y	z	Occupancy	B_{eq}(nm^2)
Pr8	4a	0	0	0	1	0.008010
Co1	96k	0.1667(4)	0.1667(4)	0.2293(6)	1	0.008142*
Co2	96k	0.0739(8)	0.0739(8)	–0.0165(1)	0.50(3)	0.008142*
Co3	32f	0.3942(6)	0.3942(6)	0.3942(6)	1	0.008142*
Co4	32f	0.3055(5)	0.3055(5)	0.3055(5)	1	0.008142*
Co5	4b	1/2	1/2	1/2	1	0.008142*
Co6	32f	0.057(12)	0.057(12)	0.057(12)	0.50(2)	0.008142*
Sn1	96k	0.0738(2)	0.0738(2)	0.3243(3)	1	0.008097
Sn2	96k	0.1092(2)	0.1092(2)	0.2399(2)	1	0.009422
Sn3	48i	0.2081(3)	0.2081(3)	1/2	1	0.010599
Sn4	48h	0.1454(3)	0.1454(2)	0	1	0.006082
Sn5	48g	1/4	1/4	0.141(4)	1	0.006716
Sn6	32f	0.1469(3)	0.1469(2)	0.1469(3)	1	0.008581
Sn7	24e	0.1068(5)	0	0	1	0.003817
Sn8	24e	0.2168(52)	0	0	1	0.000949
Sn9	32f	0.4399(3)	0.4399(3)	0.4399(3)	1	0.019127

*Temperature factors were fixed.

The calculation pointed out that the shortest interatomic distance in the structure of $Tb_{117}Fe_{52}Ge_{112}$ is the shortest interatomic distance between Fe and Ge atoms ($d_{Fe\text{-}Ge}$ = 0.24023 nm), while in the structure of $Pr_{117}Co_{57}Sn_{112}$ the shortest interatomic distance is between Co and Co atoms ($d_{Co\text{-}Co}$ = 0.23820). The shortest interatomic distance between Co and Sn in the latter structure is $d_{Co\text{-}Sn}$ = 0.24396 nm [53].

5. Lattice Thermal Expansion of Rare Earth Compounds

5.1. $CuSb_2Ti_5$

The lattice thermal expansion property of a compound can be studied by the HTXRD technique. The lattice parameters at different temperatures can be calculated from the X-ray diffraction patterns recorded at the corresponding temperatures. Once the lattice parameters at different temperatures are obtained, the thermal expansion coefficient (or the lattice thermal expansion property) of the measured compound can be determined.

The thermal expansion coefficient, α_p, of a given crystalline parameter p can be determined by using the relation expressed as:

$$\alpha_p = \frac{1}{p}\frac{dp}{dT} \tag{5-1}$$

Then, the coefficients of average lattice thermal expansion along various axes and the average coefficient of unit cell volume thermal expansion in a certain temperature range can be obtained using the equation (5-1).

The HTXRD patterns of $CuSb_2Ti_5$ recorded at different temperatures in the range of 298–773 K are given in Figure 30 [29]. These patterns are similar to each other but with a little shift of Bragg reflections toward lower angle side. These results indicated that there is no phase transformation but expansion of the lattice within the selected temperature range. The typical variations of the unit cell parameters of $CuSb_2Ti_5$ with different temperatures are shown in Figure 31.

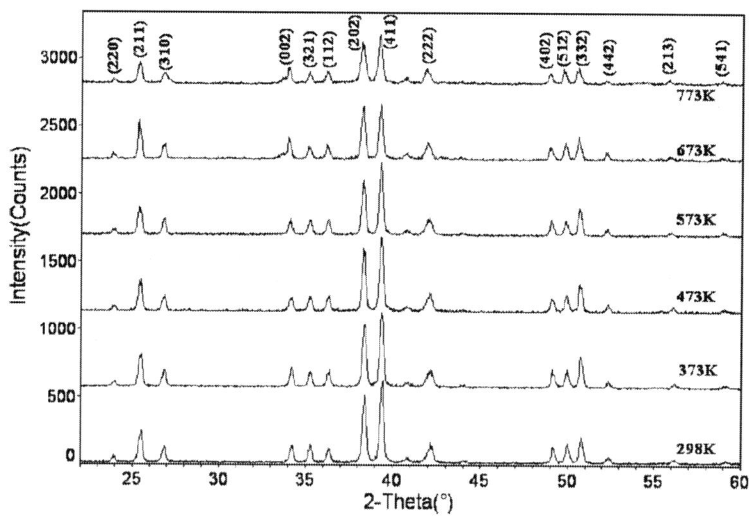

Figure 30. The HTXRD patterns of $CuSb_2Ti_5$ recorded at different temperatures [29].

The variations of the unit cell parameters show that the unit cell parameters increase with the elevated temperature. The variations of the unit cell parameters of $CuSb_2Ti_5$ with different temperatures can be represented by the following polynomial functions:

$$a\ (\text{nm}) = 1.0403(2) + (3.183\times10^{-5})T - (2.017\times10^{-8})T^2, \tag{5-2}$$
$$c\ (\text{nm}) = 0.5267(2) - (1.514\times10^{-5})T + (2.289\times10^{-8})T^2, \tag{5-3}$$
$$V(\text{nm}^3) = 0.5689(2) + (1.927\times10^{-5})T + (2.247\times10^{-9})T^2 \tag{5-4}$$

The thermal expansion coefficient, α_p, of a given crystalline parameter, p, can be determined by the Equation (5-1). Thus, the thermal expansion coefficients along various axes were $\alpha_a = 9.6\times10^{-6}$ K^{-1} and $\alpha_c = 18\times10^{-6}$ K^{-1}, and the bulk thermal expansion coefficient (α_V) is 38×10^{-6} K^{-1} in the temperature range of 298–773 K. One can notice that the average of coefficients obey the law of $\overline{\alpha}_V = 2\overline{\alpha}_a + \overline{\alpha}_c$ for the tetragonal structure [29].

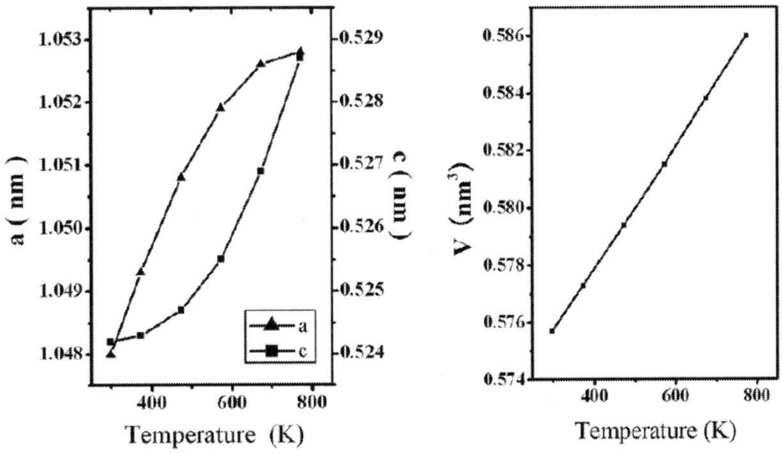

Figure 31. Variations of the lattice parameters and unit cell volumes of $CuSb_2Ti_5$ with different temperatures [29].

5.2. $Dy_{1.2}Fe_4Si_{9.8}$

The lattice thermal expansion property of a new compound $Dy_{1.2}Fe_4Si_{9.8}$ was investigated. The HTXRD patterns of compound $Dy_{1.2}Fe_4Si_{9.8}$ were recorded at different temperatures. The results reveal that these patterns are similar to each other, but with a little shift of Bragg reflections toward the lower angle side. The unit cell parameters at elevated temperature are given in Table 31. The variation of the unit cell parameters shows that the unit cell parameters increase with the elevated temperature. The density of $Dy_{1.2}Fe_4Si_{9.8}$ decreases with the elevated temperature. The typical variations of lattice parameters, unit cell volume and density within the temperature range of 298K to 1013K are shown in Figure 32 and Figure 33[50]. The variations of the unit cell parameters and volumes with different temperature can be expressed by the following polynomial functions:

$$a\,(\text{nm}) = 0.39971 - (3.7046\times10^{-5})T + (1.0259\times10^{-7})T^2 - (1.0463\times10^{-10})T^3 + (3.8436\times10^{-14})T^4, \qquad (5\ 5)$$

$$c\,(\text{nm}) = 1.52457 + (2.52427\times10^{-5})T - (8.5366\times10^{-9})T^2, \qquad (5\text{-}6)$$

$$V(\text{nm}^3) = 0.2080 - (1.6259\times10^{-5})T + (6.3268\times10^{-8})T^2 - (6.8963\times10^{-11})T^3 + (2.65818\times10^{-14})T^4. \qquad (5\text{-}7)$$

The coefficients of average lattice thermal expansion along various axes in the temperature range of 298 to 1013K are obtained by Equation (5-1): $\overline{\alpha}_a = 1.42\times10^{-5}K^{-1}$, $\overline{\alpha}_b = 9.17\times10^{-6}K^{-1}$, and the average coefficient of unit cell volume thermal expansion is $\overline{\alpha}_V = 3.78\times10^{-5}K^{-1}$. This indicates that the thermal expansion is slightly anisotropic in the tested temperature range. In addition, the average coefficients obey the relationship of $\overline{\alpha}_V = 2\overline{\alpha}_a + \overline{\alpha}_c$ for the hexagonal structure [50].

Table 31. The unit cell parameters, volumes and densities at different temperatures [50]

T(K)	a(nm)	c(nm)	V(nm^3)	Density (nm^3)
298	0.39519(2)	1.5312(6)	0.2071	5.562
323	0.39522(2)	1.5317(6)	0.2072	5.559
373	0.39567(2)	1.5325(6)	0.2078	5.543
453	0.39594(2)	1.5341(6)	0.2083	5.530
533	0.39601(2)	1.5366(8)	0.2087	5.518
613	0.39709(2)	1.5366(7)	0.2099	5.488
693	0.39728(2)	1.5373(8)	0.2101	5.482
768	0.39789(3)	1.5387(9)	0.2110	5.460
853	0.39808(1)	1.5390(6)	0.2112	5.453
933	0.39856(2)	1.5418(8)	0.2121	5.430
1013	0.39920(2)	1.5412(5)	0.2127	5.415
α (K^{-1})	$\overline{\alpha}_a = 1.42\times10^{-5}$	$\overline{\alpha}_b = 9.17\times10^{-6}$	$\overline{\alpha}_V = 3.78\times10^{-5}$	

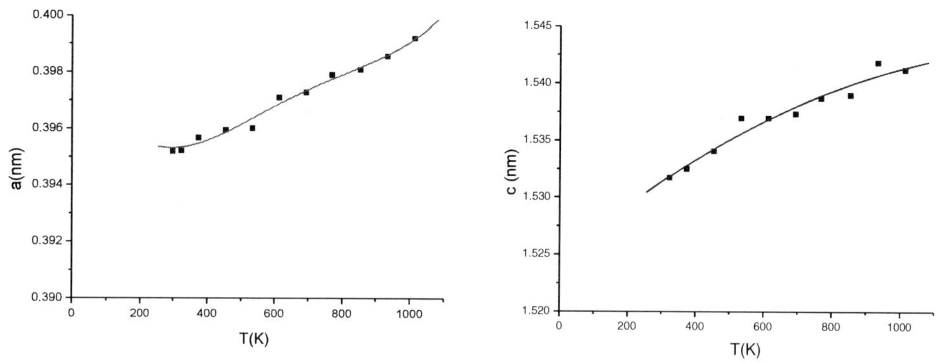

Figure 32. Temperature dependence of the lattice parameters of $Dy_{1.2}Fe_4Si_{9.8}$ [50].

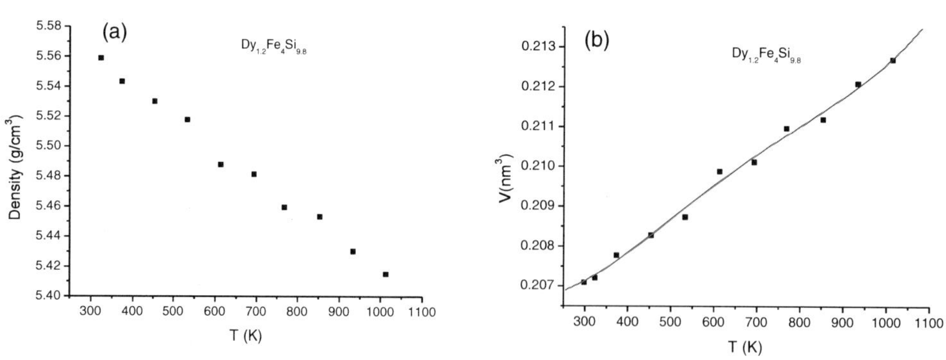

Figure 33. Variation of unit cell volume (a) and density (b) of $Dy_{1.2}Fe_4Si_{9.8}$ with different temperature [50].

5.3. $Al_2Er_{11}Ge_8$

Investigation on the lattice thermal expansion property of new compound $Al_2Er_{11}Ge_8$ was carried out using HTXRD technique. The HTXRD patterns of the new compound $Al_2Er_{11}Ge_8$ were recorded at different temperatures in the range of 293 – 743 K, as shown in Figure 34. A shift of Bragg reflections towards the lower angle was observed. No phase transformation occurred in the selected temperature range, while the compound exhibited a thermal expansion in the lattice. The unit cell parameters at different temperatures are given in Table 32[49]. The variations of lattice parameters and unit cell volume with different temperatures, as shown in Figure 35, can be expressed by the following polynomial functions:

$$a\ (\text{nm}) = 1.10109 - 5.23758 \times 10^{-6}T + 1.78788 \times 10^{-8}T^2 \quad (5\text{-}8)$$

$$c\ (\text{nm}) = 1.61734 + 1.98921 \times 10^{-5}T + 3.0303 \times 10^{-10}T^2 \quad (5\text{-}9)$$

$$V\ (\text{nm}^3) = 1.96086 + 6.28788 \times 10^{-6}T + 6.31818 \times 10^{-8}T^2 \quad (5\text{-}10)$$

where T is in Kelvin.

The percentage of the mean linear thermal expansion is given as:

$$\text{MLTE}(\%) = 100\left\{\frac{a_T - a_{298}}{a_{298}}\right\} \quad (5\text{-}11)$$

The calculated values of the average lattice thermal expansion coefficients and lattice parameters at different temperatures are listed in Table 32.

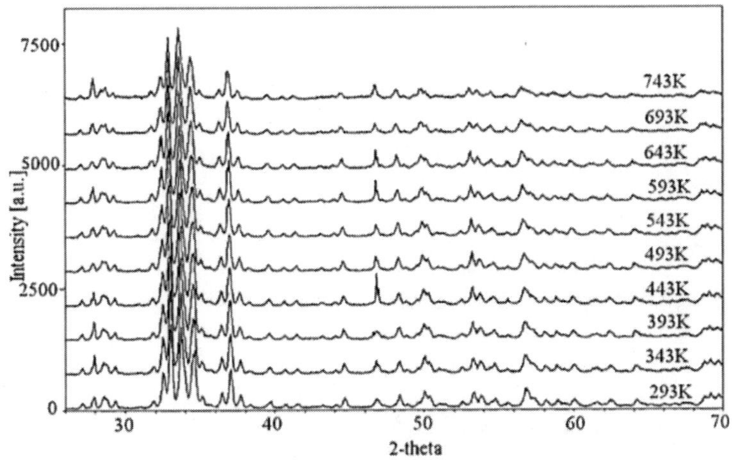

Figure 34. High temperature X-ray diffraction patterns of $Al_2Er_{11}Ge_8$ recorded at different temperatures [49].

Table 32. The temperature dependences for a, c, and V, including both measured and fitted α_a, α_a, and α_V, and the MLTE for the compound $Al_2Er_{11}Ge_8$ [49]

T (K)	a (nm)	a (nm) fitted	α_a $10^{-5}K^{-1}$	MLTE (%)	c (nm)	c (nm) fitted	α_c $10^{-5}K^{-1}$	MLTE (%)	V(nm^3)	V(nm^3) fitted	α_V $10^{-5}K^{-1}$	MLTE (%)
293	1.1011	1.1011			1.6228	1.6232			1.9676	1.9681		
343	1.1013	1.1014	0.539	0.03	1.6246	1.6242	1.22365	0.06	1.9704	1.9705	2.39	0.12
393	1.1019	1.1018	0.629	0.06	1.6254	1.6252	1.22459	0.12	1.9744	1.9731	2.54	0.25
443	1.1024	1.1023	0.713	0.11	1.6261	1.6262	1.2249	0.18	1.9763	1.9760	2.69	0.40
493	1.1026	1.1029	0.796	0.16	1.6272	1.6271	1.22506	0.25	1.9782	1.9793	2.85	0.57
543	1.1036	1.1035	0.878	0.22	1.628	1.6281	1.22515	0.31	1.9829	1.9829	3.01	0.75
593	1.1043	1.1043	0.960	0.29	1.6291	1.6291	1.22521	0.37	1.9868	1.9868	3.17	0.95
643	1.105	1.1051	1.042	0.37	1.6305	1.6301	1.22526	0.43	1.9908	1.9910	3.33	1.16
693	1.1062	1.1060	1.123	0.45	1.6311	1.6311	1.22529	0.49	1.9959	1.9956	3.49	1.40
743	1.1070	1.1071	1.205	0.54	1.6324	1.6321	1.22532	0.55	2.0004	2.0004	3.65	1.64

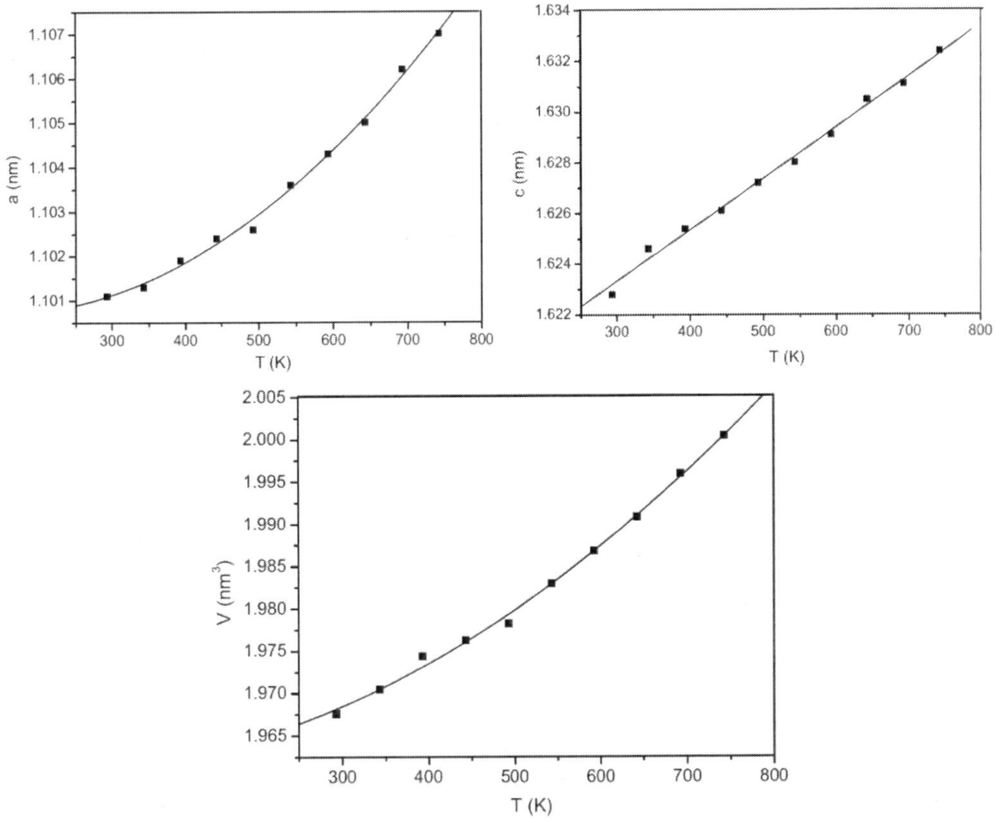

Figure 35. Variations of lattice parameters and unit cell volume with different temperatures for $Al_2Er_{11}Ge_8$ [49].

As seen from Figure 35, the lattice parameters (a, c and volume) of $Al_2Er_{11}Ge_8$ increase progressively with the increase of temperature in the range of 293–743 K. The average relative thermal expansion coefficients along a and c axes and the average coefficient of unit cell volume thermal expansion were $\overline{\alpha}_a = 0.88\times10^{-5}K^{-1}$, $\overline{\alpha}_c = 1.22\times10^{-5}K^{-1}$, and $\overline{\alpha}_V = 3.01\times10^{-5}K^{-1}$, respectively. These results indicate that the thermal expansion of the compound is slightly anisotropic with temperature, and the average coefficients obey the relationship of $\overline{\alpha}_V = 2\overline{\alpha}_a + \overline{\alpha}_c$ for the tetragonal structures [49].

ACKNOWLEDGMENTS

The author would like to express her gratitude to Professor Lingmin Zeng for his assistance and helpful suggestions in writing this chapter. This work is supported by the Natural Science Foundation of China (No. 51261002) and Guangxi Natural Science Foundation (No. 2011GXNSFD018005).

REFERENCES

[1] B.G. Shen, J.R. Sun, F.X. Hu, H.W. Zhang, Z.H. *Cheng, Adv. Meter.*, 21 (2009) 4545-4564.

[2] J. Fidler, J. Bernardi, T. Schrefl, *Scripta Metallurgica et Materialia*, 33 (1995) 1781-1791.

[3] W. He, J. Zhang, L. Zeng, *Journal of Alloys and Compounds,* 424 (2006) 105-107.

[4] W. He, L. Zeng, J. Zhang, C. Li, *Journal of Alloys and Compounds*, 455 (2008) 221-224.

[5] Jade5.0, XRD Pattern Processing Materials, Data Inc., Livermore, California, USA, 1999.

[6] H.M. Rietveld, *Acta. Crystallogr.* , 22 (1967) 151.

[7] H.M. Rietveld, *J. Appl. Crystallogr.*, 2 (1969) 65.

[8] A.K. Gheetham, Ab initio structure solution with powder diffraction data, in: R.A. Young (Ed.) The Rietveld Method, IUCr. Oxford University Press, Oxford, 1995, pp. 276-292.

[9] A.K. Gheetham, A.P. Wilkinson, *Journal of Phys. Chem. Solids,* 52 (1991) 1199.

[10] R.A. Young, Introduction to the Rietveld method, in: R.A. Young (Ed.) The Rietveld Method, Oxford University Press, Oxford, New York, 1993.

[11] G.S. Smith, P.L. Snyder, *J.Appl.Cryst.,* 12 (1979) 60.

[12] P. Villars, Pearson's Handbook, Crystallographic Data for Intermetallic Phases, ASM International, Materials Park, Ohio 1997.

[13] W. He, J. Zhang, L. Zeng, M. Yu, X. Xia, *Rare Metals,* 27(6) (2008) 623-626.

[14] H.E. Wei, Z. Jiliang, Z. Lingmin, *Chinese rare earths,* 27(5) (2006) 33-35.

[15] H. Wei, Z. Jiliang, Z. Lingmin, *Journal of Rare Earths*, 24 (2006) 78-81.

[16] W. He, X. Xia, J. Zhang, J. Huang, Y. Liu, *Rare Metals,* 28(5) (2009) 491-493.

[17] Pcpdfwin Powder Diffraction File (PDF release 2002), International Centre for Diffraction Data, Pennsylvania, USA, 2002.

[18] W. Kraus, G. Nolze, in, Federal Institute for Materials Research and Testing,, Berlin, Germany, 2000.

[19] P. Qin, Y. Chen, J. He, W. He, L. Nong, L. Zeng, *Materials Chemistry and Physics*, 109 (2008) 515-518.

[20] C. Kranenberg, D. Johrendt, A. Mewis, *Solid State Sciences,* 4 (2002) 261-265.

[21] C. Kranenberg, D. Johrendt, A. Mewis, R. Pottgen, G. Kotzyba, C. Rosenhahn, B.D. Mosel, *Solid State Sciences,* 2 (2000) 215-222.

[22] G. Demchenko, J. Konczyk, P. Demchenko, V. Kuprysyuk, R. Gladyshevskii, *Acta Crystallogr.,* E62 (2006) i55–i57.

[23] P. Qin, Z. Xu, Z. Qin, D. Li, L. Zeng, W. He, *Journal of Alloys and Compounds*, 509 (2011) 4484-4494.

[24] L. Zeng, P. Qin, H. Qin, J. Zhang, *Materials Letters*, 61 (2007) 300-303.

[25] L. Zeng, G. Cai, X. Xu, *Journal of Alloys and Compounds*, 395 (2005) 104-107.

[26] G. Cai, J. Zhang, W. He, P. Qin, L. Zeng, *Journal of Alloys and Compounds*, 421 (2006) 42-44.

[27] S. Rundqvist, F. Jellinek, *Acta Chem. Scand.*, 13 (1959) 425–432.

[28] G. Melnyk, W. Tremel, *Journal of Alloys and Compounds*, 349 (2003) 164-171.

[29] X. Wei, H. Liu, W. He, L. Zeng, *Materials Letters*, 62 (2008) 615-618.
[30] W. He, L.M. Zeng, Y.H. Zhuang, Z.Q. Wu, *Journal of Alloys and Compounds*, 298 (2000) 173-176.
[31] L. Zeng, Z. Sun, W. He, Y. Zhuang, *Journal of Alloys and Compounds*, 366 (2004) 187-190.
[32] Y. Zhao, J.K. Liang, G.H. Rao, *J. Alloys Comp.*, 241 (1996) 191.
[33] W.H. Tang, J.K. Liang, Y. Zhao, *J. Alloys Comp.*, 226 (1995) 139.
[34] H. Flandorfer, D. Kaczorowski, J. Grobner, P. Rogl, R. Wouters, C. Godart, A. Kostikas, *Journal of Solid State Chemistry*, 137 (1998) 191-205.
[35] S. Pukas, Y. Lutsyshyn, M. Manyako, E. Gladyshevskii, *Journal of Alloys and Compounds,* 367 (2004) 162-166.
[36] A.R. Young, A.C. Larson, C.O. Paiva-Santos, in, R. School of Physics, Georgia Institute of Technology, Atlanta, Georgia, 1999.
[37] S. Miraglia, J.L. Hodeau, F.d. Bergevin, M. Marezio, G.P. Espinosa, ACTA CRYSTALL. , 43B (1987) 76-83.
[38] Y. Chen, B. He, J. He, W. He, L. Zeng, L. Nong, Powder Diffraction, 23 (2008) 26-30.
[39] L. Zeng, J. He, P. Qin, X. Wei, *Powder Diffraction,* 23 (2008) 60-62.
[40] P. Qin, Y. Chen, H. Liu, W. He, L. Nong, L. Zeng, *Materials Letters*, 62 (2008) 4393-4395.
[41] L. Zeng, P. Qin, Y. Chen, H. Liu, W. He, L. Nong, *Materials Chemistry and Physics*, 112 (2008) 934-937.
[42] P. Qin, Y. Chen, H. Liu, L. Nong, L. Zeng, *Journal of Alloys and Compounds*, 463 (2008) 34-37.
[43] O.P. Bodak, E.I. Gladyshevskii, *Kristallografi*ya 14 (1970) 990.
[44] L. Zeng, H.F. Franzen, *Journal of Alloys and Compounds*, 267 (1998) 86-89.
[45] V.K. Pecharsky, O.Y. Mruz, M.B. Konyk, P.S. Salamakha, M.F. Starodub, O.I. Bodal, *J. Struct. Khim., 30* (5) (1989) 6.
[46] W. He, H. Zhong, H. Liu, J. Zhang, L. Zeng, *Journal of Alloys and Compounds*, 467 (2009) 6-9.
[47] O. Tougait, D. Kaczorowski, H. Noel, *Journal of Solid State* Chemistry, 178 (2005) 3639-3647.
[48] G. SMITH, Q. Johnson, A.G. Tharp, *Acta Crystall.*, 23 (1967) 640-644.
[49] L. Zeng, W. He, W. Shen, M. Lu, *Journal of Alloys and Compounds*, 529 (2012) 158-162.
[50] W. He, L. Zeng, J. Huang, J. Zhang, *Journal of Alloys and Compounds*, 458 (2008) 83-87.
[51] S. Noguchi, K. Okuda, T. Adachi, T. Yoshida, *Physica B,* 237 (1997) 612-615.
[52] S. Noguchi, T. Sakon, H. Nojiri, M. Motokawa, *Physica B-Condensed Matter*, 346 (2004) 183-186.
[53] W. He, J. Zhang, J. Yan, Y. Fu, L. Zeng, *Journal of Alloys and Compounds*, 491 (2010) 49-52.
[54] P. Salamakha, O. Sologub, G. Bocelli, S. Otani, T. Takabatake, *Journal of Alloys and Compounds,* 314 (2001) 177-180.
[55] O.I.B. V.K. Pecharskii, V.K. Belskii, P.K. Stardub, I.R.Mokra and E.I. Gladyshevskii, Kristallografiya, 32 (1987) 334.

In: X-Ray Diffraction: Structure, Principles and Applications
Editor: Kaimin Shih
ISBN: 978-1-62808-591-4

Chapter 5

Application of Quantitative X-Ray Diffraction in Geoenvironmental Problems: Overview and Case Studies

Maria Chrysochoou[*]
University of Connecticut,
Department of Civil and Environmental Engineering,
Storrs, CT, US

Abstract

There are several methods developed by material scientists, geochemists and soil scientists to quantitatively analyze the mineralogy of earth and other materials; quantitative XRD is by far the most widely applied one.

This chapter presents an overview of the application of quantitative XRD and specifically the Rietveld method to selected geoenvironmental problems: a) characterization of alternative cementitious stabilizers such as cement and lime kiln dust that can be used to improve the current empirically driven approach for mix design of stabilized soils; b) characterization of clay soils, which can be used to develop correlations with mechanical properties and improve predictive capabilities in geotechnical applications; and, c) assessment of swell potential from ettringite induced heave in sulfate-bearing stabilized clays. The principles of the method, specific examples of its use and limitations and challenges are discussed.

1. Introduction

Spectroscopic techniques such as X-ray diffraction (XRD), scanning electron microscopy (SEM) and optical microscopy, electron probe microanalysis, Fourier-transform infrared (FTIR) and X-ray absorption spectroscopy (XAS) are very mature and frequently applied in several fields neighboring mainstream geotechnical engineering including geochemistry,

[*] Corresponding author: Tel: +1-860-486-3594. E-mail address: mchrysoc@engr.uconn.edu.

materials science, cement and concrete research, and, more recently, geo-environmental site characterization. Material properties can be evaluated on numerous scales ranging from the macroscale (bulk properties, mm scale) to the nanoscale (discrete chemical structures, nm scale), allowing for an integrated and systematic approach that can provide insight into geotechnical media spanning the measurement continuum. For example, while measurement of the grain size distribution of a soil can provide information with regard to its large-scale geotechnical behavior (e.g., clay vs. sand), micro/nano-scale characterization can provide insight on the soil geochemical properties (e.g., reactivity) that will in turn influence its mechanical properties.

Micro/nano-structural characterization techniques can be used in a qualitative or quantitative mode, the former case being the rule in the geotechnical and geoenvironmental literature. A common approach has been to use, for example, a combination of SEM images and corresponding XRD patterns to indicate purity or composition of various media, such as clays and fly ash in a qualitative or semi-quantitative fashion [1, 2, 3]. An author/researcher may also show complimentary total metal concentrations or major oxide chemistry based on X-ray fluorescence (XRF) in an effort to use multiple lines of evidence to substantiate observations central to their work, whether it be a metal sorption study, ash applications, a soil heave problem involving ettringite or expansive clays [4], or to demonstrate heavy metals immobilization using stabilization/solidification (S/S) technology [1, 5].

The qualitative application of micro/nanoscale characterization techniques has enabled a more sophisticated approach to environmental materials characterization and geotechnical forensic work. The next step is to deploy these techniques *quantitatively* in a predictive design mode. Once this threshold is crossed and the aforementioned techniques become increasingly applied and gain consensus in geotechnical engineering (as they have in other fields), the stage is set to ultimately develop the complementary quality/control quality assurance (QA/QC) approaches that can be implemented in real time at the field scale. Such efforts have already begun to a limited extent. For example, the American Society for Testing and Materials (ASTM) recently developed a standard for the Determination of the Proportion of Phases in Portland Cement and Portland-Cement Clinker Using X-Ray Powder Diffraction Analysis (ASTM C1365-06).

Moreover, the American Association of State Highway and Transportation Officials developed a test protocol for the evaluation of polymer content of polymer-modified emulsified asphalt residue and asphalt binders (T302-05) using FTIR. Although neither of the two methods is directly applicable to common geotechnical problems, both have been developed by related bodies and in neighboring fields. Similarly, environmental science and engineering has also shifted from the exclusive use of wet chemical methods to the application of spectroscopic techniques to investigate the chemical structure and behavior of pollutants.

There is an overall shift of the focus of the scientific and engineering community from the macroscopic description of system behavior to the microscopic understanding of the underlying mechanisms that are manifested in macroscopic phenomena. Quantitative XRD is particularly useful in several geotechnical and geoenvironmental engineering problems. This chapter will provide an overview of the method and examples of selected geoenvironmental applications.

2. Background on Quantitative XRD

There are two main approaches to performing quantitative XRD analysis: the internal standard method that utilizes individual reflections and the Rietveld method that uses the entire XRD pattern and is thus described as a whole pattern fitting (WPF) method.

The internal standard method relies on the relationship $\frac{I_{ia}}{I_{js}} = K\frac{X_a}{X_s}$ (1) where I_{ia} is the intensity of line i of phase α, I_{js} is the intensity of line j of phase s, X_a, X_s the corresponding amounts and K a proportionality constant. The phase s is the internal standard, a crystalline compound that is added to the sample in known amount. To establish a calibration curve, $X_s\frac{I_{ia}}{I_{js}}$ is plotted versus X_a [6]. This method is rather cumbersome in that it may require several calibration constants if multiple analytes and internal standard lines were to be chosen for analysis. A more general method that avoids this is the Reference Intensity Ratio (RIR) method that relies on relative intensity. Most often, corundum is used as the internal standard and its most intense line I(113) to determine the RIR [7] in which case Equation (1) becomes $X_a = \left(\frac{X_{cor}}{RIR_{cor}}\right)\cdot\left(\frac{I_{(hkl)a}}{I_{(113)cor}}\right)$ (2). The RIR_{cor} values are published in the Powder Diffraction Files (PDFs) of pure mineral phases that are included in the database of the International Centre for Diffraction Data (ICDD), even though they are often subject to experimental errors [6]. For a given corundum spike, a measurement of the two reflection intensities yields the amount of the unknown phase. The two main weaknesses of the RIR method are that the chosen reflections should not overlap with the reflections of any other minerals, which may be challenging in a multi-phase mixture, and the published RIRs are typically inadequate to produce accurate quantitative results [6]. The Rietveld method can overcome both because it utilizes crystal structures as input and it considers the entire pattern.

Quantitative XRD analysis is essentially a by-product of the Rietveld method [8]. Rietveld first developed a mathematical model in order to determine crystal structures from any kind of diffraction data, i.e. X-ray, electron or neutron diffraction. Rietveld constructed a mathematical model that produced a synthesized diffraction pattern based on the physics of diffraction, which was then matched as closely as possible to the experimental diffraction pattern. The function to be minimized is a least-squares function [9]:

$$S_y = \sum_i w(i)(y_i - y_{ic})^2 \qquad (1)$$

where w(i) is the weighting factor equal to $1/y_i$, y_i is the observed intensity and y_{ic} the calculated intensity at the i_{th} step, summed over all data points. The objective of the model is to minimize the least-squares function by optimizing the various parameters that are used to calculate the diffracted intensities of all crystals, also termed as *refinable* parameters. This way, crystal structure parameters (interatomic distances, atomic positions) are refined to obtain the best structure that produced the observed diffraction pattern. A key advantage of this approach is that the production of a synthesized pattern is possible for complex mixtures having significant peak overlap between different mineral phases; thus, Rietveld is the only method that enables quantification of the mineralogy of complex soil and soil-like media. Rietveld originally created the method in order to come up with the best crystal structures that produced a pattern; however, in quantitative analysis the crystal structure is considered known (an input to the model) and the quantities of the observed minerals are the unknown. If one has a fairly good idea which crystal structures produce the experimental pattern, the Rietveld model can be used to conduct quantitative analysis instead of mere structure refinement. For example, Portland cements are known to consist of four mineral phases (alite, belite, tricalcium aluminate and brownmillerite) and gypsum; the Rietveld model is in this case used to determine the relative quantities of the four minerals, which have high impact on the reactivity and the setting characteristics of cement. A variation of the Rietveld method is to utilize diffraction patterns of pure minerals as input instead of crystal structures (a.k.a. observed pattern method [6]); WPF is utilized as the generic term to describe both approaches.

Figure 1 shows an example of WPF for a clay soil obtained from Texas (sample courtesy of A. Puppala). The solid line represents the actual XRD pattern, the dashed line the calculated pattern and the difference between them is shown at the top. A perfect flat line for the difference plot represents the ideal refinement, which is never achieved, because the calculation cannot represent noise. However, the absence of large positive or negative spikes in the difference plot indicates that there is very close agreement between the experimental and calculated pattern. The success of the fit is also evaluated by means of statistical parameters, such as the goodness-of-fit $\frac{S}{E} = \left[\frac{N-P}{S_y}\right]^{-1/2}$ (2), where S_y is the function shown in equation (1); N is the number of fitted data points; and P the number of refined parameters. The goodness of fit S/E is close to one in an ideal refinement and is an indicator of how trustworthy the obtained values are for quantification. Young (1993) [9] suggests a value for 1.3 for a good refinement and a threshold of 1.7 to re-evaluate the choice of crystal structures used in the model. It should be stressed, however, that these values are proposed for high-resolution XRD patterns that aim at the determination of precise crystal structures for isolated compounds or simple mixtures. A mixture of five to ten phases (most soils) is unlikely to approach these values. The S for the refinement shown in Figure 1 was 1.8; in the experience of the author, S values up to 2.0 are still reasonable for highly noisy patterns with broad peaks.

Figure 1 shows that zincite (ZnO) was identified and quantified in the XRD pattern. Zincite was not a soil phase, but was added at 10 wt.% to the sample as internal standard in order to quantify the amorphous content of the soil. WPF provides a *relative quantification* of the phases that are used in the model; thus, the presence of phases that are too disordered or too low in amount to observe by XRD is not captured and leads to an overestimation of all the observed phases in the model. To estimate the **amorphicity** of a specimen and obtain **true**

phase quantification, the XRD powder sample is spiked with a known mass of a substance of very high crystallinity (an internal standard) [10, 11, 12]. The choice of internal standard depends on the properties of the analyzed sample. Ideally, an internal standard should not be part of the sample, should not interact with it and should have similar properties with the sample minerals in terms of absorption coefficient and scattering power. Corundum (α-Al_2O_3) has been used successfully in recent geoenvironmental studies [13, 14]. The amorphous content is $A = \dfrac{W_{s,o} - W_s}{W_{s,o}(1 - W_s)}$ where $W_{s,o}$ is the amount of the spike calculated by the WPF model and W_s the actual amount. All phases can then be normalized using the amorphous content to obtain true weight percentages.

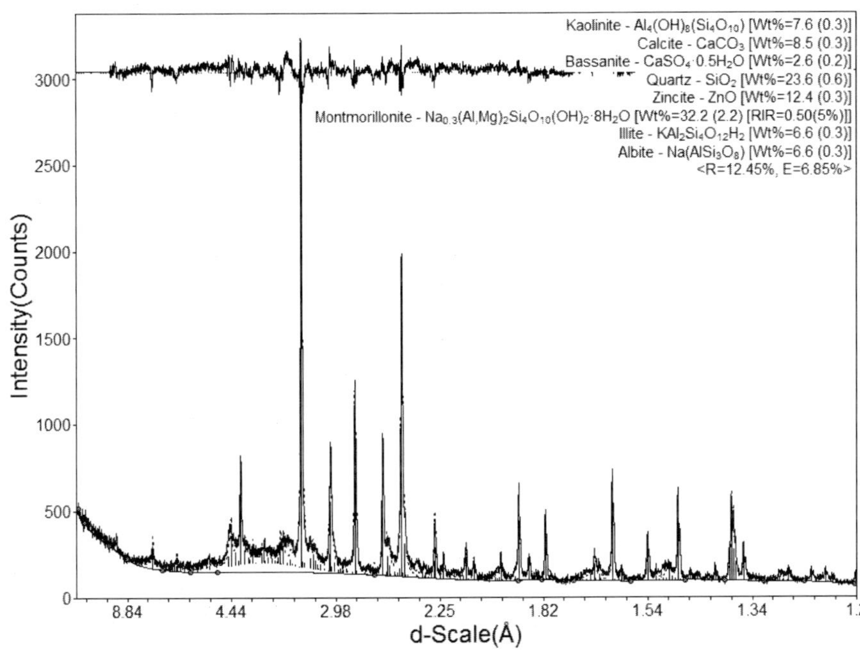

Figure 1. Example of clay soil analysis using the WPF.

There are various software codes, both free (GSAS, FullProf, Rietica) and commercial (Jade, Topas, Siroquant), that have been developed to implement the Rietveld method. Several of which also enable quantitative analysis. These codes are primarily used by the materials science community and have varying levels of complexity and user-friendliness. Even though the principle of the method is the same regardless of the application, there are several factors that influence the accuracy and precision of the method. The choice of software/model and the specific application determine which set of parameters is relevant and what the desired and attainable accuracy and precision of the method are. For example, if the desired end result is the precise determination of a crystal structure, parameters such as temperature and anisotropic strain may influence the results significantly and a high level of accuracy is desired. However, if the desired end result is the quantitative composition of a soil sample within a marginal error of several percentage points, the input to the model is more important and the tolerance for residual error in the least-squares function will be different.

The parameters that can be refined in the Rietveld method generally fall under two categories: global parameters that apply to the whole pattern and phase parameters that address each mineral phase loaded in the model. Global parameters include fitting of the background, angular adjustments that correct for shift of peak positions due to experimental aberrations, absorption, fitting of amorphous humps and the profile-shaping function. Phase parameters are the lattice constants of the unit cell, atom position coordinates and site occupancies, the scale factor that adjusts peak height and parameters related to the Full-Width at Half Maximum (FWHM that refers to the peak width at 50% of the intensity). Preferred orientation coefficients correct for non-random orientation of crystals; platy crystals tend, for example, to orientate horizontally on the sample holder, producing abnormally intense reflections. Temperature corrections and other effects such as crystallite size and extinction are also refinable parameters for each phase.

It should be stressed that the above listed parameters are possible rather than required refinable parameters. The analyst should choose which parameters to input in the model based on a series of criteria:

- *Pattern complexity:* a pattern with high noise and low peak resolution requires a minimum set of parameters, in order to maintain control over the physical meaning of individual parameter shifts and avoid curve fitting. McCusker et al. (1999) [15] suggest a minimum of 3 for the ratio of individual reflections observed vs. the number of parameters refined.
- *Purpose of refinement:* the determination of crystal structures generally requires a more complex model that includes parameters other than the crystal structure. In this way, peak shifts and intensity differences due to changes in crystal structure can be modeled more accurately.
- *Experience of analyst:* the analyst has to be in position to evaluate the physical meaning of calculated values for the input parameters. This is essential in order to avoid curve fitting that is not related to the reality.

Obtaining high quality WPF data requires good understanding of one's system and how the various refinable parameters and potential pitfalls that may be applicable, depending on the sample nature and composition. Several studies report on the results of round robin studies and QXRD competitions that contain invaluable information on successful strategies and sources of error [16-19]. McCusker et al. (1999) [15] provided detailed guidelines for successful Rietveld analysis and Kleeberg (2004) [20] provided important input on the specific challenges associated with complex systems, which typically applies to geoenvironmental samples.

3. Case Studies

3.1. Characterization of Recyclable Cementitious Materials

Stabilization/Solidification (S/S) of hazardous waste and contaminated soils typically relies on the use of cementing agents to produce a monolithic solid of low leachability.

Additionally, expansive clays are also stabilized for their mechanical properties using lime and Portland cement. Thus, the use of cementitious agents is common in geoenvironmental applications. Lime and Portland cement are frequently partially or completely replaced by industrial by products with cementitious or pozzolanic properties, in order to reduce costs and promote sustainability. Such materials are cement kiln dust (CKD), lime kiln dust (LKD), fly ash, blast furnace slag (BFS), steel slag and rice husk ash [21-26]. In general, all by products have to be used in dosages higher compared to lime and cement, because they contain lower amounts of the "active" minerals, i.e. calcium oxide and calcium silicates. Treatability studies that employ 28-day unconfined compressive strength (UCS) or leaching tests are typically employed to determine the optimum material blend for mix design. In both cases, the tests employed are rather lengthy and laborious, given that the benchmark dosages 3, 5 and 10% that are common for Portland cement and lime are not necessarily applicable. Additionally, these materials do not comply with any specifications, so that comparison of different sources is difficult to establish. Several studies [21, 27, 28] employ X-ray fluorescence (XRF) analysis as a measure of comparison of various stabilizers, however, as will be demonstrated, such comparisons are not very useful because reactivity is controlled by mineralogy, and not chemistry.

Table 1. QXRD analysis results for samples of various cementitious stabilizers

Mineral	Chemical formula	CKD-1[1]	CKD-2	Quick-lime	LKD-1	LKD-2	Class C fly ash	Blast furnace slag
Lime	CaO	5.1	7.7	90.2	17.6		0.5	
Portlandite	$Ca(OH)_2$			1.9		47.0		
Calcite	$CaCO_3$	37.7	33.0	2.3	47.7	15.1		2.0
Dolomite	$(Ca,Mg)CO_3$	5.7	4.6		12.0			
Anhydrite	$CaSO_4$	2.7	4.7					
Gypsum	$CaSO_4{\cdot}2H2O$							2.1
Larnite	Ca_2SiO_4	4.5	6.6			4.5		
Gehlenite	$Ca_2Al_2SiO_7$						3.7	
Melilite	$(Ca_{1.7}Na_{.2}K_{.1})Si_2Mg_{.5}Al_{.4}O_7$							49.8
Srebrodolskite	$Ca_2Fe_2O_5$						5.4	
Periclase	MgO		1.6				2.8	
Magnetite	Fe_3O_4							2.1
Quartz	SiO_2	6.8	6.5				5.5	3.5
Others+Amorphous		37.5	32.6	5.6	22.8	33.4	82.1	41.9

[1] All data in percent dry weight (wt.%). Blanks indicate not detected.

QXRD may be used as a preliminary test to determine the reactive ingredients in these materials, which may be compared to the properties of lime and Portland cement to establish a basis for developing a mix design. This will be ultimately verified using the traditional performance-based mechanical and/or leaching tests. This method has both advantages and limitations, as will be illustrated in the following examples. Table 1 shows the QXRD results of various types of stabilizers that the author has analyzed in the frame of soil stabilization studies. The two CKD samples were obtained from different cement production facilities. The results show that in both cases, the majority of the calcium is bound in calcite, which does not contribute to cementitious reactions. There are two reactive ingredients in the two CKDs, lime and larnite, so approximately 10% of the total mass constitutes of reactive crystalline ingredients. A limitation of the method is that it is not possible to assess the reactivity of the amorphous content, given that its properties are unknown. However, the amount of reactive crystalline compounds is a more meaningful parameter to show in stabilization studies in order to provide with a reference with regard to the nature and quality of the stabilizer used. In contrast, XRF analyses only provide the total oxide contents of each material. Relating the reactive crystalline percentage to performance as the stabilizer is not presently possible, because there is no published information on the relationship between CKD properties and performance. However, it is proposed that using QXRD as an instrumental tool can help establish such relationships.

In order to compare the QXRD values of the two samples, it is also necessary to have a measure of the error associated with the weight percentages. There are several factors that affect the values, including the accuracy and precision of the method itself and sample variability. The accuracy and precision of the Rietveld method depends primarily on sample preparation and proper analysis strategy [16, 17]. The lowest bias obtained in a recent QXRD competitions was 11% (approximately ±1 wt.% maximum absolute error) [19]; Hillier (2000) [7] provides ±3 wt.% at the 95% confidence level as a general accuracy for QXRD methods. It should be stressed that these values require that the analyst has optimized sample preparation and performed no systematic error during the analysis. Assuming that this was the case for the results presented in Table 1, it follows that samples CKD-1 and CKD-2 are practically indistinguishable in terms of composition, given that the difference between mineral percentages approaches the maximum accuracy of the method.

Conversely, the comparison of the two LKD samples shows that they present substantial differences in mineralogy. One sample was obtained from a U.S.-based lime production plant, while the second one was from a plant in Nicaragua. Table 2 shows the XRF analysis results for the two samples, along with XRF of the quicklime sample included in Table 1. The XRF shows that the two LKDs are very similar chemically, with the Nicaragua sample (LKD-2) having slightly less magnesium and more silica, but otherwise equivalent in terms of calcium and organic content. Conversely, the XRD analysis indicates that there is a lot more reactive calcium in the form of portlandite in LKD-2, whereas the U.S. sample LKD-1 has most Ca present as calcite. This points out a more efficient limestone burning process in the U.S. plant, which yields a lower grade LKD compared to the Nicaragua plant. A soils stabilization design relying on the XRF results would consider the two materials of similar reactivity, but, in fact, they are not.

Table 2. XRF analysis for major metals and organic content of LKD and quicklime samples

	Quicklime	LKD-1	LKD-2
CaO	95.65	61.60	63.23
Al_2O_3	0.32	2.19	2.35
SiO_2	1.08	4.25	7.01
MgO	1.11	2.27	0.80
Loss on Ignition (LOI)	2.14	26.01	23.87

All data in wt.%

The evaluation of the class C fly ash and the blast furnace slag samples is more difficult to perform on the basis of XRD. No Ca-bearing crystalline compound was identified in the fly ash sample, and over 80% of the material was observed to be amorphous, which is a typical feature of fly ashes. Thus, neither qualitative nor quantitative XRD are very useful in this case. Similarly, the blast furnace slag comprised 50% melilite and about 50% amorphous, neither of which belongs to an obvious class of cementitious minerals. However, the fact that approximately 50% of the slag is crystalline is indicative of the process it was derived from, namely a slow cooling process. Slags derived from slow cooling process are in general less reactive compared to completely amorphous slags derived from fast-cooling processes [29]. If the slag source and properties are not otherwise known, QXRD can aid in determining the amorphous content and thus the reactivity of the slag.

A major limitation of the proposed approach is that it lacks a wealth of empirical data, on the basis of which models and predictions could be made. We require a more in depth understanding of the relationship between the mineralogy, chemical reactivity and mechanical behavior of various cementitious products, similar to the one that the cement industry obtained over time for various cement types, including blended cements. Given that soil stabilization studies continue to be performed worldwide, it is plausible to propose that obtaining such data is feasible with relatively limited resources.

3.2. Analysis of Clay Minerals in Soils

Quantitative analysis of clay minerals in geologic samples with XRD has been performed for many years in the fields of geology and geochemistry [7, 19, 30, 31]. A detailed treatment of the issues involved is beyond the scope of this section, which will focus on the benefits of such analyses in geo-environmental problems. Briefly, the main challenges associated with XRD quantification of clay minerals are preferred orientation along the basal planes, turbostratic disorder that leads to broad peaks that cannot be easily modeled and wide variability in composition that can cause substantial error in the input model and the results [7].

The behavior of soft and expansive soils is largely controlled by the concentration and type of clay minerals present. Mitchell and Soga (2005) [32] discussed the influence of clay mineralogy on various engineering soil properties, including strength, compressibility, hydraulic conductivity and swell potential. Several other studies documented the influence of clay mineralogy on engineering properties: Jang et al. (2011) [33] showed that the erosion potential of levees depended on the amount and type of clay minerals; Mitchell (2009) [34] detailed how mineralogy could explain unanticipated soil behavior in Mt. St. Helens; Lin and Cerato (2011) [35] stressed the importance of surface phenomena of clay minerals to predict expansive behavior; Likos (2008) [36] related vapor adsorption to expansion potential on the basis of the properties of pure clay minerals; Little et al. (2010) [37] highlighted the role of clay minerals in sulfate-induced heave in lime-stabilized clays.

However, clay mineralogy is typically unknown in most geotechnical research and design problems. The proxy by which geotechnical engineers characterize clay soils is the plasticity chart. In general, there is a broad qualitative relationship between Atterberg limits and soil mineralogy. In turn, the plasticity chart and soil classification are used to draw general conclusions about the nature and engineering behavior of fine-grained soils. While the plasticity chart is an invaluable tool for geotechnical engineers, it faces several limitations. Casagrande (1948) [38] stated in the paper that outlined the Unified Soil Classification System (USCS): "It is not possible to classify all soils into a small number of groups to represent the many divergent problems of applied soil mechanics". This means that the system is not refined enough for quantitative prediction of soil properties or design; it is always necessary to conduct specialized mechanical tests. Attempts to correlate Atterberg limits with soil properties such as the compression index or swell potential are typically confined to a small group of soils and have large uncertainties; for example, Terzaghi and Peck (1967) [39] give a reliability range of 30% for the general correlation between liquid limit and compression index. Although this can be partially attributed to factors such as the influence of soil fabric, the liquid limit is undoubtedly too crude a parameter to accurately capture the variability of soil composition and its influence on mechanical properties.

In addition, the USCS can be misleading even for the general, qualitative evaluation of clay soil properties. Plasticity is a function of several factors, including the amount and type of each mineral, and the organic content and the presence of non-clay minerals that can influence soil plasticity. For example, Shamrani et al. (2010) [40] reported that a CL soil had higher swell potential than a CH soil obtained from two regions in Saudi Arabia. The high iron oxide content of the CL soil "masked" the real plasticity associated with the clay minerals present in this soil. Chrysochoou et al. (2010) [41] observed the opposite effect, in which dredged sediment characterized as CH soil only contained chlorite as clay mineral and the plasticity was attributed to the high organic content of the soil. Attempts to stabilize the dredged material with lime were thus largely ineffective. Moreover, the presence of rare soils, for example volcanic ash clays, cannot be established on the basis of USCS alone.

While the importance of mineralogical composition on soil and clay properties is recognized, there is very little information and practically no application of correlations of mineralogy and soil properties.

Mitchell and Soga (2005) [32] detailed the relationship between soil composition and various properties, including soil compression index and coefficient of consolidation, swelling potential, hydraulic conductivity and strength and summarized the results of several studies that attempted to correlate fundamental soil properties with their mineralogy. Yet, this approach has not gained popularity in geotechnical practice. Recently, Chittoori and Puppala (2011) [42] proposed a semi-quantitative method to determine the amount of montmorillonite, kaolinite and illite in expansive clays from Texas. The method utilized the specific surface area, cation exchange capacity and potassium content of a mixed clay sample to extrapolate the relative percentage of the three minerals based on their intrinsic properties. This analysis may then be used to assess the swell potential of the soil on the basis of the swell behavior of the pure minerals.

This approach is semi-quantitative in that it assumes all three minerals are present with a specific composition; however, it constitutes an important departure from the empirical approaches of measuring the swell potential of each individual soil. QXRD can be used to further refine it and Chrysochoou et al. (2012) [43] utilized a combination of QXRD and the method developed by Chittoori and Puppala (2011) [42] for this purpose. Figure 2 shows the results obtained with the two methods.

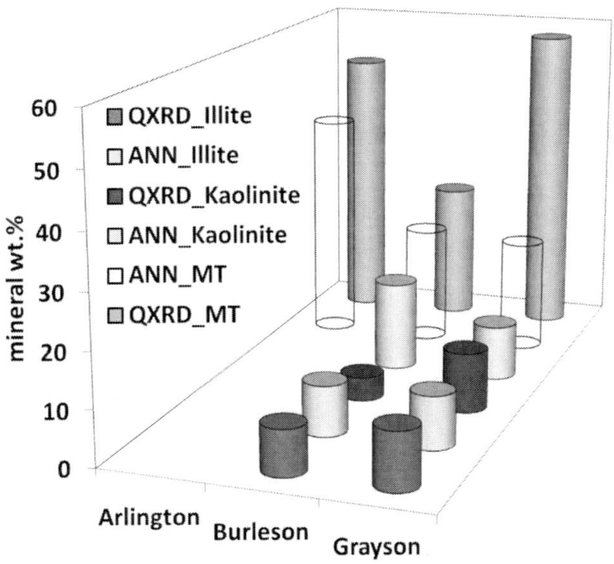

Figure 2. Comparison of soil mineralogy determined by QXRD and Artificial Neural Network (ANN) analysis (reproduced from Chrysochoou et al. 2012).

The results showed that the two methods had very good agreement for illite, with difference that was within one standard deviation of the average concentration predicted by QXRD. Good agreement was obtained for kaolinite for one soil, but not for another. This was because the ANN does not take into account the presence of other minerals in the soil, and thus it assigns all low surface area minerals to kaolinite. The Burleson soil was observed to have substantial amounts of quartz and calcite, and both of which were present as very fine silty particles. For soils with low SSA, the ANN assumes that the major mineral is kaolinite, which has the lowest SSA among the three clay minerals it includes (montmorillonite, illite and kaolinite), and is unable to take into account the contribution of other phases. This led to the overestimation of kaolinite in the Burleson soil.

The worst agreement between the ANN and QXRD was in the concentration of montmorillonite, which was not surprising given the difficulties associated with quantification of bentonites [31]. Specifically, QXRD systematically predicted higher montmorillonite concentrations compared to the ANN. This may due to several sources of error. The available structural models for montmorillonite yielded very poor fits with the observed phases, so that an observed pattern was used instead. The chosen observed pattern had no published Reference Intensity Ratio and an RIR of 1.0 was used, which is comparable to other clay minerals. However, there is considerable uncertainty with regard to its value. Additionally, the RIR is affected by preferred orientation, which was not taken into account in the analysis. Thus, the RIR is considered to be the primary source of error in the quantification of montmorillonite in this analysis. It should be stressed here that the true weight percentages of the minerals are not known, and the two methods are only compared against each other. In addition, the QXRD analysis was not optimized by utilizing additional techniques, such as oriented aggregates or complementary analyses. These are generally required in order to achieve high accuracy in QXRD [17, 19].

However, this was the first study presented to a geotechnical/geoenvironmental engineering audience and the ongoing research efforts are now focusing on establishing the correlation t between the obtained results and the soils' swell behavior to refine the obtained data. An important aspect is to understand the desired level of precision and accuracy in order to obtain reasonable predictions of soil behavior. If a ±1 wt.% accuracy level is required, then the application of QXRD will be more cumbersome, given that this level of accuracy can only be obtained using multiple methods and the substantial experience of the analyst. It will be necessary to address questions such as the impact of the presence of mixed-layer clays on mechanical behavior. This has proven to be a challenge for quantitative analysis [19]. Finally, it will be necessary to systematically explore the relationship between the mineralogy and various mechanical properties in order to develop meaningful models that would, ideally, be combined with constitutive models.

3.3. Forensic Analysis of Ettringite-Induced Heave

Certain cement or lime-stabilized clays still undergo swelling primarily due to the formation of the mineral ettringite ($Ca_6Al_2(SO_4)_3(OH)_{12}\cdot 26H_2O$). This heave is known as ettringite-induced heave and has caused damage to civil infrastructure in various states, including Texas, Iowa, California, Nevada, Oklahoma and Louisiana [44-49]. All these areas present naturally occurring sulfate-bearing clays. While ettringite had been known for years to cause cement cracking [50], it was not demonstrated that it was an active expansion mechanism in lime- and cement-stabilized clayey subgrades until 1984, when pavement heave distress was noted on the Stewart Ave., Las Vegas, Nevada. In a classic keynote paper, Mitchell (1986) demonstrated using qualitative X-ray diffraction (XRD) analyses that ettringite and the isostructural mineral thaumasite ($Ca_3Si(CO_3)(SO_4)(OH)_6\bullet 12(H_2O)$) were associated with surface heaving. Mitchell and Dermatas (1992) [51] and Dermatas (1995) [4] followed up with the studies that showed that the rates of ettringite formation and thus the rate of heave manifestation were found different in kaolinite and montmorillonite clays. However, they were able to demonstrate this only semi-quantitatively by comparing the X-ray diffraction (XRD) intensities over time.

Since then, the role of ettringite formation in the swell of stabilized clays has been widely studied both in forensic investigations and in laboratory-controlled swell tests [45-48, 52, 53, 54]. The mechanism of ettringite formation in stabilized clays is chemically well-known. Alumina dissolves from clay minerals at high pH and reacts with the Ca source of lime/cement and ground sulfates to form the ettringite mineral. Ettringite is the thermodynamically favored cement phase at high pH, i.e., its formation is inevitable when sulfate pre-exists or enters the clay system at soluble concentrations as low as 15 mg/kg [55]. It has also been hypothesized that ettringite expands in the presence of water, so that hydration instead of mineral formation contributes to the induced pressures [56]. Little et al. (2010) [37] concluded that both mechanisms are likely responsible for ettringite-induced expansion and that external water intrusion into the matrix is a necessary precondition for the manifestation of heave.

There are several factors that affect swell or cracking phenomena as a result of ettringite formation: the amount of sulfates and sulfate type; crystal size (affected by temperature); the rate of ettringite formation; soil moisture content and dry unit weight; confining pressure; and others. Given the complexity of the problem, it is difficult to develop defensible guidelines when designing the stabilization of sulfate-rich subgrades. Several studies have been published to this end, and despite a convergence towards a lower threshold of 2,000-3,000 mg/kg sulfate as the highest acceptable concentration [37], most authors conclude that the soil susceptibility to ettringite formation should be evaluated on a case-by-case basis, as reaction rates could not be captured. There are several other complex chemical and mechanical questions that still need to be addressed, such as: How much ettringite formation corresponds to certain strain magnitudes? What is the rate of ettringite formation in different clay types? Is sulfate all converted to ettringite, and if no, why? To answer these questions, a quantitative mineralogical tool has to be employed. Little et al. (2010) [37] used Differential Scanning Calorimetry (DSC) to quantify the amount of ettringite in three Colorado soils and compared the results with the predicted amount of ettringite from the thermodynamic phase diagram approach. The observation indicated that: a) the amount of ettringite formed was lower than the amount of ettringite predicted, but it could not be established whether the soils had in fact reacted to equilibrium; b) soil mineralogy affected the overall amount of ettringite formed because of the different availability of soluble Al. DSC only quantifies one mineral at a time, so that no insight could be obtained into the reactions of other minerals in the soils. QXRD may also be used in a similar fashion and potentially offer some additional insight since it is able to quantify all minerals.

Table 3 shows the QXRD results for two representative samples recovered from a site, in which substantial foundation distress was observed, due to ettringite-induced heave of the underlying stabilized fill. One of the samples was recovered from the stabilized mix and one from the native soil layer beneath. It should be noted that several samples from different locations were subjected to the same analysis in order to obtain a statistically representative picture of material properties. Only one is shown here to illustrate the methodology followed. Figure 3 shows the original and residual plots of the QXRD analyses for the two patterns.

Table 3. QXRD and mass balance analyses of the stabilized and native samples obtained from an ettringite-impacted area

Stabilized soil		Native soil	
Ettringite *	26.7	Kaolinite	4.8
Gypsum	3.5	Montmorillonite	6.6
Calcite	7.6	Quartz	70.4
Quartz	25.2	Amorphous	18.2
Amorphous	37.0		
S in ettringite	2.05	% Al in clay minerals	1.65
S in gypsum	0.65	% potential ettringite	38.3
S by XRF	3.02	Al by XRF	n.a.**
% reaction by S	68	% reaction by Al	70

* All data in wt.%, except the % reaction.

** Not available.

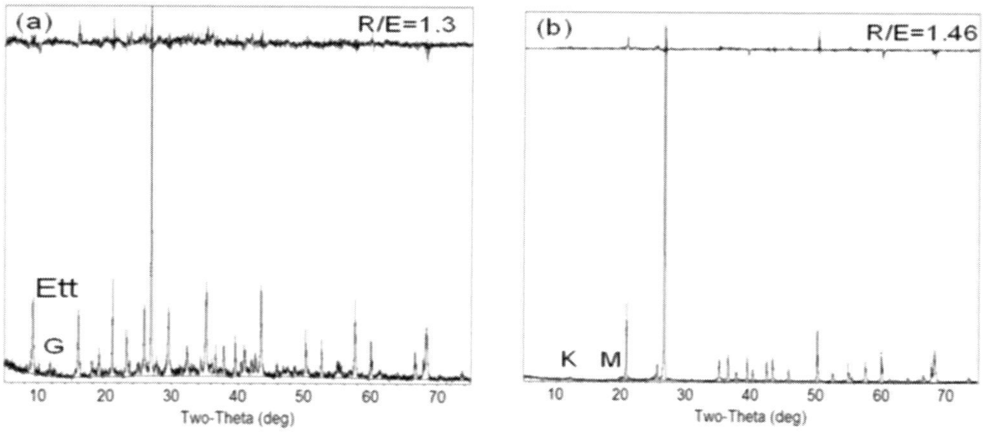

Figure 3. QXRD original and residual plots for the stabilized (a) and native (b) soils.

In the stabilized soil, quartz, calcite, ettringite and gypsum were the only minerals identified. No clay mineral was detected. The concentration of ettringite was found to be 26.7%, which corresponds to a sulfur concentration of 2.05% using the molecular weight of the ettringite structure ($Ca_6(Al(OH)_6)_2(SO_4)_3(H_2O)_{25.7}$ with MW 1248.6). Additional 0.65% of sulfur was observed as gypsum ($CaSO_4{\cdot}2H_2O$ with MW 172), so that the total sulfur content estimated by QXRD was 2.7%. The sulfur concentration measured by XRF was 3.0, which is very close to the mass balance given the uncertainty associated with both methods. Comparing the S bound in ettringite and the total available S measured by XRF, we can estimate the reaction of sulfur conversion to ettringite is 68% complete. The next question to address is whether there is sufficient aluminum available to form additional ettringite (assuming that calcium is not the limiting factor in a stabilized soil). Again, it is possible to evaluate this using an XRF measurement and/or a parallel evaluation by assessing the available Al that is bound in clay minerals in the native soil. The QXRD analysis showed that approximately 5% kaolinite and 6.6% montmorillonite were present in the native soil which yielded a total Al concentration of 1.65% bound in the two clay minerals using their chemical formulas and molecular weights (516 for kaolinite and 555 for montmorillonite). It should be noted here that the XRD pattern was dominated by quartz (Figure 3b), which renders the uncertainty associated with these concentrations relatively large; XRF analysis is necessary to complement these results. The maximum ettringite that could be formed using all the Al bound in the two minerals is 38.3%. Again, the percentage ettringite formed vs. the maximum ettringite on the basis of the Al content is approximately 70%. This indicates that there is enough Al available to convert all sulfates to ettringite and based on both elements the reaction is approximately 70% complete. In this case, the site owner was interested in knowing whether additional heave was anticipated and in what time frame. Based on this analysis, it is anticipated that there is an additional 30% of reaction that can happen, and thus there is still swell potential in the material. Whether the reaction will proceed to equilibrium and in what time frame is of course not addressed by this type of analysis; these questions require actual swell testing. However, the QXRD analysis indicates that such swell testing is necessary and sets an evaluation framework for swell tests.

Conclusion and Outlook

Quantitative XRD analysis can be used to provide important insights into several geoenvironmental problems, including:

- Evaluation of reactivity of cementitious compounds to improve the predictive design for chemical stabilization of soils and hazardous waste.
- Evaluation of the geochemical properties of recyclable and hazardous waste.
- Improved characterization of clay soils and the use of such characterization to predict mechanical properties of clays.
- Evaluation of reactions in chemically active soils, such as chemically stabilized, sulfate-bearing clays.
- Evaluation of immobilization mechanisms of heavy metals, e.g. Pb in pyromorphite or Cr(VI) in cementitious compounds.

QXRD is successful employed for characterization and design purposes in several other disciplines, such as materials science, geochemistry and the cement industry. Geotechnical and geoenvironmental engineering can be benefited from the established QXRD analysis methods, provided that the community invests in combining QXRD with traditional testing approaches to understand the correlation between mineralogy and mechanical properties in a quantitative manner. There are certainly limitations for QXRD, including the fact that it requires substantial expertise to minimize errors (a statement that is true for any experimental method and any model) and it does not offer much insight into reactions occurring in non-crystalline compounds. In addition, there are still difficulties associated with the accurate quantification of clay minerals that need to be resolved. However, it is currently possible to substantially improve current the characterization and design approaches within these limitations, as illustrated by the examples shown in the chapter.

References

[1] Kamon, M.; Nontananandh, S. *J. Geotech. Eng*. 1991, 117(1), 1-17.

[2] Wartman, J.; Riemer, M. F. *Int. J. Phys. Model. Geotech.* 2005, 2, 17-29.

[3] Santagata, M.; Bobet, A.; Johnston, C.T. and Hwang J. (2008). One-dimensional compression behavior of a soil with high organic matter content, *J. Geotech. Geoenv. Eng*. 134(1), 1-13.

[4] Dermatas, D. *Appl. Mech. Rev*. 1995, 38(10), 659-673.

[5] Dermatas, D.; Meng, X. *Eng. Geol.* 2003, 70, 377-394.

[6] Bish, D.L.; Post, J.E. Modern Powder Diffraction; Reviews in Mineralogy; Mineralogical Society of America: Washington, DC, 1989; Vol. 20.

[7] Hillier, S. *Clay Miner*. 2000, 35, 291-302.

[8] Rietveld, H.M. *J. Appl. Crystallogr.* 1969, 2, 65-71.

[9] Young, R.A. The Rietveld Method, *Oxford University Press*, New York, NY, 1993.

[10] Jones, R.C.; Babcock, C.J.; Knowlton; W.B. *Soil Sci. Soc. Am. J.* 2000, 64, 1100-1108.

[11] De La Torre, A.G.; Bruque, S.; Aranda M.A.G. *J. Appl. Crystallogr.* 2001, 34, 196-202.

[12] Whitfield, P.S.; Mitchell, L.D. *J. Mat. Sci.* 2003, 38(21), 4415-4421
[13] Dermatas, D.; Chrysochoou, M.; Moon, D.H.; Grubb, D.G; Wazne, M.; Christodoulatos, C. *Environ. Sci. Technol.* 2006, 40(18), 5786-5792.
[14] Chrysochoou, M.; Dermatas, D. *J. Haz. Mat.* 2007, 141(2), 370-377.
[15] McCusker, L.B.; von Dreele, R.B.; Cox, D.E.; Louer, D.; Scardi, P. *J. Appl. Crystallogr.* 1999, 32, 36-50.
[16] Madsen, I.C.; Scarlett, N.V.Y.; Cranswick, L.M.D.; Lwin T. *J. Appl. Crystallogr.* 2001, 34, 409-426.
[17] Scarlett, N.V.Y.; Madsen, I.C.; Cranswick, L.M.D.; Lwin, T.; Groleau, E.; Stephenson, G.; Aylmore, M.; Agron-Olshina, N. *J. Appl. Crystallogr.* 2002, 35, 383-400.
[18] Stutzman, P. *Powder Diffr.* 2005, 20(2), 97-100.
[19] Omotoso, O.; McCarty, D.K.; Hillier, S.; Kleeberg, R. *Clays Clay Miner.* 2006, 54(6), 748-760.
[20] Kleeberg, R. (1991). Applied Mineralogy, Pecchio et al; 2004 ICAM-BR, São Paulo, BR, 2004, 129-132
[21] Baghdadi, Z.A.; Fatani, M.N; Sabban, N.A. *J. Mat. Civ. Eng.* 1995, 7(4), 218-222.
[22] Petry, T.M.; Little, D.N. *J. Mat. Civil Eng.* 2002, 14(6), 447-460.
[23] Edil, T.B.; Acosta, H.A.; Benson, C.H. *J. Mat. Civ. Eng.* 2006, 18(2), 283-294.
[24] Daita, R.K.; Drnevich, V.P.; Kim, D.; Chen, R. *J. Trans. Res. Board.* 2006, 1952, 101–109.
[25] Poh, H.Y.; Ghataora, G.S.; Ghazireh, N. *J. Mat. Civil Eng.* 2006, 18(2), 229-240.
[26] Hossain, K.M.A. *J. Mat. Civ. Eng.* 2011, 23(9), 1320-1327.
[27] Miller, G.A.; Azad, S. *Constr. Build. Mat.* 2000, 14, 89-97.
[28] Sreekrishnavilasam, A.; Rahardja, S.; Kmeta, R.; Santagata, M. *Constr. Build. Mat.* 2007, 21, 318-327.
[29] Taylor, H.F.W. Cement Chemistry, Second Edition, Thomas Telford Ltd., London, UK, 1997.
[30] Taylor, J.C.; Matulis, C.E. *Powder Diffr.* 1994, 9, 119-123.
[31] Ufer, K.; Stanjek, H.; Roth, G.; Dohrmann, R.; Kleeberg, R.; Kaufhold, S. *Clays Clay Miner.* 2008, 56(2), 272-282.
[32] Mitchell, J.K.; Soga, K. Fundamentals of Soil Behavior. Third Edition, *John Wiley & Sons,* Hoboken, NJ, 2005.
[33] Jang, W.; Song, C.R.; Kim, J.; Cheng, A.H.-D.; Al-Ostaz A., J. Geotech. Geoenviron. Eng. 2011, 137(4), 398-404.
[34] Mitchell; J.K. *J. Geotech. Geoenviron. Eng.* 2009, 135, 998-1010.
[35] Lin, B.; Cerato, A.B. In Geo-Frontiers 2011: Advances in Geotechnical Engineering Conference Proceedings; Han J.; Alzamora D.A.; Geotechnical Special Publication 211, American Society of Civil Engineers: Reston, VA, 2011, pp. 397-422.
[36] Likos, W.J. *J. Geotech. Geoenviron. Eng.* 2008, 134(7), 1005-1010.
[37] Little, D.N.; Nair, S.; Herbert, B. *J. Geotech. Geoenviron. Eng.* 2010, 136, 110-118.
[38] Casagrande, A. Trans. *ASCE.* 1948, 113, 901-930.
[39] Terzaghi, K.; Peck, R. Soil Mechanics in Engineering Practice, 2nd edition, John Wiley & Sons, Hoboken, NJ, 1967.
[40] Shamrani, M.A.; Mutaz, E.; Puppala, A; Dafalla, M.A. In Geoflorida 2010: Advances in analysis, model and design; Fratta, D.O.; Puppala, A.J.; Muhunthan, B. Geotechnical

Special Publication 199, American Society of Civil Engineers, Reston, VA, 2010, pp. 78-85.

[41] Chrysochoou, M.; Grubb, D.G.; Drengler, K.; Malasavage, N. *J. Geotech. Geoenv. Eng.* 2010, 136(8), 1037-1050.

[42] Chittoori, B.; Puppala, A.J. *J. Geotech. Geoenviron. Eng.* 2011, 137(11), 997-1008.

[43] Chrysochoou, M.; Puppala, A.; Chittoori, B. In GeoCongress 2012: State of the Art and Practice in Geotechnical Engineering; Hryciw, R.D.; Athanasopoulos-Zekkos, A.; Yesiller, N.; Geotechnical Special Publication 225, American Society of Civil Engineers: Reston, VA, 2012, pp 1165-1174.

[44] Mitchell, J.K. *J. Geotech. Eng.* 1996, 112(3), 255-289.

[45] Little, D.; Graves, R. Investigation of Sulfate Levels and Expansive Properties of Subgrade Soils along Route 86 Project, Imperial County, CA, 1993.

[46] LTRC (Louisiana Transport Research Center). Stabilization techniques for reactive aggregate in soil-cement base course: technical summary, Report number 366, LA, 1993.

[47] Lee, H.; Cody, R.D.; Cody, A.M.; Spry, P.G. *Cem. Concr. Res.* 2004, 35, 332-343.

[48] Chen, D.-H.; Harris, P.; Scullion, T.; Bilyeu, J. *J. Perf. Constr. Facil.* 2005, 19, 324-330.

[49] Puppala, A.J.; Cerato, A. Geo-Strata, March/April, 2009.

[50] Taylor, H.F.W.; Famy, C.; Scrivener, K.L. *Cem. Concr. Res.* 2001, 31, 683-693.

[51] Mitchell, J. K.; Dermatas, D. In Innovations and uses for lime, ASTM Special Publication 1135, American Society for Testing and Materials, West Conshohocken, PA, 1992, pp. 41-64.

[52] Puppala, A.J.; Griffin, J.A.; Hoyos, L.R.; Chomtid, S.J. *Geotech. Geoenviron. Eng.* 2004, 130(4). 391-402.

[53] Puppala, A.J.; Intharasombat, N.; Vempati, R.K. *J. Geotech. Geoenviron. Eng.* 2005, 131(3), 325-337.

[54] Yoon, S.; Balunaini, U.; Prezzi, M. *Trans. Res. Rec.* 2007, 2026, 9-17.

[55] Hunter, D. Geochemistry of Lime Induced Heave in Sulfate Bearing Clay Soils, Ph.D. Dissertation, University of Reno, NV, 1989.

[56] Mehta, P.K.; Wang, S. *Cem. Concr. Res.* 1982, 12, 121-122.

In: X-Ray Diffraction: Structure, Principles and Applications ISBN: 978-1-62808-591-4
Editor: Kaimin Shih

Chapter 6

QUANTITATIVE X-RAY DIFFRACTION FOR REVEALING THE THERMAL INCORPORATION BEHAVIOR OF LEAD

Xingwen Lu and Kaimin Shih*
Department of Civil Engineering, The University of Hong Kong, Hong Kong

ABSTRACT

X-ray diffraction (XRD) has become one of the most powerful techniques for crystal structure studies and phase composition identifications. Our progress in identifying lead incorporation behavior suggests an opportunity of using quantitative XRD (QXRD) techniques to assist in the development of reliable engineering strategies to incorporate hazardous lead pollutants into ceramic matrix. The removal of lead from wastewater commonly generates a large amount of lead-laden sludge as a hazardous waste and requires additional treatments. Metal stabilization strategies have been sought to replace the traditional disposal methods by safely reusing the waste metal sludge. In this chapter, the fundamentals and the developmental background of QXRD will be introduced. To demonstrate the unique capability of QXRD in monitoring lead incorporation behavior, alumina was used to react with lead oxide to simulate the thermal reactions between the aluminum-rich ceramic precursor and the lead-laden sludge during the sintering process. The influences of Pb/Al molar ratio, temperature, and treatment time on lead incorporation efficiency were successfully revealed by QXRD. A predominant $PbAl_2O_4$ phase was found at temperatures of 950-1000 °C while sintering lead oxide with γ-Al_2O_3 for 3 hours. The $PbAl_{12}O_{19}$ phase was first observed at 950 °C and was effectively formed at 1000 °C in the Pb/Al = 1/12 system. Moreover, the influence of silica on lead stabilization was analyzed by blending amorphous SiO_2 and quartz with γ-Al_2O_3 as the precursors. The results reveal both silica precursors could crystallochemically incorporate lead into the lead feldspar ($PbAl_2Si_2O_8$) structure in significant quantities. By adding CaF_2 as the internal standard, amorphous contents in the products were quantified using QXRD, and the weight percentage of lead feldspar was around 30-40% after 3 hours of sintering at 1000°C. Finally, by sintering clay-based precursors with lead oxide,

* Corresponding author. Tel: +852-28591973; Fax: +852-25595337. E-mail address: kshih@hku.hk.

$Pb_4Al_4Si_3O_{16}$ and a polymorph of lead feldspar were detected at temperatures within 700-900 °C, and a complete lead incorporation occurred at 950 °C. The quantitative evaluation of thermal incorporation behavior of lead has demonstrated the powerful method capability of QXRD in contributing to the advancements of both material and environmental technologies.

1. INTRODUCTION

1.1. Basic Principles of X-Ray Diffraction

In 1895, Roentgen discovered X-rays. X-rays have very high frequencies, and their electromagnetic spectra are situated between gamma rays and ultraviolet radiation. In 1911, characteristic X-ray radiation was recognized by Charles G. Barkla. X-rays emit from heavy elements when their electrons make transitions between lower atomic energy levels. The emitted characteristic X-rays generally have specific wavelengths from ~0.01 nm to ~1 nm, which are very comparable to the radius of atoms and ions in the order of 0.1 nm in size. Thus, characteristic X-rays can be used to investigate crystal structures by X-ray diffraction technique. With the performance of the first X-ray diffraction experiment on a single crystal in 1912, X-ray crystallography has been of major importance in natural sciences.

Crystal structure information of solids can be determined with the use of Bragg's law in a Bragg spectrometer, known as,

$$n\lambda = 2d_{hkl}\sin\theta \tag{1.1}$$

where: d is the distance between two crystal planes; λ represents the wavelength of the X-ray; n refers an integer; and θ is the Bragg angle. However, Bragg's law is a necessary, but insufficient, tool for diffraction study, because the diffraction pattern includes information related to both peak positions and intensity. Peak positions indicate the crystal structure and symmetry of the contributing phase, whereas peak intensities reflect the scattering from each plane in the phase's crystal structure and are also directly dependent on the distribution of particular atoms in the structure. Some Bravais lattices are with lattice units only at unit cell corners, such as simple cubic structure, in which all of the atoms positions within the unit cell are fully constrained by symmetry. However, crystal structures with non-primitive unit cells have atoms at additional lattice sites located along a unit cell edge. This may occur in a unit cell face or in the interior of the unit cell as in face-centered cubic structure or body-centered cubic structure [1]. The extra scattering centers can cause out-of-phase scattering to occur at certain Bragg angles; that means some of the diffraction lines predicted by Bragg's law do not occur. The relationship between the fractional coordinates of the atoms and the diffracted intensity of a reflection indexed as H = hkl is represented by the structure-factor equation [1, 2]:

$$F_{hkl} = \sum_{l}^{j} f_j e^{2\pi i(hu_j + kv_j + lw_j)} \tag{1.2}$$

Besides structure factor, there are many other factors that can affect the diffracted intensity including: polarization factor, Lorentz factor, multiplicity factor, temperature factor, and absorption factor. Intensities are ultimately related to both the structure and the composition of a phase. Thus, the expression for the diffraction intensity of a Bragg reflection in a θ/2θ scan can be summarized below [3, 4]

$$I_{(hkl)\alpha} = \frac{I_0 \lambda^3}{64\pi r}\left(\frac{e^2}{m_e c^2}\right)^2 \frac{M_{(hkl)}}{V_\alpha^2}\left|F_{(hkl)\alpha}\right|^2 \left(\frac{1+\cos^2(2\theta)\cos^2(2\theta_m)}{\sin^2\theta\cos\theta}\right)_{hkl} \frac{v_\alpha}{\mu_s} \tag{1.3}$$

where: $I_{(hkl)\alpha}$ = reflection intensity of the *hkl* plane in phase α; I_0 = incident beam intensity; λ = X-ray wavelength; r = distance from specimen to detector; $(e^2/m_ec^2)^2$ = square of classical electron radius; M_{hkl} = multiplicity of reflections from *hkl* planes of phase α; V_α = volume of the unit cell of phase α; $F_{(hkl)\alpha}$ = structure factor for the reflection of *hkl* plane of phase α (i.e., the vector sum of scattering intensities of all atoms contributing to that reflection); $2\theta_m$ = diffraction angle of the monochromator; the fraction in parentheses, $(\cdots)_{hkl}$, equals the Lorentz and polarization corrections for the diffractometer $(Lp)_{hkl}$, including a correction for the diffracted beam monochromator; v_α = volume fraction of phase α; and μ_s = linear absorption coefficient of the specimen.

1.2. Phase Identification of Crystalline Phases

The X-ray diffraction pattern of a pure substance is like a fingerprint of the substance, and thus, the powder diffraction method is ideally suited for the identification and characterization of polycrystalline phases. The diffraction pattern of an unknown sample can be measured to compare with known standards for phase identification. The International Centre for Diffraction Data (ICDD), formerly known as Joint Committee on Powder Diffraction Standards (JCPDS), is the organization that maintains the database of inorganic and organic spectra [5, 6]. The database, known as the Powder Diffraction File database (PDF), is designed for material identification and characterization. The database is available from diffraction equipment manufacturers or from ICDD directly. Today, about 50,000 inorganic and 25,000 organic components, crystalline phases, and diffraction patterns have been collected as standards [6]. The main use of powder diffraction is to identify components/phases in the sample by pattern search and/or match procedure.

As each crystal structure generates a unique fingerprint in the form of a diffraction pattern, the crystalline phases co-existing in the materials can be identified via modern search-match software. Semi-quantitative elemental analysis using scanning electron microscopy (SEM) or X-ray fluorescence (XRF) should still be considered since unwanted elements might have been incorporated into the crystal structures and reflected in the diffraction measurement. Advanced application of powder diffraction technique can further quantify the amount of different phases and identify the spatial distribution of observed phases.

1.3. Rietveld Refinement with Fundamental Parameters Profile Fitting (FPPF)

Rietveld refinement is a whole profile fitting method and has been well-developed to quantify the phase compositions of complex mineral assemblages. It is known that many Bragg reflections can contribute to the observed intensity, $Y_{i(obs)}$, at the ith step in the XRD pattern. Thus, in the Rietveld method, the calculated intensity, $Y_{i(cal)}$, is determined by the following equation [7]:

$$Y_{i(cal)} = S_j \sum_K L_k \left|F_k\right|^2 \phi(2\theta_i - 2\theta_k) P_k A_j + Y_{bi} \tag{1.4}$$

Here, the subscript j refers to the different crystalline phases; S_j represents the Rietveld scale factor for the jth phase; k refers to the Miller indices (which are h, k, and l) for the Bragg reflections; L_k contains the Lorentz-polarization and multiplicity factor; F_k is the structure factor for kth Bragg reflection; P_k is the preferred orientation function; A_j is the microabsorption correction function that takes into account the absorption and/or texture effects; Φ_g is the reflection profile function; 2θ is the diffraction angle; and Y_{bi} represents the background intensity at the ith step. Therefore, it can be seen that the model parameters that may be refined include not only atomic positions and thermal and site-occupancy factors, but also parameters for the background, lattice, the instrumental geometrical-optical features, and the specimen reflection-profile-broadening agents. Multiple phases may be refined simultaneously with Rietveld scale factors for the phases, and such function offers what is probably the most reliable method for performing quantitative phase-composition analysis [7, 8].

In this regard, the weight fraction of the jth phase (W_j) can be obtained using the following equation:

$$W_j = S_j(ZMV)_j / \sum_j S_j(ZMV)_j \tag{1.5}$$

where: S_j is the refined scale factor; Z is the number of formula units per cell; M is the atomic weight of the formula unit; and V is the unit cell volume of the jth phase. The summation includes all crystalline and identified components [9-14]. If an internal standard phase s has been mixed into the sample prior to data collection, the weight fraction W_a pertaining to X-ray amorphous or nonidentified phases may be estimated using Equation (1.6):

$$W_a = 1 - \sum_j W_{j,c} \tag{1.6}$$

where: $W_{j,c}$ is the recalculated, actual weight fraction of phase j, which can be determined by:

$$W_{j,c} = W_j W_{s,w} / W_s(1 - W_{s,w}) \tag{1.7}$$

W_j and W_s are the refined weight fractions of phase j and the internal standard respectively, and $W_{s,w}$ is the actual added weight of the internal standard.

The fundamental parameter approach, comparing to the phenomenological approach, for Rietveld refinement analysis is to separate the diffraction signals contributed from the instrumental factors and isolate the diffraction signals derived from samples in order to more accurately reflect crystal characteristics. This technique of profile synthesis was first introduced by Leroy Alexander [15], but was only implemented as a standard fitting procedure [16-18]. Diffraction broadening can be incorporated into the profile function $I(2\theta)$ by convoluting the broadening function $B(2\theta)$ into the instrument profile function $K(2\theta)$, where $K(2\theta)$ is synthesized by convoluting the geometrical instrument function $J(2\theta)$ with the wavelength profile $W(2\theta)$ at the Bragg angle 2θ of the peak.

$$I(2\theta) = K(2\theta)\times B(2\theta)=\mathrm{W}(2\theta)\times J(2\theta)\times B(2\theta) \quad (1.8)$$

Freeware and commercial software packages [19-21] with FPPF have been used to synthesize and fit XRD data collected from different diffractometers. For Rietveld refinement programs using fundamental-parameter approach, the refining parameters may be influenced by the configuration of each diffractometer, including the angular aperture of the analyser slits, the width and axial length of the receiving slit, the angular apertures of the axial Soller slits, the length and projected width of the X-ray source, the absorption coefficient, axial length of the sample, etc. FPPF is based on a physical model of the diffractometer, and its refined parameters should be consistent with the physical dimensions of an actual diffractometer and the physical properties of samples. On this basis, it should be more reliable for refining diffraction data collected by different powder diffractometers and fitting profiles of widely different shapes [22].

2. Refinement for Lead Ceramics

2.1. Quantitative Analysis via Powder X-Ray Diffraction

Rietveld refinements for quantitative analysis of the phase compositions in sintered ceramic products containing hazardous metals were extensively employed by the authors to explore new metal stabilization strategies. For example, the successful application of the TOPAS (version 4.0) crystallographic program, which is capable of performing the Rietveld refinement with fundamental parameter approach, has contributed new quantitative understandings in the different metal incorporation efficiencies by a variety of metal hosting phases. For systems potentially containing amorphous or poorly crystalline phases (i.e., γ-Al_2O_3, amorphous SiO_2, and organics), a refinement method using CaF_2 as the internal standard was utilized to quantify the amorphous content in the samples [23-25]. Stable and reliable background refinement was achieved using the 6-coefficient polynomial function with 4 released coefficients. In addition, scale factors and instrumental parameters including zero position and systematic shift, due to the transparency and off-centering of the sample, were released. Moreover, lattice parameters, profile parameters, and structural parameters (atomic coordinates and thermal coefficients) for all crystal phases were refined. Finally, the correction of preferred orientation for some crystalline phases was taken into account using the March-Dollase algorithm.

To access the quality of matching results, the refinement of each pattern generated the pattern factor (R_p), the expected pattern factor (R_{exp}), and the goodness of fit (*GOF*) values. The value of *GOF* is defined by the following equation [26, 27]:

$$GOF = chi^2 = \frac{R_{wp}}{R_{\exp}} = \left\{ \frac{\sum \omega_i \left[Y_i(obs) - Y_i(calc) \right]^2}{N_d - N_p} \right\}^{1/2} \tag{2.1}$$

where: $Y_{i(obs)}$ and $Y_{i(calc)}$ are the observed and calculated data, respectively, at data point m; N_d is the number of data points; N_p is the number of parameters; and ω_i is the weighting given to data point m. The counting statistics is given by $\omega_i = 1/\delta(Y_{i(obs)})^2$, where: $\delta(Y_{i(obs)})$ is the error in $Y_{i(obs)}$. The R_{wp}/R_{exp} ratio (the *GOF* value) will be equal to 1 in an ideal refinement. However, in an actual situation, the background and peak profile mismatch lead to *GOF* > 1. A *GOF* value between 1.0 and 2.9 is generally considered satisfactory [28].

2.2. Incorporating Lead in Ceramics

Global production and consumption of lead have substantially increased in recent decades. Available data from the International Lead and Zinc Study Group estimates lead consumption reached more than 9 million tons in 2010. Wastewaters generated from the industries of lead batteries, oil-based paints, mining, plating, electronics, and wood production are the main lead pollution sources [29-31]. Not biodegradable, lead can easily accumulate in the vital organs of humans and animals, inducing lead poisoning [32]. Physicochemical techniques of removing lead from wastewater usually include precipitation, coagulation, reduction, ion exchange, and membrane processes [33-38]. These treatments often result in large amounts of hazardous metal-laden sludge, which is then solidified into cementitious materials and disposed of in landfills. Considering alternative solutions to better immobilize hazardous metal, thermal treatment offers the advantage of producing stable and usable products [39].

Ceramic technique is considered to be highly effective for rendering hazardous waste inert because the process destroys organic matters and immobilizes metals in the framework of glass or crystallization phases [40]. Previous studies reported the mechanisms of stabilizing nickel and copper into aluminum-rich ceramic matrices by spinel crystalline structures [41-44]. In this study, lead incorporation experiments were conducted using PbO, together with alumina- and silica-rich precursors, to simulate the phase transformation of lead in sludge under high temperature sintering conditions [45]. PbO was mixed with several different precursors by ball milling in water slurry at different Pb/Al molar ratios. The slurry samples were then dried and homogenized by mortar grinding. The derived powder was pressed into pellets at 650 MPa to ensure consistent compaction of the powder sample for the sintering process. The pellets were sintered at targeted temperatures with different treatment time and then quenched in air to room temperature.

2.3. Ceramic Raw Materials

The received PbO (Sigma-Aldrich) powder was found as a mixture of the litharge (α-PbO; ICDD PDF #77-1971) and massicot (β-PbO; ICDD PDF #05-0561) phases. The γ-Al_2O_3, which was prepared from PURAL SB powder (with an average particle size ~45 μm) fabricated by Sasol, was first used as an alumina-rich precursor to react with PbO. The phase of PURAL SB powder was identified by XRD as boehmite (AlOOH; ICDD PDF # 74-1875), and it was successfully converted to the γ-Al_2O_3 phase after heat treatment at 650 °C for 3 hours [46] as shown in Figure 1.

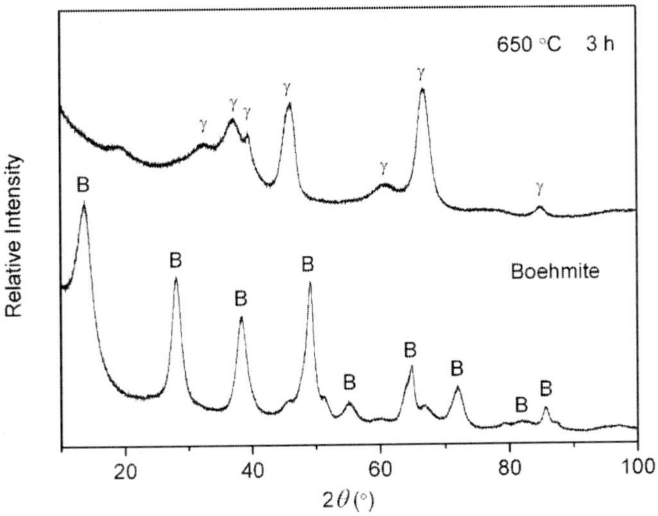

Figure 1. Phase changes of PURAL SB alumina powder after thermal treatment at 650 °C for 3 hours. The XRD pattern of PURAL SB powder was well-matched by boehmite (B), and after 650 °C and 3-hours of treatment, the XRD pattern was consistent with the γ-alumina (G) pattern.

Secondly, natural clays and silicates are usually the most common ceramic raw materials. Kaolin, which consists chiefly of kaolinite ($Al_2Si_2O_5(OH)_4$), is a common clay type, and most ceramics contain high percentages (40-65%) of kaolinite [47]. Because it is widely available, kaolinite is one of the most commonly used preliminary materials for aluminosilicate-based ceramics [48]. Mullite, the only stable binary crystalline phase in the SiO_2-Al_2O_3 system, can be produced by thermal decomposition of kaolinite [49]. Calcined kaolin is a primary byproduct generated by heating spray-dried kaolin particles to render them anhydrous through their transformation into mullite ($3Al_2O_3{\cdot}2SiO_2$) and silica (SiO_2) [50, 51]. Kaolinite powder (Sigma-Aldrich) was used as one precursor, and its XRD pattern matched that of standard kaolinite peaks (ICDD PDF #78-1996). The mullite precursor was prepared by calcining the kaolinite powder at 1350 °C for 24 h to yield mullite (ICDD PDF #79-1455) and cristobalite (SiO_2; ICDD PDF #76-0938) phases, confirmed by their XRD patterns (Figure 2).

Amorphous silica (SiO_2) and crystalline quartz (SiO_2; ICDD PDF#79-1910) were also used as Si-rich sintering precursors. In addition, water treatment sludge collected from the waterworks in Hong Kong, with a typical aluminum-based coagulation process, was also used as a sintering precursor. The dried sludge was first heated at 900 °C for 15 min to remove organic content and to facilitate the observation of metal transformation processes among different phases. The crystalline phases and the chemical compositions of the heat-treated sludge were analyzed by XRD (D8 Advance, Bruker) and XRF (JSX-3201z, JEOL) as shown in Figures 3 and 4, respectively.

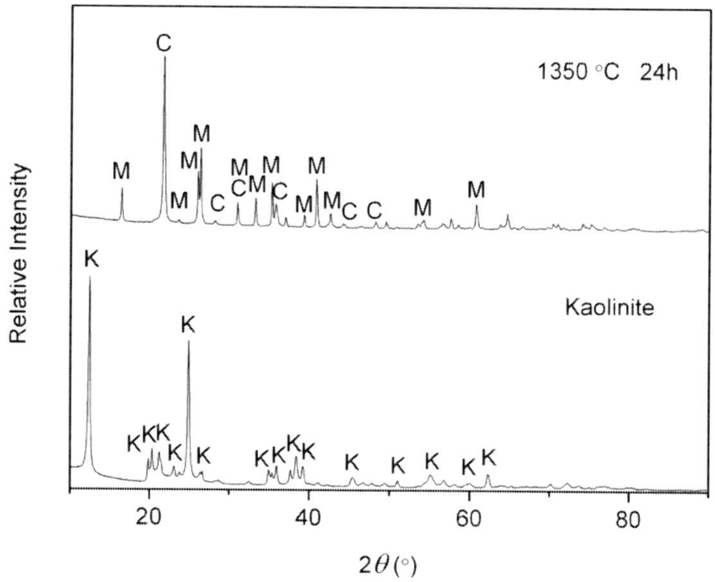

Figure 2. XRD patterns of kaolin powder used in this study and of the powder calcined at 1350 °C for 24 h. "K" represents peak positions of referenced kaolinite; "M" represents mullite phase, and "C" represents cristobalite phase.

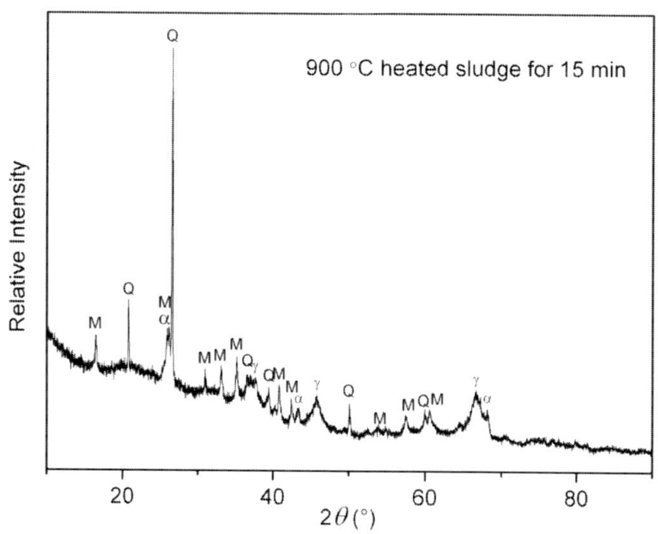

Figure 3. XRD pattern of the 900 °C and 15 min treated water treatment sludge used in this study. "M" represents peak positions of referenced mullite; "Q" represents quartz; "γ" represents γ-Al_2O_3; and "α" represents α-Al_2O_3.

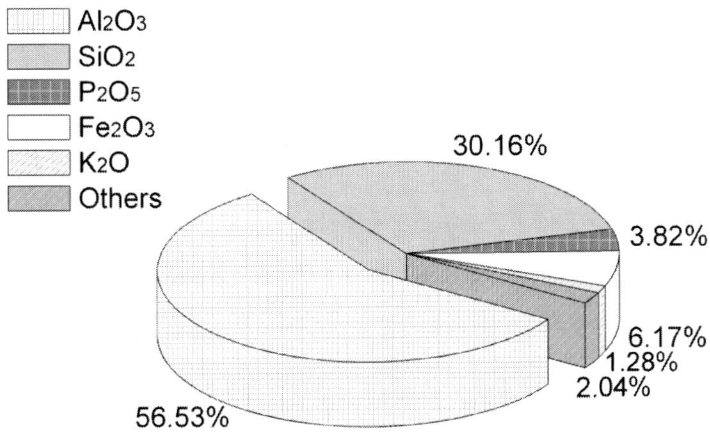

Figure 4. Major compositions of the calcined water treatment sludge used in this study. The elemental compositions of the 900 °C for 15 min calcined sludge were analyzed by X-ray fluorescence spectrometry and expressed as basic oxide forms.

3. Thermal Incorporation Behavior for Lead by Different Precursors

3.1. Formation of $PbAl_2O_4$

When γ-Al_2O_3 was used as an inexpensive ceramic precursor to react with PbO, potential phase formations are magnetoplumbite-like structure ($PbAl_{12}O_{19}$) and lead aluminate ($PbAl_2O_4$) [52]. Although the equilibrium study has indicated potential interactions between PbO and aluminum-rich precursors, the opportunity of obtaining the same product phases dominant in the time-limited sintering scheme used in industrial process is still uncertain. To explore the phases that may appear in the high-lead concentration sintering, Figure 5 demonstrates the XRD patterns of the products sintered from the Pb/Al molar ratio = 1:2 mixture of PbO + γ-Al_2O_3 powder at 600-1000 °C for 3 hours. The result shows that when sintered at 700 °C, the formation of crystalline $PbAl_2O_4$ phase described by Equation (3.1) was first observed.

$$PbO + \gamma\text{-}Al_2O_3 \rightarrow PbAl_2O_4 \qquad (3.1)$$

It was reported that $PbAl_2O_4$ can form at 600 °C, but that experiment involved 1-2 months of dwelling time [52]. Therefore, this result may suggest that the formation of $PbAl_2O_4$ at temperatures below 700 °C is limited by slow diffusion although it is thermodynamically feasible at temperatures above 600 °C. The solid-state reaction is affected by both thermodynamic constraint and kinetic process. Below 700 °C, the $PbAl_2O_4$ phase formed by the short sintering scheme might only be limited at the grain boundary of reactants, and such small quantity in the sample was not reflected in the XRD result. As the diffraction intensity of the $PbAl_2O_4$ phase increased with the rise of sintering temperature, a significantly stronger $PbAl_2O_4$ signal was achieved at 750 °C. Above 800 °C, the Bragg reflections of α-PbO or β-PbO reactants were not observable in the products. An intermediate phase, $Pb_9Al_8O_{21}$ (ICDD PDF # 73-1875), was found when the sintering temperature was around 750 °C, but its Bragg diffractions disappeared for the sample sintered at 900 °C. Above 950 °C, $PbAl_2O_4$ became the only product phase in the samples of the tested system.

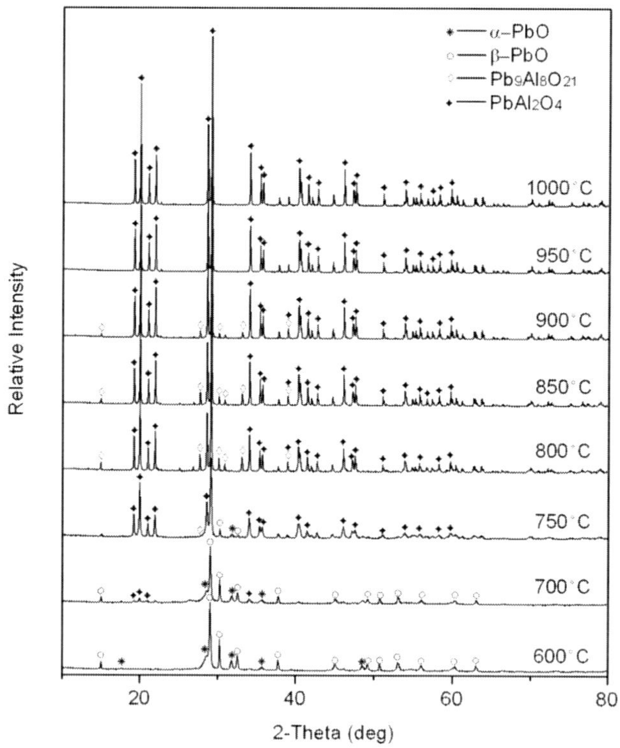

Figure 5. XRD patterns of the sintered PbO + γ-Al_2O_3 samples with a Pb:Al molar ratio of 1:2. The result shows the formation of $PbAl_2O_4$ (ICDD PDF # 85-1289) and $Pb_9Al_8O_{21}$ (ICDD PDF # 73-1875) at sintering temperatures between 600 and 1000 °C for 3 hours.

To optimize the lead incorporation efficiency during ceramic sintering at different temperatures, diffraction data, with CaF_2 as an internal standard, were analyzed by Rietveld method to quantitatively evaluate the efficiency of transforming lead in the products (Figure 6). The mass fraction of $PbAl_2O_4$ substantially increased to 79 % at 750 °C, which can be considered as a crucial temperature point to achieve Pb incorporation into the $PbAl_2O_4$ structure. However, a further increase of temperature to 800 °C led to a small amount (~8 wt.%) of $Pb_9Al_8O_{21}$ phase (ICDD PDF #73-1875) in the system. Above 950 °C, the $Pb_9Al_8O_{21}$ phase disappeared, and $PbAl_2O_4$ was the only lead-containing product phase observed at higher temperatures.

The formation of $Pb_9Al_8O_{21}$ during the sintering process was probably initiated by the reaction between PbO and γ-Al_2O_3 in a non-equilibrium system, due to insufficient sintering time at lower temperatures. The efficiency of a reaction generally depends on the encountering rate between reactant molecules [53]. The short sintering time and low temperature were not able to provide sufficient contact, and hence, the formation of $PbAl_2O_4$ is incomplete with an intermediate compound appearing in the system. At higher temperatures, intensive interaction between reactants was achieved, and $Pb_9Al_8O_{21}$ was not observed in the sintered samples. Therefore, to effectively achieve the formation of $PbAl_2O_4$, the temperature of a short sintering scheme (such as a 3-hour scheme) should be higher than 950 °C.

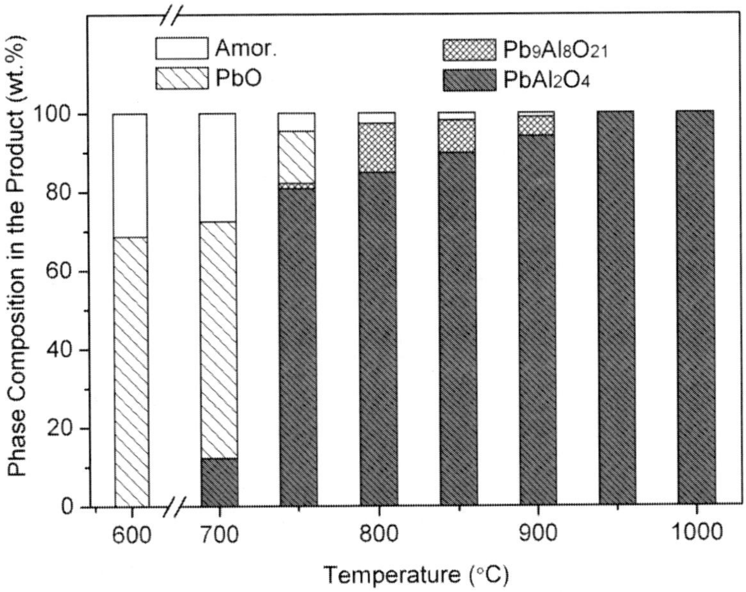

Figure 6. Quantitative comparison of phase compositions in the products of sintering PbO + γ-Al_2O_3 samples (Pb/Al molar ratio of 1:2) at temperatures between 600 and 1000 °C for 3 hours. Crystalline lead aluminates were observed to be the dominant products at high temperatures.

3.2. Formation of $PbAl_{12}O_{19}$

When the PbO molar content in the PbO-Al_2O_3 system was lower than 50%, lead dodecaaluminate ($PbAl_{12}O_{19}$; ICDD PDF # 80-1174) was the only product phase reported in equilibrium experiments [54, 55]. Therefore, a potential thermal reaction of incorporating lead by γ-Al_2O_3 precursor at lower lead levels may be described by Equation (3.2):

$$PbO + 6\,\gamma\text{-}Al_2O_3 \rightarrow PbAl_{12}O_{19} \qquad (3.2)$$

Figure 7 presents the XRD patterns of the 600-1000 °C sintered PbO + γ-Al_2O_3 mixtures with a Pb/Al molar ratio of 1:12, and they show that the peaks of the $PbAl_{12}O_{19}$ phase appeared at sintering temperatures above 950 °C. The lowest temperature at which $PbAl_{12}O_{19}$ was observed after a short sintering scheme was nearly 200 °C higher than that derived from the equilibrium experiment [54]. This discrepancy can be explained by the shorter sintering time and the potential diffusion barrier created by the newly formed $PbAl_{12}O_{19}$ layer between the PbO and γ-Al_2O_3 grains. At 1000 °C, substantial growth of $PbAl_{12}O_{19}$ phase was found, and $PbAl_{12}O_{19}$ was the only lead-containing phase in the product. However, an intermediate phase, hydrocerussite ($Pb_3(CO_3)_2(OH)_2$; ICDD PDF # 73-4362), was found to form in the products sintered at temperatures between 700 °C and 950 °C. When sintered at 1000 °C, the significant conversion to the $PbAl_{12}O_{19}$ phase completely eliminated the formation of hydrocerussite in the product, and this result may also indicate the superior stability of $PbAl_{12}O_{19}$ phase in the sintered products.

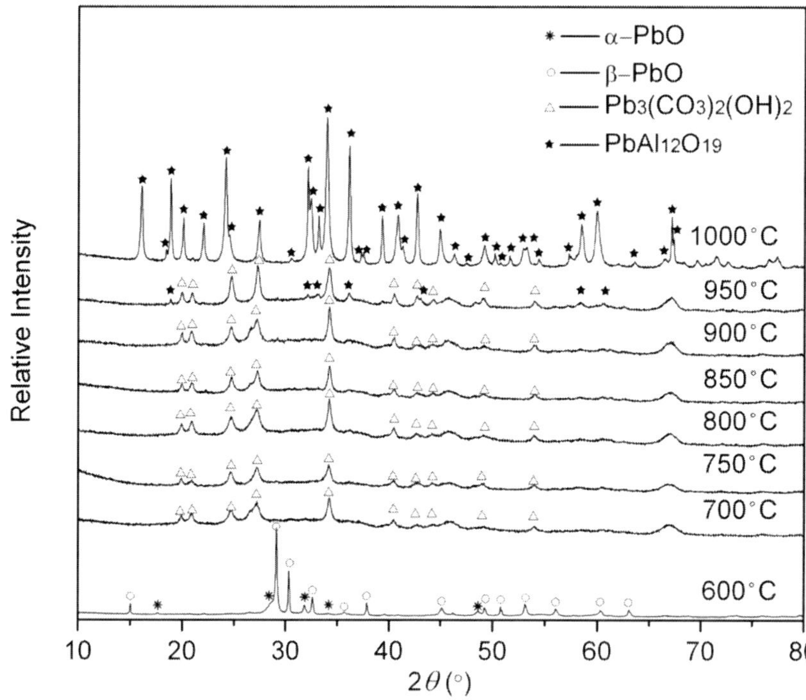

Figure 7. XRD results of products sintered from the PbO + γ-Al_2O_3 samples with a Pb/Al molar ratio of 1:12 at temperatures between 600 and 1000 °C for 3 hours.

Figure 8 presents the phase distributions of the sintered PbO + γ-Al_2O_3 samples with a Pb/Al molar ratio of 1:12 at 600-1000 °C. The figure shows that the formation of $PbAl_{12}O_{19}$ was first detected at 950 °C, and then achieved nearly full incorporation at 1000 °C. However, an intermediate phase, hydrocerussite ($Pb_3(CO_3)_2(OH)_2$), formed in the products at sintering temperatures between 700 °C and 950 °C. $Pb_3(CO_3)_2(OH)_2$ was identified as a common lead corrosion product by other studies [56,57]. Hence, the formation of poor crystalline $Pb_3(CO_3)_2(OH)_2$ phase was probably due to the instability of product phases sintered at 700-950 °C, which might be vulnerable to the attack of CO_2 and moisture in the air during sample quenching. Starting from 950 °C, the formation of small amounts of $PbAl_{12}O_{19}$ phase can be observed. Significant growth of the $PbAl_{12}O_{19}$ phase clearly occurred within the temperature range of 950-1000 °C. Nevertheless, $PbAl_{12}O_{19}$ was found to be the only lead-containing phase in the products. This result confirms the potential formation of $PbAl_{12}O_{19}$ to incorporate lead into aluminum-rich ceramics in systems with low Pb/Al ratios.

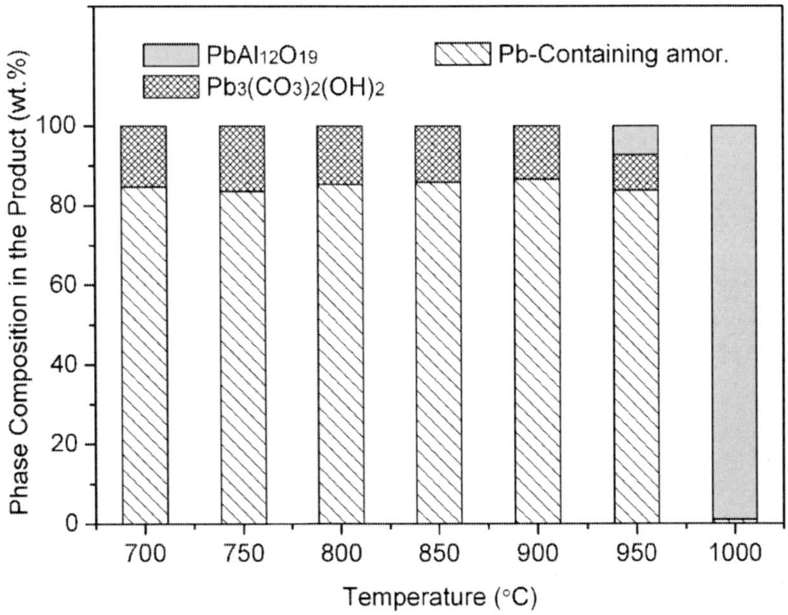

Figure 8. Weight fractions of crystalline and amorphous phases obtained from sintering PbO with γ-Al_2O_3 in a Pb/Al molar ratio of 1:12 at temperatures from 700 to 1000 °C for 3 hours.

3.3. Silica for Lead Feldspar Formation

The recently published PbO-Al_2O_3-SiO_2 equilibrium phase diagram [55] suggests that the mixture of PbO, Al_2O_3 and SiO_2 may result in glass-ceramic materials with the crystallizations of lead aluminate ($PbAl_2O_4$) and lead feldspar ($PbAl_2Si_2O_8$) phases. To achieve a successful metal stabilization strategy, a quantitative understanding on the partitioning of lead into different hosting phases under different thermal conditions is crucial. Therefore, our study first quantitatively compared the reaction efficiencies among lead oxide, Al_2O_3, and the different forms of Si-containing precursors through ceramic sintering to delineate potential lead incorporation mechanisms. The feasibility of using water treatment sludge as a material source for stabilizing hazardous metals during the sintering process was also evaluated.

The influence of silica on lead *transformation* was evaluated using γ-Al_2O_3 + amorphous SiO_2 (Figure 9) and γ-Al_2O_3 + quartz (Figure 10) as the two precursors to mix with PbO in the molar ratio of Pb:Al:Si = 1:2:2. The powder XRD data indicated that lead could be incorporated into the lead feldspar structure ($PbAl_2Si_2O_8$; ICDD PDF #89-6435) at temperatures higher than 800 °C in the γ-Al_2O_3 + amorphous SiO_2 system and 750 °C in the γ-Al_2O_3 + quartz system. Unlike the system only with γ-Al_2O_3, no $PbAl_2O_4$ phase occurred in either of these two Si-containing systems throughout the entire sintering process. When sintering with the γ-Al_2O_3 + amorphous SiO_2 precursor at 600 °C, no residual PbO was found, and two silicate phases, lead silicate (Pb_2SiO_4; ICDD PDF # 30-0723) and lead metasilicate ($PbSiO_3$; ICDD PDF #28-0540), were observed. However, their Bragg diffraction peaks disappeared and the system was completely amorphorized at temperatures between 700 and

750 °C, potentially due to the low melting points of Pb_2SiO_4 (625-720 °C) and $PbSiO_3$ (700-750 °C) [58]. Although the poorly crystalline $PbAl_2Si_2O_8$ phase began to form at 800 °C, the growth of lead feldspar generally became more prominent at temperatures above 900 °C. This phase ($PbAl_2Si_2O_8$) is confirmed to be the only lead-hosting crystal phase at high temperatures with a solid-state reaction described as follows:

$$PbO + \gamma\text{-}Al_2O_3 + 2SiO_2 \rightarrow PbAl_2Si_2O_8 \quad (3.3)$$

When using γ-Al_2O_3 + quartz as a precursor, the depletion of PbO occurred at a temperature of 650 °C, but a strong amorphorization of the system also began at this temperature. This finding indicates the important role of quartz in amorphorizing PbO under thermal conditions and suggests the potential of producing lead-silicate glass in products. The initial growth of lead feldspar ($PbAl_2Si_2O_8$) was detected at 750 °C and a slight increase of lead feldspar diffraction signal was observed at higher temperatures. This indicates a recrystallization preference of $PbAl_2Si_2O_8$ phase at high temperatures although the amorphous phase still dominates the system.

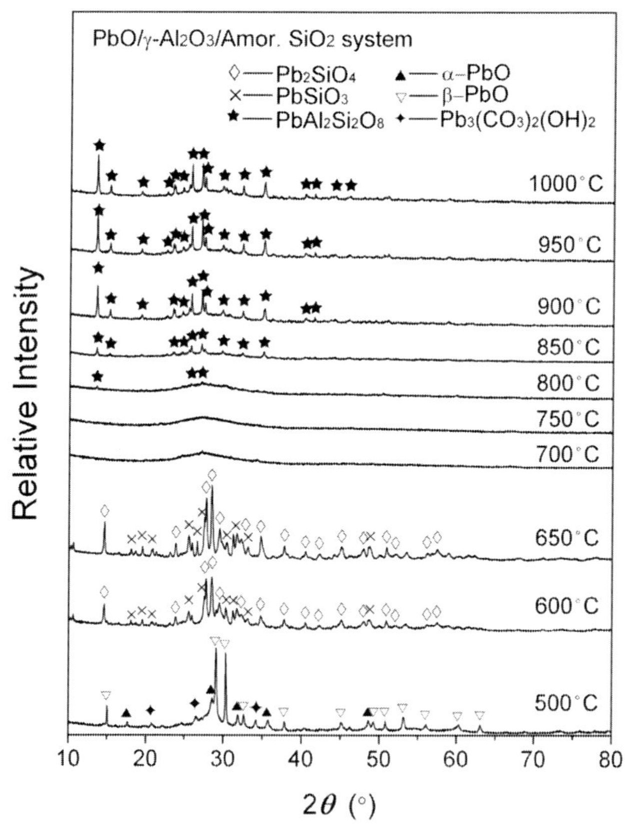

Figure 9. XRD patterns of the products sintered from the PbO + γ-Al_2O_3 + amorphous SiO_2 system (Pb:Al:Si molar ratio of 1:2:2) at temperatures between 500 °C and 1000 °C for 3 hours.

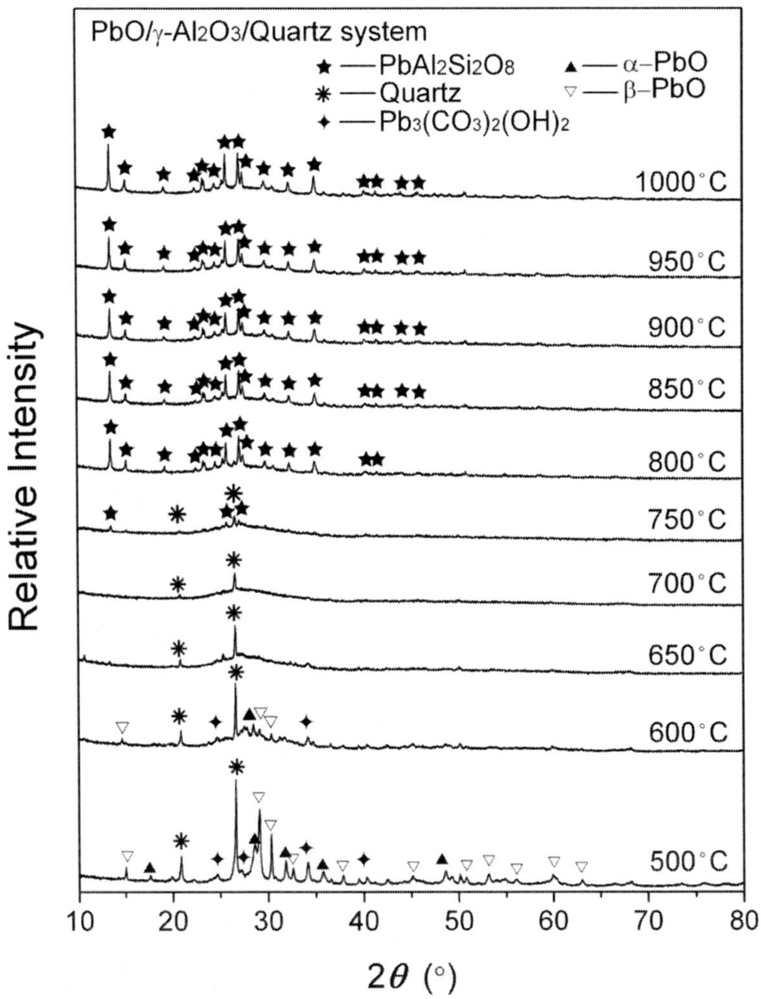

Figure 10. XRD patterns of the products sintered from the PbO + γ-Al_2O_3 + quartz system (Pb:Al:Si molar ratio of 1:2:2) at temperatures between 500 °C and 1000 °C for 3 hours.

By adding 20 wt.% CaF_2 as an internal standard, the quantities of crystalline and amorphous phases in each product sintered by γ-Al_2O_3 + amorphous SiO_2 or γ-Al_2O_3 + quartz precursor can be further revealed. Figure 11 summarizes the quantitative outcomes of lead distribution. For the reactions of using γ-Al_2O_3 + quartz as a precursor, the weight percentage of $PbAl_2Si_2O_8$ increased from 17 to 30 wt.% at temperatures from 750 to 1000 °C. While in the PbO + γ-Al_2O_3 + amorphous SiO_2 system, an increase of lead feldspar phase from 6 to 38 wt.% occurred at temperatures from 800 to 1000 °C. Although the growth of $PbAl_2Si_2O_8$ could be initiated at a lower temperature by γ-Al_2O_3 + quartz precursor, potentially due to the enhancement of PbO amorphoriziation by quartz, a slightly higher transformation to the lead feldspar phase was achieved by amorphous SiO_2 at the highest temperature (1000 °C). The incomplete recrystallization of $PbAl_2Si_2O_8$ indicates that the solid-state reaction of this tertiary-component system probably occurs with a significant overlap of nucleation and crystal growth rate curves [59]. Therefore, a glass-ceramic material is the product of sintering PbO with the aluminum-rich precursor, if SiO_2 is also a part of the precursor material.

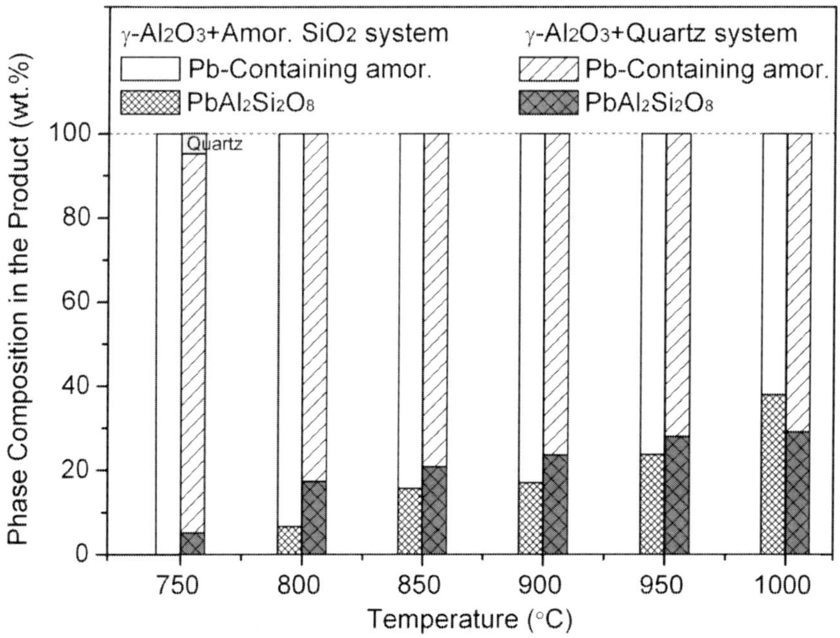

Figure 11. Quantities of lead-containing phases in the products of sintering PbO with γ-Al_2O_3 + amorphous SiO_2 and γ-Al_2O_3 +quartz precursors at temperatures between 750 °C and 1000 °C for 3 hours. A glass-ceramic material containing lead feldspar ($PbAl_2Si_2O_8$) was the final form of sintered products.

3.4. Formation of Lead Feldspar from Kaolinite Precursor

To investigate the effect of temperature on initiating lead reaction with clay-based precursors (e.g. kaolinite, mullite) during the ceramic sintering process, we carried out a short (3-hour) sintering scheme at temperatures ranging from 500 to 950 °C. The XRD patterns revealed that the sample sintered at 700 °C started to develop a distinguishable lead feldspar phase as shown in Figure 12. The chemically bound water of kaolinite is generally removed at temperatures above 550 °C with the formation of amorphous metakaolin ($Al_2Si_2O_7$) [50,51]. XRD patterns of products sintered at 700 °C showed the crystalline incorporation of lead by a recrystallization process when sintering PbO with metakaolin:

$$PbO + Al_2Si_2O_7 \rightarrow PbAl_2Si_2O_8 \quad (3.4)$$

The diffraction intensity of the lead feldspar phase increases with the intensification of sintering temperature and demonstrates a very strong lead feldspar signal at 800 °C. By high-temperature XRD technique (temperature increase rate of 5 °C min^{-1}), the formation of lead feldspar was also observed at this temperature [60]. Together with our experimental results, both outcomes indicate a strong and effective formation of lead feldspar under this thermal condition. Lead feldspar was the only lead-containing phase in the system when sintering temperatures reached 950 °C. However, two intermediate phases, namely lead aluminum silicate ($Pb_4Al_4Si_3O_{16}$; ICDD PDF #32-0505) and a polymorph of lead feldspar (ICDD PDF #14-0326), were also observed at temperatures of 700-900 °C. The appearance of these two intermediate phases suggests that the reaction between PbO and metakaolin was not in equilibrium at low temperatures (700-900 °C) after a 3-hour sintering scheme. The short sintering time and low temperature did not allow sufficient contact for this solid-state reaction and resulted in intermediate compounds and an incomplete formation of lead feldspar. More intensive interaction between reactants can be achieved by higher temperatures, and the two intermediate silicates were not observed in products sintered at higher temperatures.

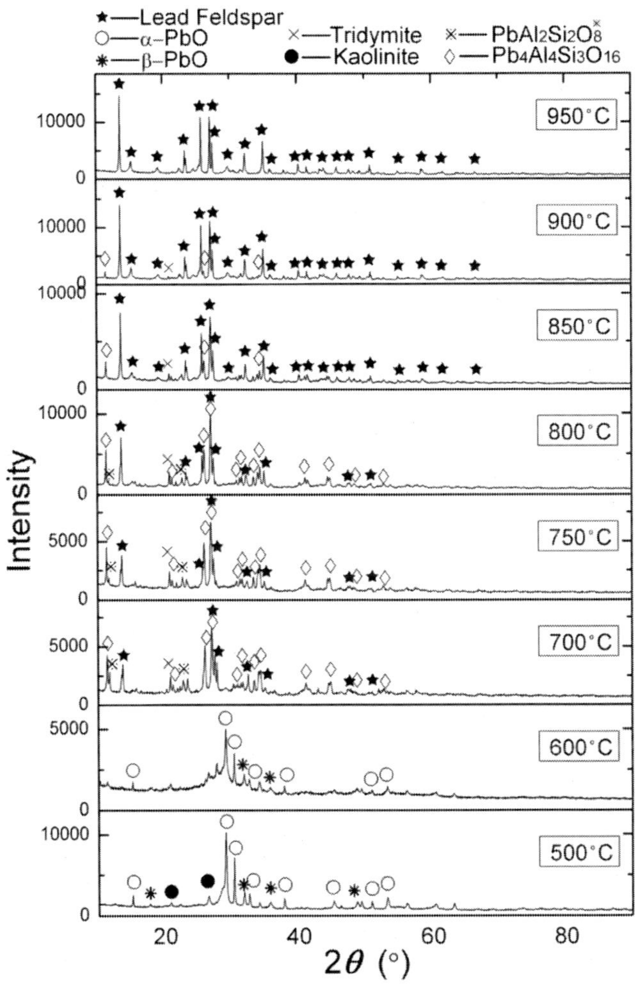

Figure 12. XRD patterns of the sintered PbO + kaolinite samples (Pb/Al molar ratio of 1:2) show the formation of lead feldspar ($PbAl_2Si_2O_8$, ICDD PDF #89-6435) at temperatures between 500 °C and 950 °C after the 3-hour treatment.

According to the powder diffraction data of lead feldspar phase, its two major peaks are located at $2\theta = 13.58°$ and 25.75°, corresponding to the diffraction planes (0 2 0) and (2 2 0), respectively. Therefore, two 2θ ranges ($2\theta = 13.25$-13.95° and 25.55-26.00°) were selected to further observe the formation of lead feldspar at elevated temperatures (insets of Figure 13). The peak intensities of lead feldspar were weak in the 700 °C sintered sample, and substantial formation of lead feldspar was not detected until 800°C which represents the energy needed for overcoming the diffusion barrier in the sample. Figure 13 shows relative intensities of the peaks at $2\theta = 13.58$ (a) and 25.75° (b), and the continuing formation of lead feldspar was found at temperatures between 700 and 950 °C.

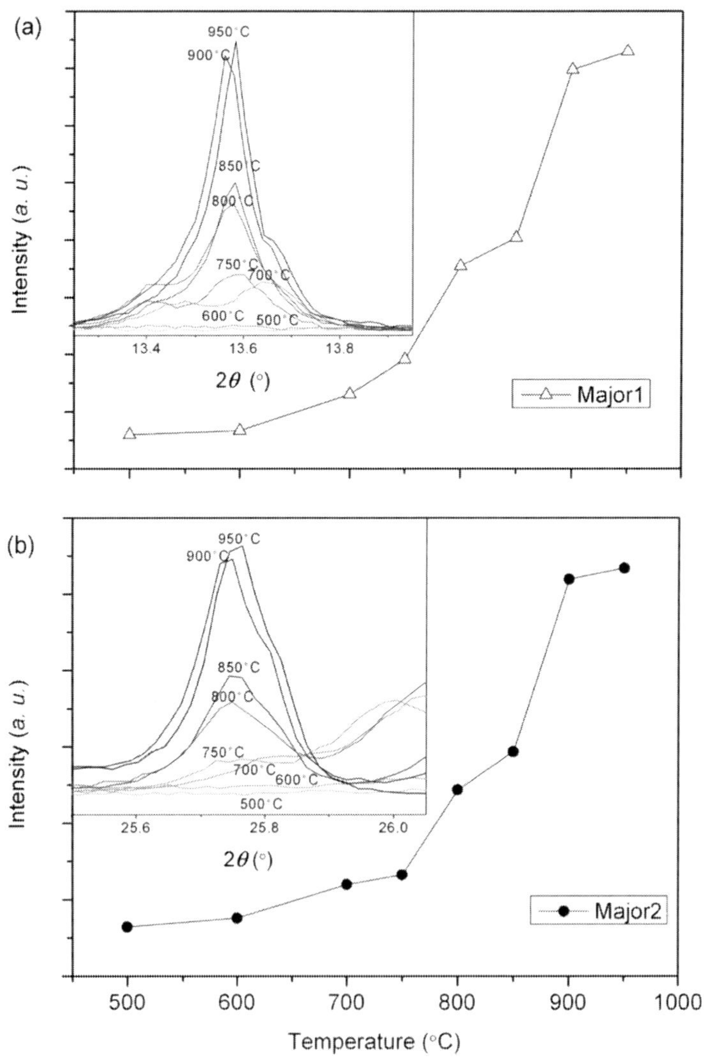

Figure 13. XRD patterns and peak intensities between (a) $2\theta = 13.25°$ and 13.95°, and (b) $2\theta = 25.55°$ and 26.00° for the PbO + kaolinite samples (Pb/Al molar ratio of 1:2) sintered at 500-950 °C for 3 hours. Maximum lead feldspar formation occurred at 950 °C. The top-right insets show the relative peak intensities at $2\theta = 13.58°$ (a) and $2\theta = 25.75°$ (b).

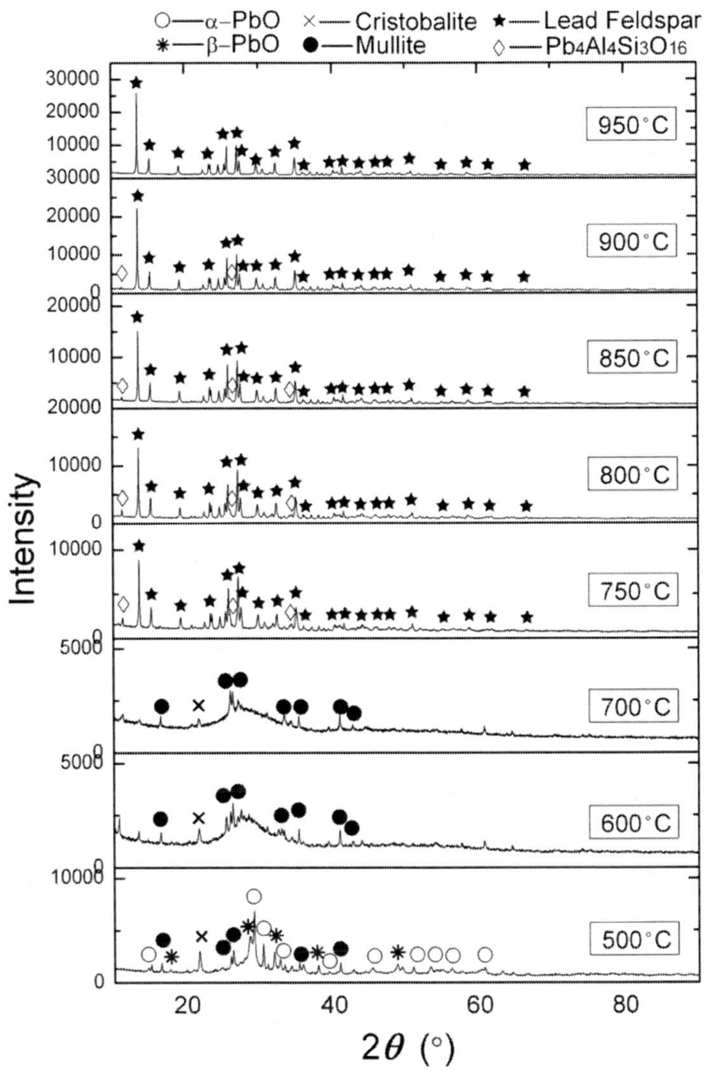

Figure 14. XRD patterns of the PbO + mullite samples (Pb/Al molar ratio of 1:2) sintered at temperatures of 500-950 °C for 3 hours show amorphization and recrystallization of lead-containing phases.

3.5. Formation of Lead Feldspar from Mullite Precursor

Phase transformation through thermally treating kaolinite has been extensively studied, and the reaction sequence is known as the kaolinite-to-mullite series. At 1000 °C, mullite begins to form, with excess cristobalite converted from amorphous silica at approximately 1300 °C [50, 51]. For the PbO + mullite system, Figure 14 shows the powder XRD patterns of products sintered between 500 and 950 °C for 3 hours. A distinguishable lead feldspar signal was detected at temperatures above 750 °C, and no reactants were detected at 950 °C. The formation of lead feldspar initiated by the mullite precursor can thus be described as:

$$3PbO + 3Al_2O_3{\cdot}2SiO_2 + 4SiO_2 \rightarrow 3PbAl_2Si_2O_8 \quad (2)$$

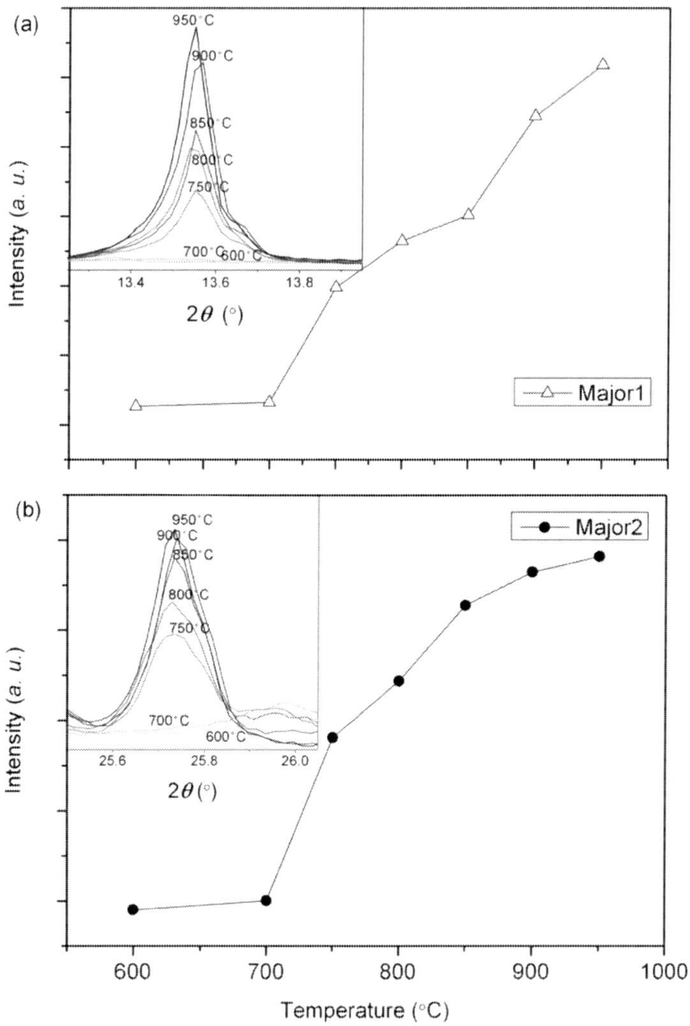

Figure 15. XRD patterns and peak intensities between (a) $2\theta = 13.25°$ and 13.95°, and between (b) $2\theta = 25.50°$ and 26.00° for PbO + mullite samples (Pb/Al molar ratio of 1:2) sintered at 600-950 °C for 3 hours. Effective formation of lead feldspar began at 750 °C and continuously increased at higher sintering temperatures.

In Figure 14, three stages of phase transformation were observed, representing different lead incorporation behavior at corresponding temperature ranges. In the first stage (below 500 °C), no reaction between the mullite precursor and lead oxide was detected by XRD. The XRD patterns at 500 °C show diffraction peaks of PbO, mullite, and cristobalite phases which were the reactants in the system. At sintering temperatures between 600 and 700 °C (the second stage), the peaks of PbO disappeared, and cristobalite substantially decreased. This finding indicates PbO amorphization initiated by silica under thermal condition and suggests the existence of a lead-silicate glass phase in the product. In the third stage, a significant increase of lead feldspar in the system strongly indicates a recrystallization of the lead-containing phase through reactions with mullite at temperatures of 750-950 °C. However, in the third stage, an intermediate phase of $Pb_4Al_4Si_3O_{16}$ was detected in samples sintered below 900 °C, which can also be explained by the insufficient mass transfer of the formation of lead feldspar. At 950 °C, more intensive interaction between reactants was possible, and lead feldspar was found to be the only lead-containing phase in the sintered product.

The formation of the lead feldspar phase is demonstrated in more detail by the increase of two major peaks located at 2θ = 13.58° and 25.75°, which are attributed to diffraction planes (0 2 0) and (2 2 0), respectively (Figure 15). The increase of both peak intensities indicates a strong formation of lead feldspar starting at 750 °C, even with the existence of $Pb_4Al_4Si_3O_{16}$ intermediate in the system.

3.6. Formation of $PbAl_2Si_2O_8$ from Water Treatment Sludge Precursor

On a global scale, water treatment systems generate approximately 10,000 tons of water treatment (*water*works) sludge per day [61]. The costs of handling the enormous amount of *water* treatment sludge *are likely to multiply due to increasingly stringent regulations [62].* This type of sludge typically contains the residues of treatment chemicals used as coagulants, commonly aluminum or iron salts [63]. Recently, several studies evaluated the feasibility of using water treatment sludge as a raw material for ceramic production [64-66], and the results of the study suggest the success of incorporating harmful metals in crystal structures to largely reduce the risk of their environmental hazard. Recycling water treatment sludge for quality ceramic production can encourage a more sustainable use of natural resources. Moreover, if the application of waste sludge in ceramics can effectively stabilize harmful materials, such as hazardous metals, it is also a promising waste-to-resource strategy to reduce the burden of managing environmentally hazardous waste.

The potential of using water treatment sludge as the sintering precursor to form $PbAl_2Si_2O_8$ was investigated using a mixture of PbO and heat-treated sludge in a molar ratio of Pb/Si = 1:2 (the Al was stoichiometrical excess in the system) for the 3-hour sintering scheme at temperatures of 500-1000 °C. The XRD patterns (Figure 16) illustrate the success of incorporating lead into the $PbAl_2Si_2O_8$ structure. At 600°C, the peak intensity of PbO significantly decreased, showing the amorphorization effect observed in the previous PbO-Al_2O_3-SiO_2 system. When sintered at 700 °C, no Bragg reflection of PbO phase was detected, and the $PbAl_2Si_2O_8$ phase began to crystallize. Above 750 °C, $PbAl_2Si_2O_8$ was the dominant crystalline phase in the system and continuously increased at higher temperatures. However, with the influence of minor elements (such as K and P) in the system, two minor Pb-containing phases, dundasite ($PbAl_2(CO_3)_2(OH)_4{\cdot}H_2O$; ICDD PDF #71-0805) and lead potassium phosphate ($Pb_8K_2(PO_4)_6$; ICDD PDF #72-0888), were also found in the sintered products. A small amount of dundasite was discovered as a low-temperature (700-750 °C) intermediate, and the formation of dundasite was probably due to the attack of CO_2 and moisture in the unstable product phase(s) during sample quenching. At temperatures above

800 °C, the significant increase of $PbAl_2Si_2O_8$ phase completely eliminated the formation of dundasite, suggesting the superior stability of $PbAl_2Si_2O_8$ phase at high temperatures.

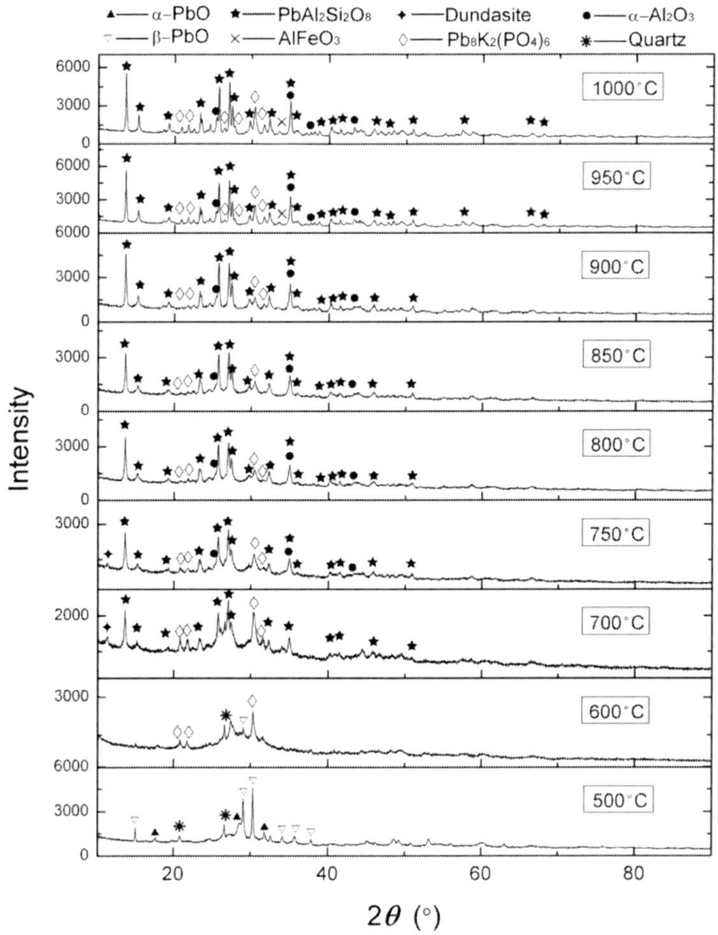

Figure 16. XRD patterns of products sintered from using water treatment sludge precursor for lead incorporation (Pb/Si molar ratio of 1:2). The results show the amorphorization and the recrystallization processes of lead-containing phases at temperatures between 500 and 1000 °C with a 3-hour sintering scheme.

To further evaluate the lead incorporation efficiency, quantitative compositions of different phases in the products were delineated by Rietveld refinement analysis with an internal standard (i.e., 20 wt.% CaF_2) to also quantify the amorphous content. Figure 17 summarizes the weight fractions of crystalline and amorphous phases at different sintering temperatures. The results reveal a key feature of $PbAl_2Si_2O_8$ increase from 10 wt.% to 46 wt.% at temperatures from 700 °C to 1000 °C, with a corresponding decrease of amorphous content from 77 wt.% to 30 wt.%. Furthermore, the weight fractions of two minor phases, $Pb_8K_2(PO_4)_6$ (~5.5 wt.%) and $AlFeO_3$ (~3.6 wt.%), were generally not affected by the temperatures in this range. Above 950 °C, the weight fraction of $PbAl_2Si_2O_8$ in the product stabilized at around 45%, and thus, the incorporation of lead by waterworks sludge system can be optimized by using a 950 °C sintering temperature.

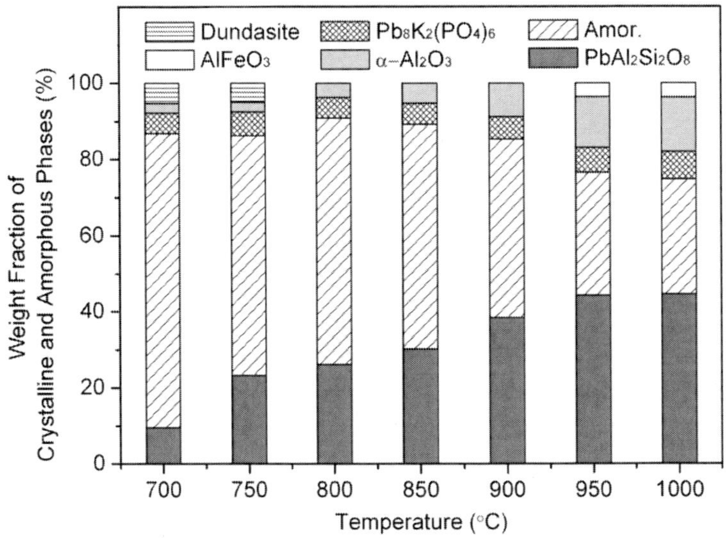

Figure 17. Weight fractions of crystalline and amorphous phases observed in the 3-hour sintered products of PbO and water treatment sludge samples. A complete (100%) transformation of the lead into $PbAl_2Si_2O_8$ phase is corresponding to a 67 % weight fraction (dashed line) of $PbAl_2Si_2O_8$ in the product.

With a complete (100 %) transformation of lead into $PbAl_2Si_2O_8$ structure, the weight fraction of lead feldspar in the product can reach 67 %, based on the elemental composition and stoichiometry information, in the system. The highest weight fraction (45%) of $PbAl_2Si_2O_8$ represents a nearly 69 % transformation of lead into the $PbAl_2Si_2O_8$ phase, and this result indicates a substantially enhanced $PbAl_2Si_2O_8$ crystallization in the water treatment (waterworks) sludge system, comparing to the $Al_2O_3 + SiO_2$ systems. The enhancement effect is likely due to the existence of impurities, such as K_2O and P_2O_5, in the water treatment sludge precursor. Capable of forming liquid phase(s) with silica and alumina, K_2O is commonly used as a fluxing agent in ceramic sintering to enhance the particle contact frequency and ceramic density [67-69]. In addition, P_2O_5 can also act as a nucleating agent in glass-ceramic systems to promote the crystallization of ceramics [70, 71]. Therefore, as a potential waste-to-resource strategy, the beneficial use of water treatment sludge can effectively incorporate lead by the formation of $PbAl_2Si_2O_8$ structure. The transformation efficiency can also be higher than that from a precursor with only γ-Al_2O_3 and SiO_2, potentially due to the additional fluxing and nucleation effects.

CONCLUSION

Hazardous lead was successfully incorporated into the $PbAl_2O_4$ and $PbAl_{12}O_{19}$ structures using γ-Al_2O_3 precursor. Competitive formations of $PbAl_2O_4$ and $PbAl_{12}O_{19}$ occurred in the

PbO + γ-Al_2O_3 system at 950 °C and 1000 °C for 3 hours, respectively. The reaction sequences performed differently depending on the thermal treatment conditions. The occurrence of intermediate $Pb_9Al_8O_{21}$ and $Pb_3(CO_3)_2(OH)_2$ phases was mainly caused by the incomplete reaction of PbO and γ-Al_2O_3 in the system. However, with the addition of SiO_2 to the system, the formation of $PbAl_2Si_2O_8$ dominates the incorporation of lead in glass-ceramics with a maximum incorporation efficiency of 30-40% found at 1000 °C. As a comparison to sintering PbO with kaolinite and mullite precursors, the results indicate that lead feldspar formation can be initiated at a lower temperature range of 700-750 °C, and it is the only lead-containing phase in the product when sintered at 950 °C. However, attention should be paid to the formation of $Pb_4Al_4Si_3O_{16}$ and the polymorph of lead feldspar as intermediate phases at temperatures between 700 and 900 °C in the PbO + kaolinite system. In the PbO + mullite system, a lead-silicate glass phase appeared between 600 and 700 °C, and $Pb_4Al_4Si_3O_{16}$ is the only intermediate phase in the temperature range of 750-900 °C. Finally, the mechanism of predominate formation of $PbAl_2Si_2O_8$ in a glass-ceramic matrix was successfully confirmed by using the waste sludge generated from water treatment plants (waterworks) as a precursor material. A high lead incorporation of 69% into the $PbAl_2Si_2O_8$ phase was found using the water treatment sludge precursor, potentially due to the K_2O and P_2O_5 impurities acting as fluxing and nucleating agents in the precursor. With a crucial assist from quantitative X-ray diffraction technique, a promising waste-to-resource strategy of using water treatment sludge as a precursor to stabilize hazardous metals in glass-ceramics was successfully demonstrated as a viable and dependable environmental technology.

REFERENCES

[1] Wang, Y.; Uchida, T.; Robert, V. D.; Nozawa, A.; Funakoshi, K.; Nishiyama, N.; Kaneko, H. High-pressure angle-dispersive powder diffraction using an energy-dispersive setup and white synchrotron radiation; Chen, J.; Wang, Y.; Duffy, T. S.; Shen, G.; Dobrzhinetskaya, L. F. Ed.; *Elsevier:* Amsterdam, 2005, pp 339-352.

[2] Louër, D. Powder X-Ray diffraction, applications; Lindon, J. Ed.; *Academic Press:* Oxford, 1999, pp 2253-2262.

[3] Jenkins, R.; Snyder, R. L. Introduction to X-ray Powder Diffractometry; *Wiley:* New York, 1996.

[4] Fujio, I. The Rietveld method and its applications to synchrotron X-ray powder data; Saisho, H.; Gohshi, Y. Ed.; *Elsevier,* 1996, Vol.7, pp 405-452.

[5] Chatterjee, A. K. X-Ray Diffraction; Ramachandran, V. S.; Beaudoin, J. J. Ed.; William Andrew Publishing: *Norwich,* NY, 2001, pp 275-332.

[6] Sharma, R.; Bisen, D.P.; Shukla U.; Sharma, B. G. X-ray diffraction: a powerful method of characterizing nanomaterials. *Recent Research in Science and Technology.* 2012, 4(8), 77-79.

[7] Young, R. A. The Rietveld Method; Oxford University Press: Oxford, 1993.

[8] Canton, P.; Fazzini, P.F.; Meneghini, C.; Benedetti, A.; Pozzi, G. Nanoscale Characterization of Metal Nanoclusters by Means of X-Ray Diffraction (XRD) and Transmission Electron Microscopy (TEM) Techniques; Corain, B.; Schmid, G.; Toshima, N. Ed.; Elsevier: Amsterdam, 2008, pp 129-147

[9] Lutterotti, L. (2002). Maud version 1.85, <http://www.ing.unitn.it/~luttero/maud>.
[10] Rietveld, H. M. *Acta Crystallogr.* 1967, 22, 151.
[11] Rietveld, H. M. *J. Appl. Crystallogr.* 1969, 2 65.
[12] Young, R. A.; Wiles, D. B. *J. Appl. Cryst.* 1982, 15, 430.
[13] Ortiz, A. L.; Sánchez-Bajo, F.; Padture, N.P.; Cumbrera, F. L.; Guiberteau, F. *J. Eur. Ceram. Soc.* 2001, 21(9), 1237-1248.
[14] Wiles, D. B.; Young, R. A. *J. Appl.* Cryst. 1981, 14, 149.
[15] Alexander, L. E. *J. Appl. Phys.* 1954, 25, 155.
[16] Cheary, R. W.; Coelho, A. A. *J. Appl. Cryst.* 1992, 25, 109.
[17] Cheary, R. W.; Coelho, A. A. *J. Appl. Cryst.* 1998, 31, 851-861.
[18] Cheary, R. W.; Coelho, A. A. *J. Appl. Cryst.* 1998, 31, 862-868.
[19] Cheary, R. W.; Coelho, A. A. (1996). Software: Xfit-Koalariet CCP14 Library <http://www.ccp14.ac.uk>.
[20] Bergmann, J.; Kleeberg, R. (1998). Software: BGMN. <http://www.bgmn.de>.
[21] Kern, A. A.; Coelho, A. A. *Allied Publishers Ltd.,* 1998, ISBN 81-7023-881-1, 144-155.
[22] Cheary, R. W.; Coelho, A. A.; Cline, J. P. *J. Res. Natl. Inst. Stand. Technol.* 2004, 109, 1-25.
[23] de la Torre, A. G.; Bruque, S.; Aranda, M. *J. Appl. Cryst.* 2001, 34, 196-202.
[24] Magallanes-Perdomo, M.; Carrodeguas, R. G.; Pena, P.; DeAza, P. N.; De Aza, S.; De Aza, A. H. *Key Engineering Materials,* 396-398, 2009, 127-130.
[25] Rendtorff, N. M.; Conconi, M. S.; Aglietti, E. F.; Chain, C. Y.; Pasquevich, A. F.; Rivas, P. C.; Martínez, J. A.; Caracoche, M. C. *Hyperfine Interact.* 2010, 198(1-3), 211-218.
[26] Manik, S. K.; Pradhan, S. K. *Mater. Chem. Phys.* 2004, 86(2-3), 284-292.
[27] Lutterotti, L. (2002). Maud version 1.85. <http://www.ing.unitn.it/~luttero/maud>.
[28] Fansuri, H.; Zhang, D. K.; French, D.; Elcombe, M.; Studer, A. (2004). An X-ray and neutron diffraction study of the structure of alpha- $Bi_2Mo_3O_{12}$ as a catalyst for partial oxidation of propylene to acrolein [online]. In: Chemeca 2004: 32nd Australasian Chemical Engineering Conference: Sustainable Processes. Barton, A.C.T.: Engineers Australia, 153-160. <http://search.informit.com.au/documentSummary;dn= 92731 53 22346188;res=IELENG> ISBN: 9781604233278.
[29] Jalali, R.; Ghafourian, H.; Asef, Y.; Davarpanah, S. J.; Sepehr, S. J. *Hazard. Mater.* 2002, 92(3), 253-262.
[30] Gupta, V. K.; Gupta, M.; Sharma, S. *Water Res.* 2001, 35(5), 1125-1134.
[31] Conrad, K.; Bruun Hansen, H. C. *Bioresource Technol.* 2007, 98(1), 89-97.
[32] Gupta, V. K.; Agarwal, S.; Saleh, T. A. *J. Hazard. Mater.* 2011, 185(1), 17-23.
[33] Husein, M. M.; Vera, J. H.; Weber, M. E. *Sep. Sci. Technol.* 1998, 33(12), 1889-1904.
[34] Lin, S. W.; Navarro, R. M. F. *Chemosphere* 1999, 39(11), 1809-1817.
[35] Petruzzelli, D.; Pagano, M.; Tiravanti, G.; Passino, R. Solvent. *Extr. Ion Exc.* 1999, 17(3), 677-694.
[36] Saeed, A.; Iqbal, M.; Akhtar, M. W. *J. Hazard. Mater.* 2005. 117(1), 65-73.
[37] Doyurum, S.; Çelik, A. *J. Hazard. Mater.* 2006, 138(1), 22-28.
[38] Ali, I.; Gupta, V.K. *Nat Protoc.* 2007, 1(6), 2661-2667.

[39] Basegio, T.; Berutti, F.; Bernardes, A.; Bergmann, C.P. *J. Euro. Ceram. Soc.* 2002, 22(13), 2251-2259.

[40] Zoua, J. L.; Xua, G. R.; Lia, G. B. *J. Hazard. Mater.* 165, 2009, 995-1001.

[41] Shih, K.; White, T.; Leckie, J. O. *Environ. Sci. Technol.* 2006, 40, 5077-5083.

[42] Shih, K.; White, T.; Leckie, J. O. *Environ. Sci. Technol.* 2006, 40, 5520-5026.

[43] Tang, Y.; Shih, K.; Chan, K. *Chemosphere* 2010, 80, 375-380.

[44] Hu, C. Y.; Shih, K.; Leckie J. O. *J. Hazard. Mater.* 2010, 181, 399-404.

[45] Yao, H.; Naruse, I. P. *Combust. Inst.* 2009, 32, 2685-2691.

[46] Zhou, R. S.; Snyder, R. L. *Acta Crystallogr. B-struct. Sic.* 1991, 47(5), 617-630.

[47] Velde, B. Introduction to Clay Minerals; Chapman & Hall: London. 1992.

[48] Sainz, M. A.; Serrano, F. J.; Amigo, J. M.; Bastida, J.; Caballero, A. *J. Eur. Ceram. Soc.* 2000, 20, 403-412.

[49] Lee, W. E.; Souza, G. P.; McConville, C. J.; Tarvornpanich, T.; Iqbal, Y. *J. Eur. Ceram. Soc.* 2008, 28, 465-471.

[50] MacKenzie, K. J. D.; Hartman, J. S.; Okada, K. *J. Am. Ceram. Soc.* 1996, 79, 2980-2982.

[51] Chen, C. Y.; Tuan, W. H. *J. Am. Ceram. Soc.* 2002, 85, 1121-1126.

[52] Geller, R. F.; Bunting, E. N. *Journal of research of the National Bureau of Standards* 1943, 31(5), 255-270.

[53] Kukukova, A.; Aubin, J.; Kresta, S. M. Chem. Eng. Res. Des. 2009, 87(4), 633-647.

[54] Kuxmann, U. Fischer P. *Erzmetall* 1974, 27(11), 533-537.

[55] Chen, S.; Zhao, B.; Hayes, P. C.; Jak, E. *Metall. Mater. Trans. B.* 2001, 32(6), 997-1005.

[56] Kim, E. J.; Herrera, J. E.; Huggins, D.; Braam, J.; Koshowski, S. *Water Res.* 2011, 45(9), 2763-2774.

[57] Kim, E. J.; Herrera, J. E. *Environ. Sci. Technol.* 2010, 44(16), 6054-6061.

[58] Ott, W. R.; McLaren, M. G. *J. Am. Ceram. Soc.* 1970, 53(7), 374-375.

[59] Rawlings, R. D.; Wu, J. P.; Boccaccini, A. R. *J. Mater. Sci.* 2006.41 (3), 733-761.

[60] Sorrell, C. A. *Am. Mineral.* 1962, 47, 291-309.

[61] Dharmappa, H. B.; Hasia, A.; Hagare, P. *Water Sci. Technol.* 1997, 35(8), 45-56.

[62] Babatunde, A. O.; Zhao, Y. Q. *Crit. Rev. Env. Sci.* 2007, 37(2), 129-164.

[63] Petruzzelli, D.; Volpe, A.; Limoni, N.; Passino, R. *Water Res.* 2000, 34(7), 2177-2182.

[64] Lin, D.F., Weng, C.H. *J. Environ. Eng.* 2001, 127(10), 922-927.

[65] Vicenzi, J.; Bernardes, A. M.; Bergmann, C. P. *Ind. Ceram.* 2005, 25(1), 7-16.

[66] Xu, G.; Zou, J.; Li, G. *Environ. Sci. Technol.* 2008, 42 (19), 7417-7423.

[67] Kingery, W. D.; Bowen, H. K.; Uhlmann, D. R. Introduction to Ceramics, second ed.; *John Wiley and Sons:* New York, 1975.

[68] Callister, W. D., Materials Science and Engineering: An Introduction, third ed; John Wiley and Sons: New York, 1994.

[69] Vieira, C. M. F.; Monteiro, S. N. *Revista Matéria* 2009, 14 (3), 881-905.

[70] Mei, S.; Yang, J.; Ferreira, J. M. F. *J. Eur. Ceram. Soc.* 2000, 20(12), 2191-2197.

[71] Li, B.; Yuan, Y.; Zhang, S.; Xu, Y. *J. Mater. Sci-Mater. El.* 2011, 22(8), 924-928.

In: X-Ray Diffraction: Structure, Principles and Applications ISBN: 978-1-62808-591-4
Editor: Kaimin Shih

Chapter 7

X-RAY ANALYSIS OF METAL OXIDE-METAL CORE-SHELL NANOPARTICLES

Faruq Mohammad[1, 2, 3], Lissinda du Plessis[1] and Tanvir Arfin[4,5,*]
[1]Unit for Drug Research and Development, North-West University, South Africa
[2]Environmental Toxicology, Southern University and A&M College, Baton Rouge, LA, US
[3]Center for Advanced Microstructures and Devices (CAMD), Louisiana State University, Baton Rouge, LA, US
[4]PGM Group, Chemical Resource Beneficiation (CRB) Research Focus Area, North-West University, South Africa
[5]Department of Applied Chemistry, University of Johanneshberg, Doornforntein, South Africa

ABSTRACT

During recent years, the SPIONs containing core-shell architecture became the interest of many researchers due to their unique chemical, optical, electronic, catalytic, magnetic and biological properties. In this chapter, we synthesized two different SPIONs containing core-shell architectures by different approaches with fine-tuned properties; one with an Au shell and the other with Ag. Both the core-shell structures were extensively characterized by various X-ray based instrumental techniques such as HRTEM conjugated with EDAX, XPS, XAS-based XANES and powdered XRD. The presence of a SPIONs@Au, SPIONs@Ag core-shell structure was supported by HRTEM conjugated EDAX analysis, where the peaks corresponding to two distinct layers with different lattice constants were observed. Also from the analysis, both the particles found to be spherical, uniformly dispersed with a mean size distribution of 6.5±0.7 nm for SPIONs@Au and 10.8±0.9 nm corresponding to SPIONs@Ag. The XPS analysis of elemental oxidation states, it confirmed that the iron oxide core in both the particles is magnetite, where maghemite has a possibility for its formation through oxidation by different routes. From the synchrotron radiation-based XANES investigation, a further confirmation of magnetite core in both particles, in addition to Au and Ag in 0 and +1 oxidation states, means that no further oxidation/reduction to the core or the shell occurs.

[*] Corresponding author. Tel: +27-018-299-1576; Fax: +27-018-299-1667. Email address: tanvirarfin@ymail.com.

On the other hand, comparison of the XRD patterns of the SPIONs@Au and SPIONs@Ag particles with that of SPIONs seeds provided a proof of the formation of Au or Ag layers. From the temperature dependent XRD pattern, an increase of crystallite size and a decrease of peak intensity for both of the core-shell NPs were observed with an increase of sintering temperature. Also from the XRD analysis, an increased crystallite size with sintering temperature, resulted in a decrease of lattice parameter was observed for both the particles. Therefore, the XRD analysis of the SPIONs@Au, SPIONs@Ag particles allowed for the investigation of the stability of core-shell structure at various temperatures of up to 400°C.

1. Introduction

The term "nanotechnology and nanomaterials" needs no special introduction as interest in them during recent years sparked exponential growth in this field, not only in scientific laboratories from academia but also in industries around the world. The main drivers for this remarkable growth are new scientific understandings, unanticipated unique properties, and applications leading to possibilities for tremendous economic impact. With the unraveling of their chemical, physical and biological properties coupled with supercomputing methods, the nanomaterials of different sizes and shapes are being investigated, and the largest activity has been centered on spherical particles and carbon nanotubes (CNTs) [1]. In addition to size and shape effects, newer designs of nanostructures created using nanofabrication technologies also have begun gaining prominence and therefore provided new opportunities to tailor the properties of nanomaterials and investigate their fundamental behavior. Some of these new designs are alloy, core-shell, onion-type, multilayered, multimetallic and multifunctional nanoparticles (NPs) [2-10]. Among those different types of nanomaterials, core-shell architecture is extremely important as they offer a great opportunity to modulate the properties of both core as well as shell and therefore can present multifunctionality. The core-shell structure due to its design offers nanomaterials with unique chemical, optical, electronic, catalytic, magnetic and biological properties; therefore by making use of this, a number of possibilities and opportunities can be envisaged such as 1) *Core protection*: a shell can be utilized as a protective shield for sensitive core material, for example, in the case of air sensitive magnetic nanomaterials such as cobalt [11, 12]; 2) *Tailored properties*: the dimensions and composition of the shell can be modulated in order to influence the properties of the core and similarly the core dimensions can be modulated to influence the properties of the shell [13, 14]; 3) *Easy biofunctionalization*: the shell surface can be utilized for biofunctionalization which otherwise either difficult or not possible on the core surface, for example, gold or silica coated magnetic nanomaterials [15, 16]; 4) *Enhanced biocompatibility*: a biocompatible shell around toxic core material enables reduction in their toxicity, for example, coated quantum dots [17] or cobalt NPs [18]; 5) *Cost effective approach*: expensive nanomaterials can be formed as shells around inexpensive nanomaterials and this is particularly useful for catalysis technology [19]; and 6) *Multifunctional modality*: multiple functions such as detection and treatment modalities or multiple treatment options can be incorporated in a single core-shell particle [20]. All these different possibilities are reflected to enhance the properties of core-shell nanomaterials in a number of applications.

During recent years, the magnetic core-shell nanoclusters are gaining much more attention of researchers due to their possibility to offer opportunities for developing devices, cluster assembled materials with new and improved functions in magnetic recording media, catalysis and biomedical technology [21, 22]. In fact, the superparamagnetic iron oxide nanoparticles (SPIONs) with suitable biocompatible coatings have important implications in biomedical disciplines such as the contrast enhancing agents for magnetic resonance imaging (MRI) and hyperthermia-based cancer therapy, thermoresponsive drug delivery vehicles and are also found to be promising for the current medical paradigm of ‘‘detect and treat’’ strategy [23, 24]. In a recent study, the SPIONs containing Au metallic shell found to be effective heat releasing agents as compared against SPIONs with no shell and also acting as efficient catalysts for the CO oxidation when supported with titania [25, 26]. Hence, the SPIONs containing the core-shell structure is extremely important in the contemporary research areas of catalysis and nanomedicine due to their enhanced structural, optical, magnetic, biocompatible and surface adsorption characteristics.

Several phases of iron with different oxidation states and structures are available, which includes Fe^0, FeO (wüstite), Fe_3O_4 (magnetite), γ-Fe_2O_3 (maghemite), α-Fe_2O_3 (hematite) and FeOOH (oxyhydroxy) [27]. The most abundant and stable forms of iron are the magnetite and maghemite and when falls within a nanosize range (<15 nm), they both exhibit superparamagnetism (the property of ferromagnetic and ferrimagnetic nanomaterials whose magnetization values falls in between ferromagnetic and paramagnetism). In magnetite, Fe exhibits two different oxidation states of +2 and +3, while maghemite has only one oxidation state of +3. For applications where the magnetic properties and electron rich surface characteristics are important, magnetite is the most preferred form due to the availability of Fe^{2+} state which has a potential to further act as an electron donor. In addition, to fine tune the intrinsic magnetic properties and to increase the biocompatibility of magnetite, Fe_3O_4-Au or Fe_3O_4-Ag core-shell NPs have been prepared by many different groups and characterized extensively by various instrumental techniques. Several approaches for the synthesis of magnetite core-shell NPs have been introduced, which includes the chemical reduction in presence of strong reducing agents, reverse micelle, redox transmetallation, spray deposition, etc [16, 25, 28-30]. The key issue in all these synthesis methods is the chemical state of the core iron/shell material and whether the oxide forms during or after the synthesis process, as it will determine further the sensory, magnetic and catalytic properties of those nanostructured materials. For the iron core-metallic shell clusters, the total magnetic moment depends not just on the number of core iron atoms, but also on the number of surrounding atoms of shell, in addition to their shape and orientation [24, 30]. In spite of numerous publications on the structural, electronic and magnetic studies of iron oxide coated metal NPs, there are still only a limited number of studies with a complete characterization at the microstructural level of the core, which allows achieving a better understanding of their local electronic structure and further the magnetic, surface adsorption behavior [25, 28, 31]. Some of the most sophisticated strategies currently available for the investigation of microstructure of the core after forming core-shell architecture includes the X-ray absorption spectroscopy (XAS), electron micrographs equipped with EELS/EDAX, powdered XRD, and Mössbauer spectroscopy.

In the present study, by keeping all the discussed factors in view, two different types of SPIONs containing two different biocompatible metallic shells, Au and Ag were prepared and were characterized by various instrumental techniques that use X-rays for analysis. We discussed in this book chapter about the synthesis of thermally stable (up to 400°C) superparamagnetic iron oxide core and Au or Ag metallic shell NPs. The characterization of as-prepared SPIONs@Au and SPIONs@Ag NPs presented here are only the radiation-based techniques such as the HRTEM equipped with EDAX, XPS, powdered XRD and XAS-based XANES analysis. The main goal is to understand the microstructure and electronic states of iron oxide core before and after the formation of Au or Ag shell and also its stability by thermal treatment using those X-ray based techniques. We used the synchrotron radiation-based XANES analysis in order to determine the chemical state, local atomic structure and chemical transformations of the core iron oxide and the Au or Ag shell nanomaterials.

2. Experimental

2.1. Chemicals and Reagents

Iron (III) acetylacetonate ($Fe(acac)_3$, 99.9%), 1,2-hexadecanediol ($C_{14}H_{29}CH(OH)CH_2(OH)$, 90%), Oleylamine ($C_9H_{18}dC_9H_{17}NH_2$, 70%), Oleic acid ($C_9H_{18}dC_8H_{15}$-COOH, 99%), Diphenyl ether ($C_{12}H_{10}O$, 99%), Ferric chloride ($FeCl_3$), Ferrous chloride ($FeCl_2{\cdot}4H_2O$), Sodium borohydride ($NaBH_4$), Silver nitrate ($AgNO_3$, 99.99%) were purchased from Sigma-Aldrich. Gold acetate ($Au(OOCCH_3)_3$) was obtained from Alfa-Aesar. All chemicals were of the highest grade and used as received without any further purification.

2.2. Synthesis of SPIONs@Au NPs

The SPIONs@Au NPs were synthesized by the procedure as described by Mohammad *et al.* [25] and is the modification of the previously reported method [28]. Briefly, 0.7 g of iron acetylacetonate was mixed with 20 mL of diphenyl ether, 2 mL of oleic acid and 2 mL of oleylamine under inert atmospheric conditions with vigorous stirring. To this mixture, 2.6 g of 1,2-hexadecanediol was added and the solution was heated to around 210°C with reflux for 2 h maintaining oxygen free conditions.

The reaction mixture was cooled to room temperature and then ethanol (degassed) was added to precipitate the black colored product. The precipitate was separated out by centrifugation and then washed with a series of solvents starting with hexane, followed by a mixture of hexane and ethanol and then finally with ethanol. The product was dispersed in ethanol and separated magnetically and dried to obtain the black colored powder of SPIONs. In the following step, approximately 0.5 g of SPIONs was added to 10 mL of diphenyl ether in the flask. To this mixture, a solution containing 30 mL of diphenyl ether, 3.1 g of 1,2-hexadecanediol, 0.5 mL of oleic acid, 3 mL of oleylamine and 0.8g of gold acetate was added under inert atmosphere.

The reaction mixture was heated to around 190°C with reflux for about 1.5 h and maintaining the oxygen free conditions. After cooling to room temperature, ethanol was added and the dark purple material was separated out by centrifugation. The material was re-suspended in hexane, washed trice with ethanol and dried. By using magnetic separation technique and by suspending the particles in ethanol, the magnetic gold-coated iron oxide NPs were separated from the non-magnetic gold NPs and any unwanted carbon mass produced from oleicacid/oleylamine used during the course of the reaction.

2.3. Synthesis of SPIONs@Ag NPs

For the synthesis of SPIONs@Au NPs, we first synthesized SPIONs using the procedure from Kumar *et al.* [31]. Briefly, 1.6 g of $FeCl_3$ and 1.0 g of $FeCl_2 \cdot 4H_2O$ were weighed and mixed together in a 100 mL RB flask, which was previously purged thoroughly with nitrogen gas to maintain oxygen free conditions. To this, 25 mL of air-free nanopure water was added and stirred magnetically on a magnetic stirrer. To this solution, 29.05% NH_4OH was added until the solution becomes black completely. A black precipitate was obtained by decanting the supernatant after settling on a permanent magnet. The precipitate was separated and then heated at 80°C for 30 min using a hot plate, washed thrice with 25 mL water (oxygen free) and twice with 20 mL ethanol, and dried under nitrogen flow. In the following step, the $AgNO_3$ was reduced by $NaBH_4$ onto the surface of SPIONs further to form SPIONs@Ag NPs using the method described by Ahmad *et al.* [32]. Briefly, a 50 mL $AgNO_3$ solution (0.01 M) was added immediately into the solution containing 0.5 g of SPIONs in 10 mL of deionized water under constant stirring. Following this, a 20 mL $NaBH_4$ (0.04 M) was added drop wise to the $AgNO_3$/SPIONs mixture and an immediate color change from pale yellow to brown indicating the formation of SPIONs@Ag NPs was noted. Finally, the particles were separated by using the magnetic separation technique, washed several times with degassed ethanol and dried under inert atmospheric conditions.

2.4. Characterization of Core-Shell NPs

2.4.1. High-Resolution Transmission Electron Microscopy (HRTEM)

The sizes of SPIONs@Au and SPIONs@Ag particles were determined by HRTEM using JEOL 4000EX high-resolution electron microscope with an accelerating voltage of 200 keV. The SPIONs@Au was dispersed in hexane while the SPIONs@Ag in ethanol with ultra-sonication and later drop-casted onto a 200 mesh copper grid, followed by an air-drying of the grid for at least 24 h prior to imaging. The EDAX (Energy Dispersive Analysis of X-rays) were carried out simultaneously with HRTEM by focusing the electron beam onto the particles using the same JEOL 4000EX electron microscope attached to an EDAX analyzer.

2.4.2. X-ray Photoelectron Spectroscopy (XPS)

To determine the chemical states of Fe, Au and Ag metals present in the core-shell type NPs, the XPS analyses were performed on AXIS 165 High performance multi-technique surface analyzer which is based on a 165mm mean radius hemispherical analyzer, with an eight channeltron detection system.

2.4.3. X-Ray Absorption Near-Edge Structure (XANES)

The synchrotron radiation-based XANES data collection was performed at the double-crystal monochromator beam line of the J. Bennett Johnston, Sr., Center for Advanced Microstructures and Devices, Louisiana State University, using Ge (220) crystals. The storage ring operated with electron energy of 1.3 GeV and a current between 90 and 190 mA. Samples for XANES were prepared by spreading the powdered samples between two Kapton tapes and folding it up to 10 times until the desired thickness, corresponding to the K-, L-edge jump between 0.5-1.0 au was obtained. The graphing and data analysis was performed by using ATHENA program [33].

2.4.4. Powdered X-Ray Diffraction (XRD) Studies

The XRD patterns were recorded on a Bruker-Siemens D5000 automated powder diffractometer using Cu Kα radiation. The data were collected in the range of 20-80° in 2θ at a scan rate of 0.06° per step.

3. RESULTS AND DISCUSSION

In the present work, we synthesized two different types of core-shell NPs, SPIONs coated with Au and SPIONs coated with Ag by two different approaches. Figure 1 illustrates the synthesis method (see Experimental section for a detailed description) and a schematic view of the formation of core-shell NPs of SPIONs@Au and SPIONs@Ag. The SPIONs@Au NPs are formed by the polyol method and the synthesis process involves the decomposition of organometallics at higher temperature conditions and in the presence of a suitable reducing agent such as 1,2-hexadecanediol and surfactants of oleic acid, oleylamine. These viscous surfactants completely cover the surfaces of iron oxide NPs and protects from agglomeration and aggregated structure formations, as the magnetic NPs possess high tendency to attract themselves due to surface charges and magnetic anisotropy. However, the SPIONs@Ag particles were formed by the aqueous precipitation of iron salts to form iron oxide NPs and this was followed by a reduction reaction in the presence of strong reducing agent such as sodium borohydride to form the Ag shell.

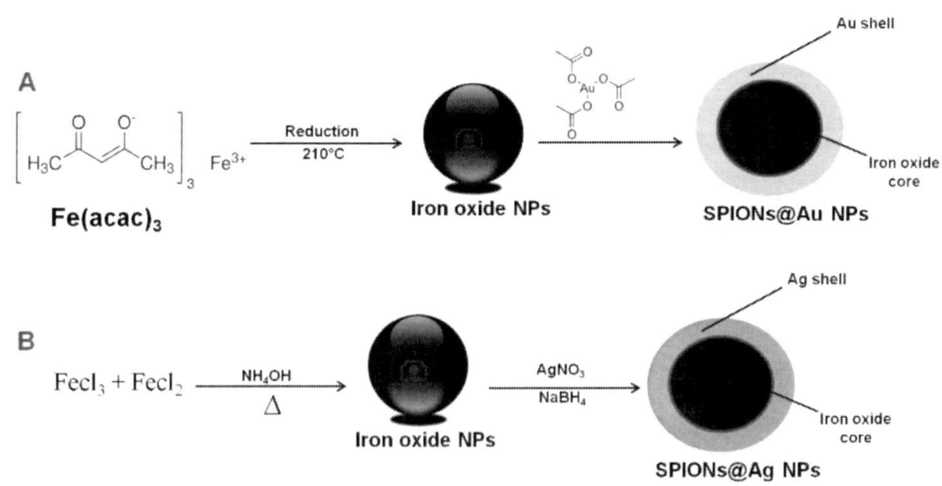

Figure 1. Schematics of the synthesis of SPIONs@Au (A) and SPIONs@Ag NPs (B).

The HRTEM images of the NPs of SPIONs@Au and SPIONs@Ag are shown in Figures 2(a) and-3(a) respectively. From the microscopic images, one can see that both the particles are uniformly dispersed on the copper grid and are found to be spherical in shape. The size distribution analysis as shown in Figure 2(b) & 3(b) gave the information about the particle sizes and the average sizes for both the particles are falling in the nanorange *i.e.* 6.5±0.7 nm for SPIONs@Au and 10.8±0.9 nm for SPIONs@Ag. The EDAX spectrums recorded simultaneously with HRTEM images gave a piece of information about the elemental composition present in each particle. From the Figure 2(c), the appearance of peaks for Au (at 2.1 keV) and Fe (6.4 keV) suggesting that the NPs contain primarily the Au-Fe bimetallic structure [34]. Similarly, the observation of peaks for Ag (at 2.7 keV), Fe (6.4 keV) in Figure 3(c) is also an indication of the presence of those corresponding elements in SPIONs@Ag sample, *i.e.* Ag-Fe bimetallic structure [29]. The presence of oxygen peak in both samples can be a primary indication that the core is in its oxidized form (*i.e.* iron oxide but not only iron). In addition to these, the appearance of peaks for Cu in both samples is due to the carbon coated copper grid used during the preparation of HRTEM samples. Therefore, from the EDAX analysis, it is confirmed that the magnetic core in both the particles is iron oxide, and this has a possibility to exist in two isomeric forms of either Fe_3O_4 or γ-Fe_2O_3. In general, both Fe_3O_4 and γ-Fe_2O_3 has similar structures of spinel ferrite and maghemite considered to be as an Fe(II)-deficient magnetite, *i.e.* Fe_3O_4 (or FeO-Fe_2O_3) when subjected to an oxidative heat treatment of 220°C, the Fe^{2+} present in it undergoes further oxidation to form γ-Fe_2O_3 [35]. Since in our case for the synthesis of both the core-shell particles, we carried the reactions at higher temperature conditions and hence there is a chance that the iron oxide formed in the first step can be the more oxidized form *i.e.* γ-Fe_2O_3. In other words, it can also be possible that the first step formed product is Fe_3O_4, but during heat treatment in the second step to form the Au or Ag shell, it is oxidizing to its more oxidized form of γ-Fe_2O_3. However, it can also be possible that the shell can even protect the core from further oxidation, if the formed shell is nonporous, strong and stable enough. In this case, the formed Fe_3O_4 can persist like that only even in SPIONs@Au and SPIONs@Au with a core of Fe_3O_4 and a shell of Au or Ag. To investigate the status of iron oxide core, we proceeded further with the studies of XPS, XANES and XRD.

To further understand the elemental composition and the oxidation states, the XPS analysis was performed on the samples of SPIONs@Au and SPIONs@Ag. The XPS analysis for the elements of Fe, Au and Ag are shown in Figure 4(A-B). From the Figure 4A, the Fe spectrum of both the samples showed a doublet containing a low energy band of $2p_{3/2}$ and a high energy band of $2p_{1/2}$ appeared at the binding energy positions of around 709.7 and 724.5 eV, respectively. These binding energy positions are in good agreement with the literature data for Fe and therefore indicating the presence of Fe in both the oxidation states of Fe(+2) and Fe(+3), *i.e.* magnetite (Fe_3O_4), but not maghemite (γ-Fe_3O_4) [36, 37]. On the other hand, the Au 4f spectrum for the SPIONs@Au sample (Figure 4B, bottom) shows a doublet containing a low energy band ($4f_{7/2}$) and a high energy band ($4f_{5/2}$) at the peak positions 81.6 and 85.3 eV, respectively. These binding energy positions are corresponding to the existence of Au in the possible oxidation states of Au(0), Au(I) and Au(III) [38, 39]. Also, one can see from Figure 4A that the binding energy positions for Ag 3d spectrum of the SPIONs@Ag sample shows a doublet with a low energy band 369 eV assigned to $3d_{3/2}$ and a high energy band at 375 eV assigned to $3d_{5/2}$. Therefore, these energy positions corresponds to the existence of Ag in the possible oxidation states of Ag(0) or Ag(I) [40, 41]. The reason that there can be a possibility for the formation of the higher oxidation states of Au and Ag shells is that, the shell either losing some of its electrons to reduce the core iron oxide or the shell itself getting oxidized due to the heat treatment.

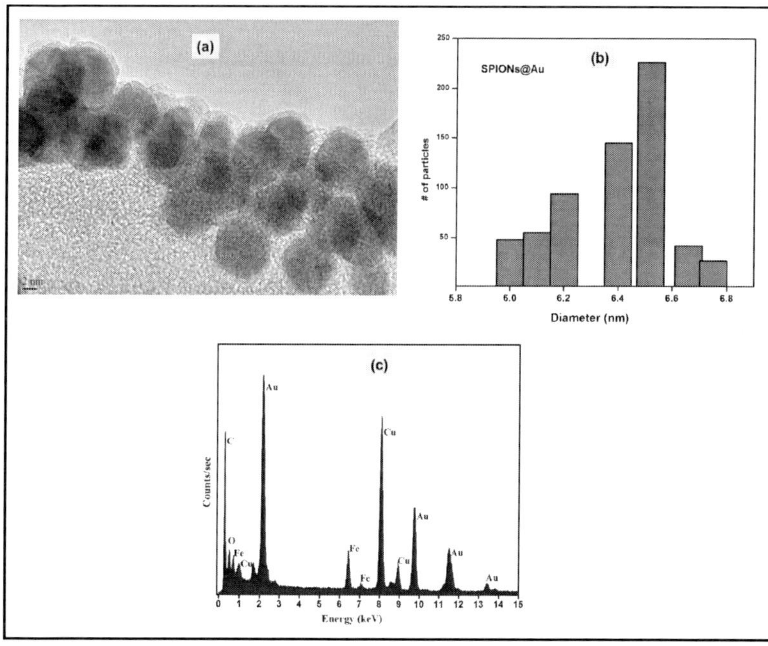

Figure 2. HRTEM image (a) with a scale of 2 nm, corresponding particle size distribution (b) and the EDAX analysis (c) for the SPIONs@Au nanomaterial.

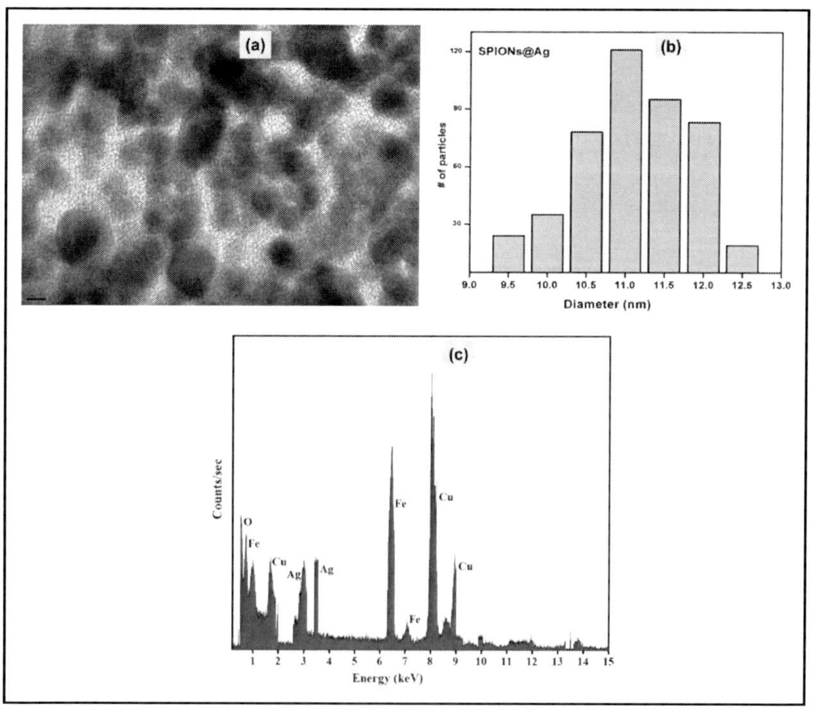

Figure 3. HRTEM image (a) with a scale of 5 nm, corresponding particle size distribution (b) and the EDAX analysis (c) for the SPIONs@Ag nanomaterial.

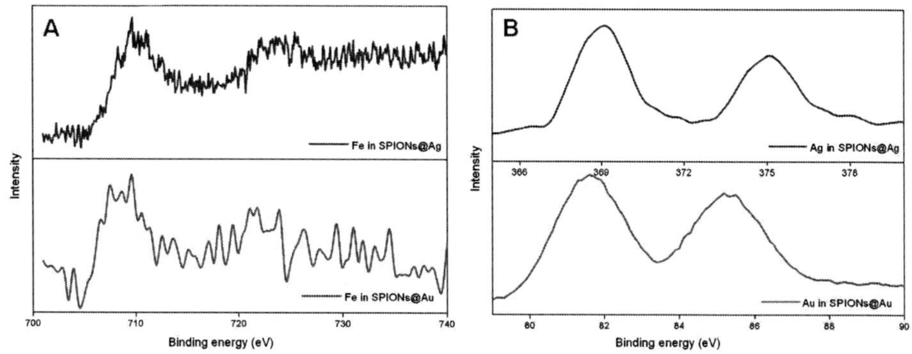

Figure 4. XPS studies of Fe (A), and Au and Ag (B) for SPIONs@Au and SPIONs@Ag NPs, respectively.

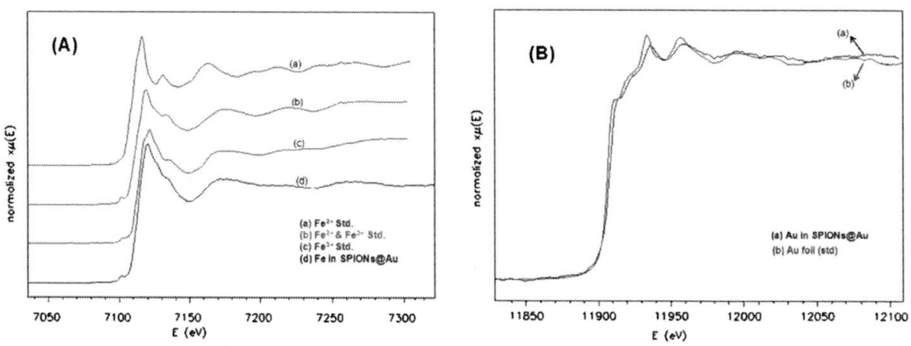

Figure 5. First derivatives of the normalized XANES spectra for Fe K-edge (A) and Au L-edge (B) of the SPIONs@Au NPs.

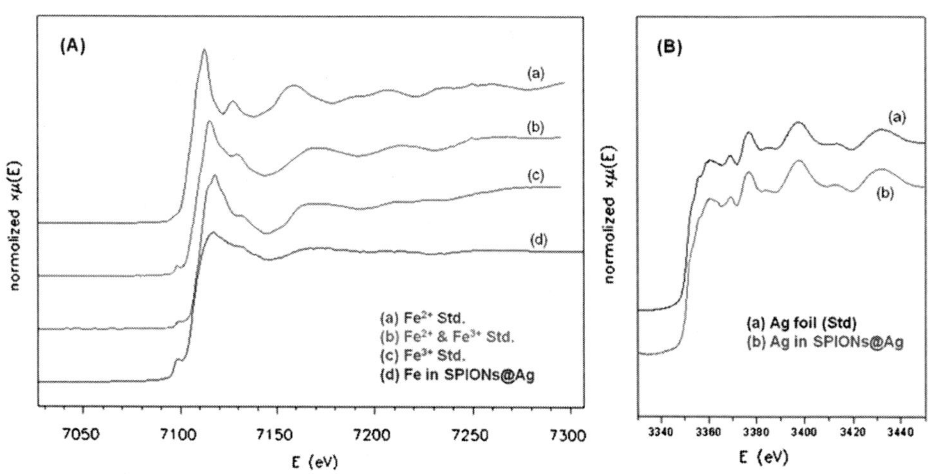

Figure 6. First derivatives of the normalized XANES spectra for Fe K-edge (A) and Ag L-edge (B) of the SPIONs@Ag NPs.

In order to fully understand the chemical state and geometric properties of two different SPIONs coated with Au and Ag metallic shells, the XAS-based XANES analysis were performed. In recent years, the XANES analysis has became a very important tool for the characterization of NPs and metal colloids, as it provides appropriate information about the local structure of absorbing atom such as its oxidation state, the effective charge of the element in different compounds, coordination geometries, and different bonding types [42]. Figures 5A and 6A shows the Fe K-edge XANES spectra for SPIONs@Au and SPIONs@AgNPs, respectively along with a series of relevant reference spectra corresponding to metallic foil (Fe^{2+}), γ-Fe_2O_3 (Fe^{3+}) and Fe_3O_4 (Fe^{2+}& Fe^{3+}). It is clear from the figure that the absorption edge of the Fe K-edge in SPIONs@Au appeared at a position of 7112 eV, while for SPIONs@Ag it is 7110 eV and can be assigned due to an electric 1s→4p dipole transition. The analysis of data further reveals that the edge positions of the core iron oxide in both types of SPIONs represents mixed oxidation states, being between the two oxide standards (+2 and +3) as similar to Fe_3O_4 but not γ-Fe_2O_3 [43]. Therefore, from the Fe-K edge spectrum of both SPIONs type and resulting electronic states, it indicates that no further oxidation to the magnetite core was occurring during the formation of Au or Ag shell, although there is a possibility for the oxidation to maghemite. Similarly, the absorption edge of the Au L-edge in SPIONs@Au at 11930 eV (Figure 5B) and Ag L-edge in SPIONs@Ag at 3355 eV (Figure 6B) are consistent with that of reference standard foils of Au or Ag, respectively. The gold shell with the edge at ~11930 eV show expected Au L-edge response corresponding to Au(0) and can be assigned to a 2p→5d shift [44], while the edge (at 3355 eV) of Ag L-edge is the response due to Ag(+1), which can be assigned to a 2p→4d dipole transition [45]. The observed oxidation states of Au and Ag also gives the information that they neither oxidized nor reduced due to the magnetite core. This, therefore, means that no inter electronic charge transfers are occurring between the Fe_3O_4 core and the metallic shells of Au or Ag.

The XRD patterns of SPIONs and SPIONs@Au NPs are compared in Figure 7. From the figure, the XRD of SPIONs showed the diffraction peaks at 2θ (Theta) of 30.1°, 35.1°, 42.5°, 52.7°, 56.5°, 62.1°, 70.5°. A closer look at each peak reveals that these peak positions are closer to those of Fe_3O_4 phase than those of γ-Fe_2O_3 phase [46]. However for SPIONs@Au NPs, the diffraction peaks are observed at 30.3°, 31.9°, 35.9, 39.6°, 44.6°, 49.5°, 58.1°, 62.9° and from the literatures, the XRD peaks for the gold NPs coated with oleic acid, oleylamine were obtained to be at 39.4°, 44.1°, 51.3°, 63.8°, 76.2° [28]. From the comparison, the pattern for SPIONs@Au displays peaks of both Fe_3O_4 and Au NPs. However, the peaks for magnetite (Fe_3O_4) in the core-shell NPs are subdued when compared to the XRD pattern of Fe_3O_4 alone. This is most likely caused by the heavy atom effect from Au[32] as a result of the formation of Au coating on the Fe_3O_4 NPs. As mentioned above that this effect has been observed previously and provides further evidence for the presence of the core-shell type of structure [28, 34, 47]. In addition, the average crystalline size of the Fe_3O_4 core particles was calculated by the Scherrer equation using the half-maximum full width of the Fe_3O_4 (311) peak and was found to be 4.8 nm, which is slightly smaller than, but in close agreement with, the 6.5 nm derived from the HRTEM images in Figure 2(b) [34].

If the particle sizes are too small, net result will be the broadening of the diffraction peaks around Bragg's angle. This phenomenon of widening the diffraction peaks is related to an incomplete, "cancelling" of small deviations from Bragg's angle in small crystallites and is known as the particle size broadening. The particle size broadening is differentiated from the normal width of diffraction peaks related to instrumental effects. In most cases, the particle size broadening will not be observed with crystallite sizes larger than 1μm. The crystallite size broadening (β_τ) of a peak can usually be related to the mean crystallite dimension (τ) by the Scherrer equation:

$$\tau = \frac{K\lambda}{\beta_\tau Cos\theta} \tag{1}$$

where β_τ is the line broadening due to the effect of small crystallites and is given by $(B-b)$, B being the breadth of the observed diffraction line at its half-intensity maximum and b is the instrumental broadening or breadth of a peak that exhibits no broadening beyond the inherent instrumental peak width. β_τ is given in radians, and K is the shape factor which typically has a value of about 0.9. λ is the radiation wavelength.

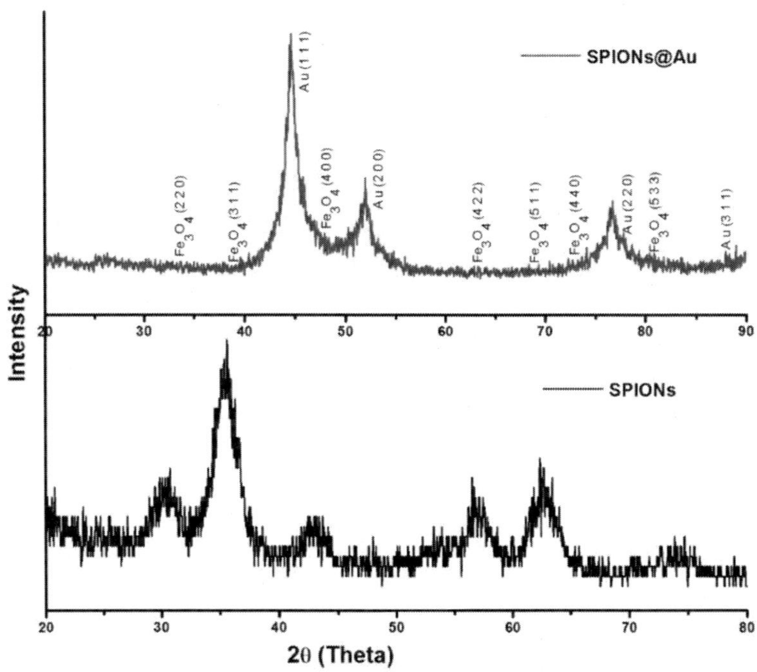

Figure 7. Comparison of the XRD patterns for SPIONs and SPIONs@Au.

Similarly, the superparamagnetic nanospheres containing Fe_3O_4 core and Ag shell were prepared by the reduction reactions between the Ag^+ and Fe^{3+} in the presence of $NaBH_4$. The crystal structure of SPIONs@Ag product was also characterized using XRD and are compared with the Fe_3O_4 NPs and is shown in Figure 8. From the figure, the observation of SPIONs@Ag peaks at corresponding reflection planes of (220), (311), (111), (400), (200), (422), (511), (440) and (220) are attributed to the face centered cubic lattice structure of SPIONs@Ag NPs [29]. As similar to SPIONs@Au NPs, the average crystalline size of Fe_3O_4 core particles calculated from Scherrer equation using the Fe_3O_4's (311) peak was found to be 8.1 nm, which is slightly smaller than, but in close agreement with 10.8 nm derived from the HRTEM images as shown in Figure 3(b).

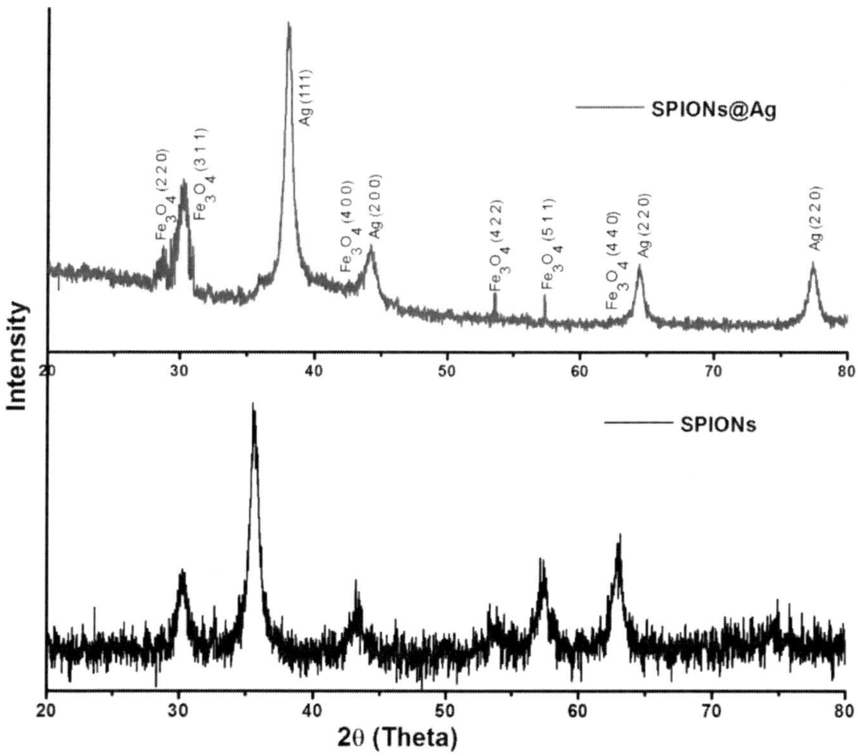

Figure 8. Comparison of the XRD patterns for SPIONs with that of SPIONs@Ag.

It has been established that the oxidation reaction for the conversion of Fe_3O_4 to γ-Fe_2O_3 can be accomplished at approximately 220°C [35]. Therefore to investigate the thermodynamic stability of core-shell structures, we subjected oxidative heat treatment to both types of NPs at two different temperatures of 200 to 400°C. Since during the oxidative heat treatment to both of SPIONs@Au and SPIONs@Ag NPs, there may be grain boundaries in the Au/Ag shell that allow for the diffusion of oxygen and, therefore, further oxidation through the metallic core. The changes in XRD patterns for SPIONs@Au and SPIONs@Ag from 200 to 400°C are shown in Figures 9 and 10, respectively. As shown in Figures 9 and 10, for both particles even after the heat treatment up to 400°C, no change of peak positions were observed and it means that no further oxidation to the cores in both of the NPs. However, it can also be seen from the figures that during the oxidative heat treatment of 200 to 400°C, the phase changes seems to be more for SPIONs@Au than for SPIONs@Ag NPs. This can be due to the decomposition of the surface ligands of oleic acid/oleylamine present on the surfaces of SPIONs@Au NPs at around 310°C [25]. However, the absence of such surface ligands on the surfaces of SPIONs@Au NPs does not show any of those phase changes. Also, the intensity of splitting peaks found to be decreased with the decrease of temperature and this situation is similar to the results reported elsewhere [48, 49].

The XRD pattern for SPIONs@Au and SPIONs@Ag powder should be produced by sufficient random orientation of crystals. Contrasted to the random orientation pattern, a significant peak reflection from the plane of (111) was observed in the samples. These results indicate that the sample type exhibited the crystallographic preferred orientation with the direction of $[111]_r$ axis being perpendicular to the sample surface. This feature of preferred orientation was same as reported elsewhere [50]. A degree of preferred orientation could be quantitatively estimated by a parameter of Γ, which is defined as $\Gamma = I(111)/I(311)$. The Γ-value of random orientation was determined to be 0.35 in the XRD pattern for SPIONs@Au, while the value of the Γ for SPIONs@Ag was 0.45. These results indicate that the degree of preferred orientation of $[111]_r$ axis was verified with the Γ values. The peak intensity for the (111) plane is increasing with an increase of temperature in both of the core-shell NPs. The values for the degree of preferred orientation indicated by the value of Γ were increased from 0.38 at the temperature of 200°C to Γ= 0.41 at 400°C for SPIONs@Au, and Γ= 0.47 at 200°C to Γ= 0.50 at 400°C for SPIONs@Ag. These results clarify the preferred orientation of $[111]_r$ axis was developed with the increase of temperature.

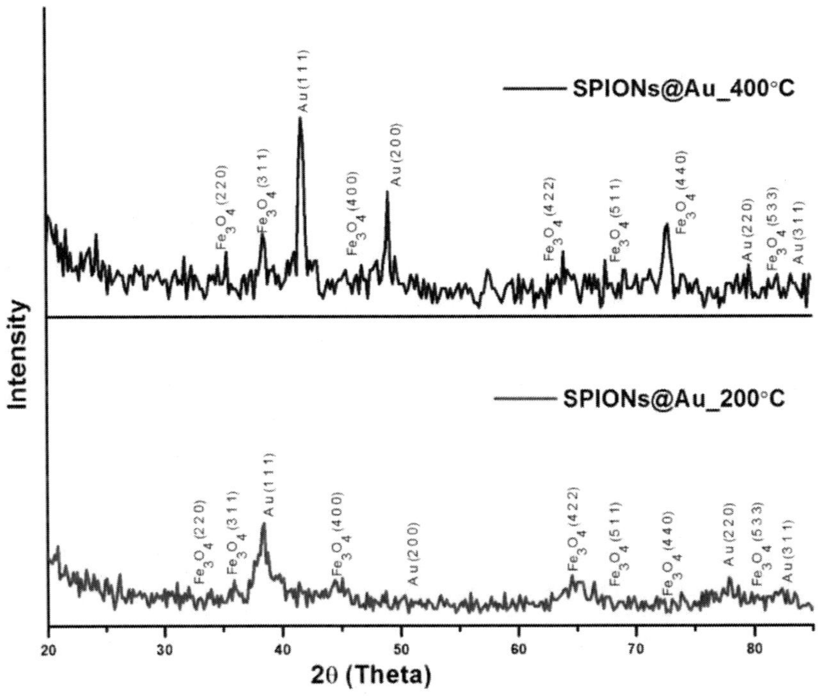

Figure 9. Comparison of the XRD patterns for SPIONs@Au after heating to two different temperatures of 200°C and 400°C.

The variation in crystalline sizes of SPIONs@Au and SPIONs@Ag NPs after the heat treatment at different temperatures were plotted in Figures 11 and 12, respectively. The crystalline sizes of the core-shell nanomaterials were determined at 200°C and 400°C using Scherrer formula and are found to be 11.4 and 12.5 nm for SPIONs@Au, 52.5 and 53.1 nm for SPIONs@Ag, respectively. Also, for both the particles it is clear that the crystallite size of the materials were increased with increase of temperature due to the smaller widths providing the larger crystalline size, which is in agreement with other results reported elsewhere [51, 52].

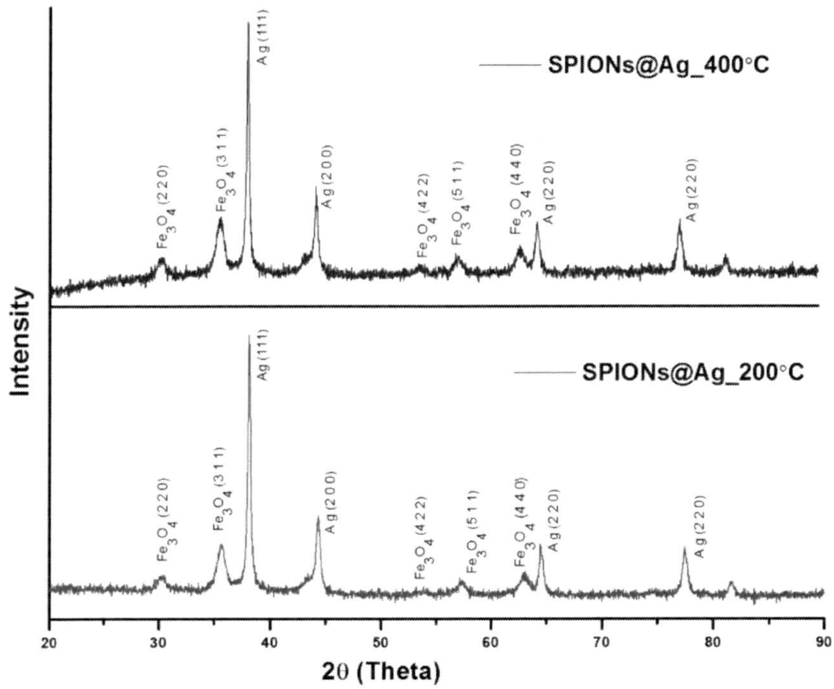

Figure 10. Comparison of the XRD patterns for SPIONs@Ag after the heating to two different temperatures of 200°C and 400°C.

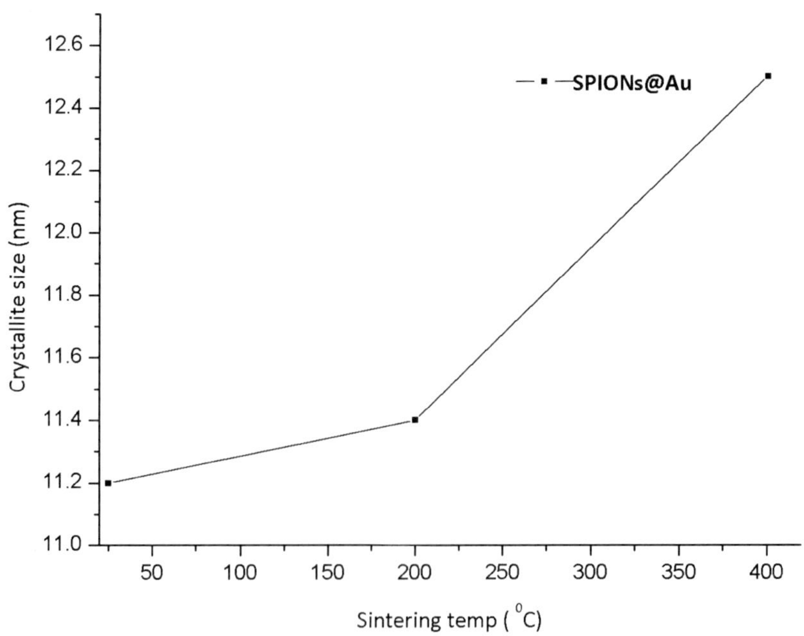

Figure 11. Crystallite size versus sintering temperature plotted for SPIONs@Au NPs when heated to a temperature of up to 400°C.

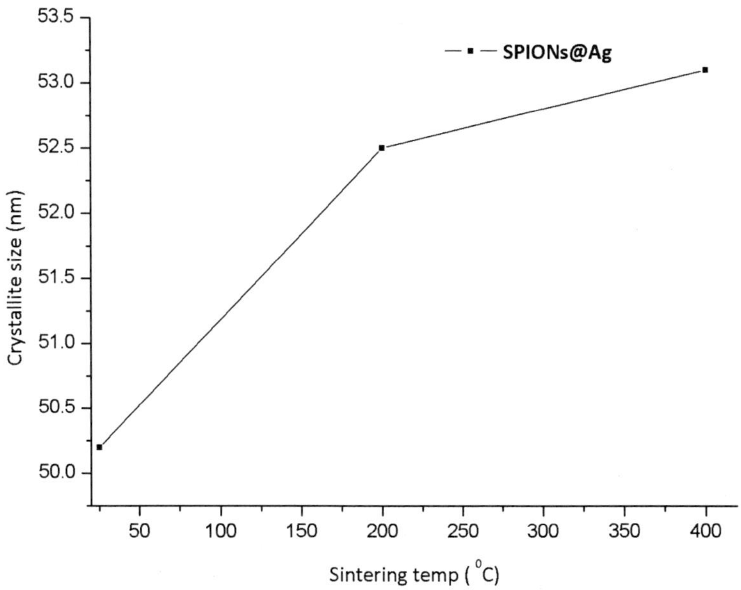

Figure 12. Crystallite size versus sintering temperature plotted for SPIONs@Ag after heating up to the temperature of 400°C.

Since the atoms are not stable at fixed points in any crystal. They are rather moving around their mean positions by thermal vibration and the amplitude of vibration increases with increasing temperature, beacuse this thermal vibration effect of atoms, a reduction in intensity of diffracted X-rays from the crystal sample are dectected. The effect of thermal vibration of atoms on the intensity of diffracted X-rays are commonly taken into consideration as a Debye-Waller factor. In practice, the temperature effect represented as e^{-2M_T} can be introduced in the following equations with the atomic scattering factor *f*.

$$f = f_0 e^{-M_T} \tag{2}$$

where f_0 corresponds to the atomic scattering factor without thermal vibration; M_T involves the amplitude *u* of thermal vibration, as well as the diffraction angle 2θ in the following form.

$$M_T = 8\pi^2 \langle u^2 \rangle \left(\frac{\sin\theta}{\lambda} \right)^2 = B_T \left(\frac{\sin\theta}{\lambda} \right)^2 \tag{3}$$

where $<u^2>$ is the square of the displacement in a direction perpendicular to the diffracted plane.

$$B_T = \frac{6h^2}{mk_B} \frac{T}{\Theta^2} \left(\phi(x) + \frac{x}{4} \right) \tag{4}$$

where T is the absolute temperature; m is the mass of the vibrating atom; h and k_B are Planck's constant and Boltzmann constant, respectively. Θ is the so-called Debye characteristic temperature of the substance given by the absolute temperature. M_T is a function of temperature.

The M_T data obtained using SPIONs@Au NPs at two different temperatures were plotted as a function of Sinθ/λ (Figure 13). From the figure, the decrease of the M_T values with the increasing Sinθ/λ is due to the θ dependence of the material, and it was observed that the M_T were low at higher temperature. Similar pattern was also obtained for SPIONs@Ag NPs and is shown in Figure 14.

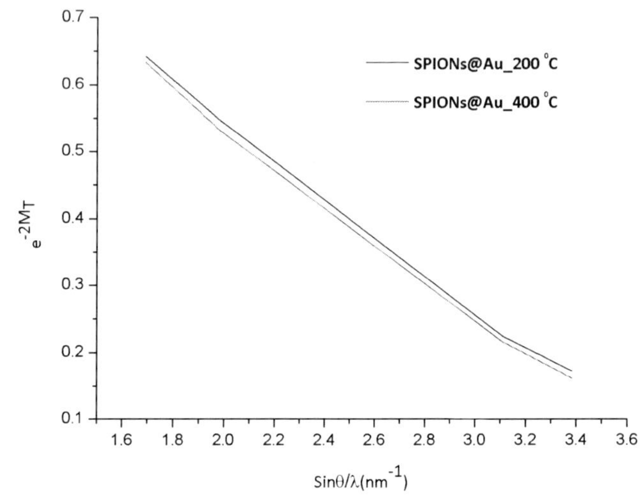

Figure 13. Variation of temperature factor for SPIONs@Au NPs as a function of Sinθ/λ.

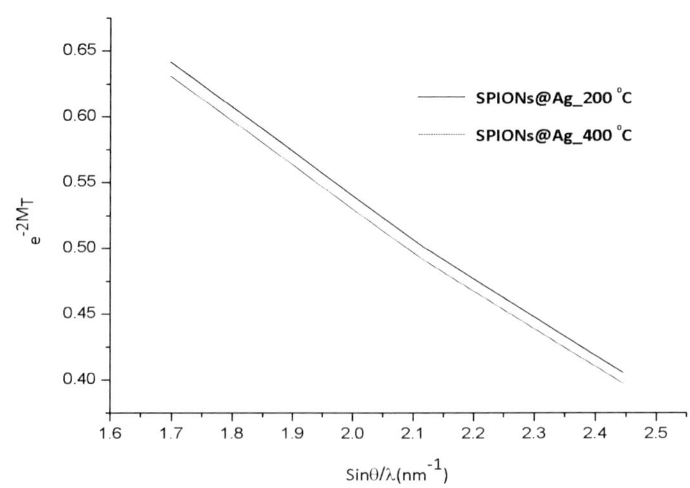

Figure 14. Variation of temperature factor for SPIONs@Ag NPs as a function of Sinθ/λ.

The cell constants can be obtained through either a single crystal or a powder X-ray diffraction experiment. For any of the diffraction to occur, the Bragg's equation must be satisfied:

$$\lambda = 2d_{hkl} \sin\theta \tag{5}$$

where d_{hkl} is the perpendicular distance between the pairs of adjacent planes (*d*-spacing); θ is the diffraction angle; λ is the wavelength of the used X-radiation.

The lattice planes are identified by their Miller indices, *hkl*, which describe their orientation relative to the unit cell and they define *d*. Each plane intersects the lattice at *a/h, b/k*, and *c/l*. Miller indices are used to identify each lattice plane. The quantities *a, b*, and *c* are called lattice parameters. The actual formula in a particular case would depend upon the crystal structure. For a cubic lattice, the relationship between the Miller indices *h, k* and *l*, the *d*-spacing, and the lattice constants *a* is:

$$d_{hkl} = \left[\frac{h^2 + k^2 + l^2}{a^2}\right]^{-1/2} \tag{6}$$

The number of formula units per unit cell is determined from the crystal structure of a solid. The variation of lattice parameter '*d*' with sintering temperature of up to 400°C for SPIONs@Au and SPIONs@Ag NPs are given in Figures 15 and 16, respectively. From the plot of lattice parameter, we found that with increasing of sintering temperature, the lattice parameter 'd′ found to be decreased for both of core-shell particles [53, 54].

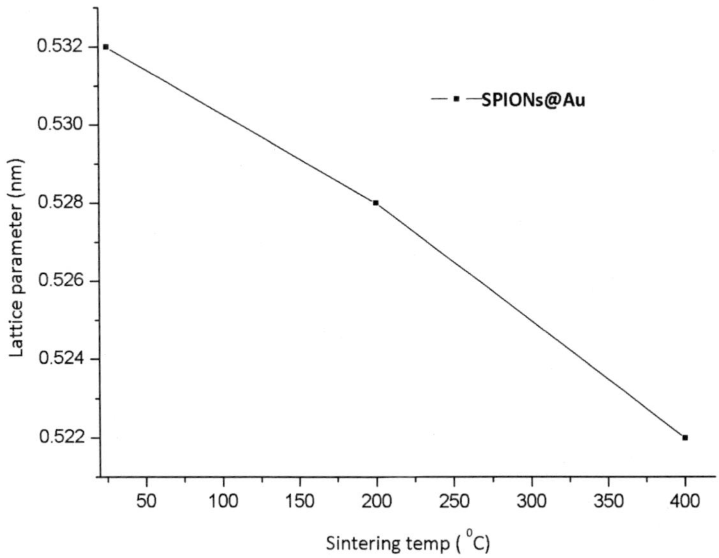

Figure 15. Lattice parameters versus sintering temperature plotted for SPIONs@Au NPs up to 400°C.

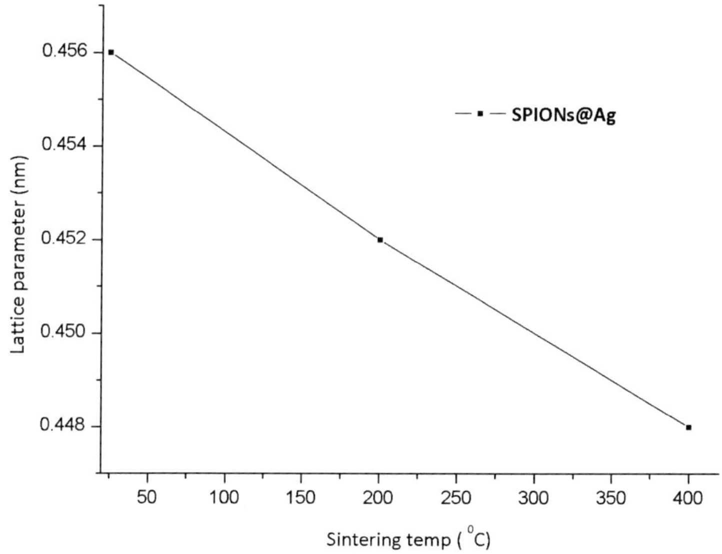

Figure 16. Lattice parameters versus sintering temperature plotted for SPIONs@Ag NPs up to 400°C.

CONCLUSION

In conclusion, we report the fabrication of core-shell superparamagnetic NPs by the reduction of gold or silver salts on the surfaces of iron oxide NPs. The gold/silver shell is anticipated to provide a versatile platform for the functionalization of NPs, in addition to fine tuning of magnetic properties. The results presented here further provide information while selecting the iron oxide core-shell nanomaterials to be used in catalysis, for the simultaneous diagnosis and treatment of cancers, and dug delivery-related applications. The data acquired from the HRTEM conjugated EDAX spectra indicated that the produced Fe-Au and the Fe-Ag particles had a spherical bimetallic structure. Further examination of the morphology by HRTEM could not discern the presence of a core-shell structure. On the other hand, the larger size and the patters generated from XRD of the SPIONs@Au and the SPIONs@Ag NPs compared to the SPIONs seeds suggested that the Au or Ag layer had successfully been added to the latter. From the temperature dependent XRD patterns, with an increase of sintering temperature, an increase of crystallite size and a decrease of peak intensity for both of the core-shell NPs were observed. It is also derived from our XRD analysis that an increased crystallite size with sintering temperature resulted in a decrease of lattice parameter. The XPS, XAS-based XANES analyses prove that both the particles contain oxidation states which can be attributed to a magnetite core, but not maghemite. Also, the presence of metallic shells of Au and Ag in 0 and +1 oxidation states means that no further oxidation/reduction to the core or the shell occurs. Finally, this report provides the use of X-ray based techniques for the analysis of metallic core-shell NPs (SPIONs@Au and SPIONs@Ag) and their local structure of absorbing atom such as the oxidation states, effective charge of element, coordination geometries, and different bonding types. The techniques discussed here are providing the sophisticated ways to see whether the formed iron oxide is magnetite or maghemite, their size and elemental composition. These data further decide the magnetic properties (for MRI, hyperthermia and drug delivery), biocompatibility and their catalytic effects [55]. In addition, such information further indicates whether the particles are biologically incompatible to cells or not, as the SPIONs itself can induce toxicity through oxidative injury of cells, if they are present in an electron deficient oxidation state (γ-Fe_2O_3, α-Fe_2O_3 and FeOOH). The preservation of particle size, elemental composition, and the core-shell structure for both particles after the thermal treatment has proven that these materials can find promising applications as quality new catalysts and also efficient heat releasing agents for magnetic fluid hyperthermia.

ACKNOWLEDGMENTS

FM thanks LSU-CAMD for the opportunity to carry out his dissertation at CAMD's nanomaterials and nanofabrication facility. Partial funding for this work from DOE's Energy Frontier Research Center (Grant No. DESC0001058) and NIH (1R01CA142842-01A1) is gratefully acknowledged.

REFERENCES

[1] Kumar, C.S.S.R.; Mohammad, F. *Adv. Drug Deliver Rev.* 2011, 63, 789-808.
[2] Daniel, M. C.; Astruc, D. *Chem. Rev.* 2004, 104, 293-346.
[3] Jun, Y. W.; Seo, J. W.; Cheon, J. *Acc. Chem. Res.* 2008, 41, 179-189.
[4] Burda, C.; Chen, X.; Narayanan, R.; El-Sayed, M. A. *Chem. Rev.* 2005, 105, 1025-1102.
[5] Saini, S.; Goundla, S.; Bagchi, B. *J. Phys. Chem.* B 2009, 113, 1817-1832.
[6] Prasad, B. L. V.; Sorensen, C. M. Klabunde, K. J. *Chem. Soc. Rev.* 2008, 37, 1871-1883.
[7] Narayanan, R.; Tabor, C.; El-Sayed, M. A. *Top. Catal.* 2008, 48, 60-74.
[8] Grzelczak, M.; Perez-Juste, J.; Mulvaney, P.; Liz-Marzan, L. M. *Chem. Soc. Rev.* 2008, 37, 1783-1791.
[9] Nehl, C. L.; Hafner, J. H. *J. Mater. Chem.* 2008, 18, 2415-2419.
[10] Lisiecki, I. *J. Phys. Chem.* 2005, 109, 12231-12244.
[11] Wang, L.; Park, H. Y.; Lim, S. I.; Schadt, M. J. Mott, D.; Luo, J.; Wang, X.; Zhong, C. J. *J. Mater. Chem.* 2008, 18, 2629-2635.
[12] Latham, A. H.; Williams, M. E. *Acc. Chem. Res.* 2008, 41, 411-420.
[13] Sweeney, C. M.; Hasan, W.; Nehl, C. L.; Odom, T. W. J. *Phys. Chem.* A 2009, 113, 4265-4268.
[14] Reiss, P.; Protiere, M.; Li, L. *Small* 2009, 5, 154-168.
[15] Burns, A.; Ow, H.; Weisner, U. *Chem. Soc. Rev.* 2006, 35, 1028-1042.
[16] Chen, J. P.; Yang, P. C.; Ma, Y. H.; Tu, S. J.; Lu, Y. J. *Int. J. Nanomedicine* 2012, 7, 5137-5149.
[17] Zhang, T.; Stilwell, J. L.; Gerion, D.; Ding, L.; Elboudwarej, O.; Cooke, P. A.; Gray, J. W.; Alivistos, A. P.; Chen, F.F. *Nano Lett.* 2006, 6, 800-808.

[18] Matoussevitch, N.; Gorschinski, A.; Habicht, W.; Bolle, J.; Dinjus, E.; Boennemann, H.; Behrens, S. *J. Mag. Mag. Mater.* 2007, 311, 92-96.
[19] Zhang, X. B.; Yan, J. M.; Han, S.; Shioyama, H.; Xu, Q. *J. Am. Chem. Soc.* 2009, 131, 2778-2779. (b) Joo, S.H.; Park, J.Y.; Tsung, C. K.; Yamada, Y.; Yang, P.; Somorjai, G.A. *Nature Mater.* 2009, 8, 126-131.
[20] Schneider, G.; Subr, V.; Ulbrich, K.; Decher, G. *Nano Lett.* 2009, 9, 636-642.
[21] Jun, Y. K.; Choi, J.; Cheon, J. W. *Chem. Commun.* 2007, 12, 1203-1214.
[22] Kim, D. K.; Zhang, Y.; Kehr, J.; Klason, T.; Bjelke, B.; Muhammaed, M. *J. Magn. Magn. Mater.* 2001, 225, 256-261.
[23] Laurent, S.; Mahmoudi, M. *Int. J. Mol. Epidemiol. Genet.* 2011, 2, 367-390.
[24] Liu, J. P.; Fullerton, E.; Gutfleisch, O.; Sellmyer, D. J. Nanoscale magnetic materials and applications; *Springer:* 2009; 1, pp 35-65.
[25] Mohammad, F.; Balaji, G.; weber, A.; Uppu, R. M.; Kumar, C. S. S. R. *J. Phys. Chem. C* 2010, 114, 19194-19201.
[26] Gaur, S.; Johansson, S.; Mohammad, F.; Kumar, C. S. S. R.; Spivey, J. J. *J. Phys. Chem.* C 2012, 116, 22319-22326.
[27] Teixeira, A. P. C.; Trisão, J. C.; Araujo, M. H.; Oliveira, L. C. A.; Moura, F. C. C.; Ardisson, J. D.; Amorim, C. C.; Lago, R. M. *J. Braz. Chem. Soc.* 2012, 23, 1579-1593.
[28] Wang, L.; Luo, J.; Fan, Q.; Suzuki, M.; Suzuki, I. S.; Engelhard, M. H.; Lin, Y.; Kim, N.; Wang, J. Q.; Zhong, C. J. *J. Phys. Chem.* B 2005, 109, 21593-21601.
[29] Jiang, W.; Zhou, Y.; Zhang, Y.; Xuan, S.; Gong, X. *Dalton Trans.* 2012, 41, 4594-4601.
[30] Cho, S. J.; Idrobo, J. C.; Olamit, J.; Liu, K.; Browning, N. D.; Kauzlarich, S. M. *Chem. Mater.* 2005, 17, 3181-3186.
[31] Kumar, C. S.; Leuschner, C.; Doomes, E. E.; Henry, L.; Juban, M.; Hormes, J. *J. Nanosci. Nanotechnol.* 2004, 4, 245-249.
[32] Ahmad, M. B.; Lim, J. J.; Shameli, K.; Ibrahim, N. A.; Tay, M. Y. *Molecules* 2011, 16, 7237-7248.
[33] Ravela, B.; Newville, M.; *J. Synchrotron Rad.* 2005, 12, 537-541.
[34] Robinson, I., Tung, L. D.; Maenosono, S.; Wälti, Thanh, N. T. K. *Nanoscal*e 2010, 2, 2624-2630.
[35] Haneda, K.; Morrish, A. H. *J. Phys. Colloques* 1977, 38, C1 321-323.
[36] Xu, C.; Sun, S. *Polym. Int.* 2007, 56, 821-826.
[37] Tan, B. J.; Klabunde, K. J.; Sherwood, P. M. A. *Chem. Mater.* 1990, 2, 186-191.
[38] Kitagawa, H.; Kojima, N.; Takahashi, H.; Mori, N. *Synthetic Metals* 1993, 56, 1726-1729.
[39] Hedman, J.; Klasson, M.; Nilsson, R.; Nordling, C.; Sorokina, M. F.; Kljushnikov, O. I, Nemnonov, S. A.; Trapeznikov, V. A.; Zyryanov, V. G. *Phys. Scripta* 1971, 4, 195-201.
[40] Rojas, T. C.; Sánchez-Lőez, J. C.; Greneche, J. M.; Conde, A.; Fernández, A. *J. Mater. Sci.* 2004, 39, 4877-4885.
[41] NIST X-ray Photoelectron Spectroscopy Database, Version 4.1 (National Institute of Standards and Technology, *Gaithersburg,* 2012); http://srdata.nist.gov/xps/.
[42] Urbina, M. C.; Zinoveva, S.; Miller, T.; Sabliov, C. M.; Monroe, W. T.; Kumar, C. S. S. R. *J. Phys. Chem.* C 2008, 112, 11102-11108
[43] Westre, T. E.; Kennepohl, P.; Dewitt, J. G.; Hedman, B.; Hodgson, K. O.; Solomon, E. I. *J. Am. Chem. Soc.* 1997, 119, 6297-6314.

[44] Simon, G.; Huang, H.; Penner-Hahn, J. E.; Kesler, S. E.; Kao, L-S. Amer. *Mineralogist* 1999, 84, 1071-1079.

[45] Miyamoto, T.; Niimi, H.; Chun, W. J.; Kitajima, Y.; Sugawara, H.; Inabe, T.; Naito, T.; Asakura, *Photon Factory Activity Report* 2005, 23, part B.

[46] Chen, R.; Song, G.; Wei, Y. *J. Phys. Chem.* C 2010, 114, 13409-13413.

[47] Mandal, M.; Kundu, S.; Ghosh, S. K.; Panigrahi, S.; Sau, T. K.; Yusuf, S. M.; Pal, T. *J. Colloid Interface Sci.* 2005, 286, 187-194.

[48] Yu, J. J.; Yang, H. J.; Kim, T.S.; Jeong, T.S.; Youn, C. J. *J. Korean Phys. Soc.* 2012, 60, 1927-1933.

[49] Lekston, Z.; Lagiewka, E. *Arch. Mat. Sci. Eng.* 2007, 28, 665-672.

[50] Kalipcilar, H.; Bowen, T. C.; Nobel, R. D.; Falcorner, J. L. *Chem. Mater.* 2002, 14, 3458-3464.

[51] Ramay, S. M.; Saleem, Murtaza.; Atiq, S.; Siddiqi, S. A.; Naseem, S.; Anwar, M. S. *Bull. Mater. Sci.* 2011, 34, 1415-1419.

[52] Mane, S. R.; Mane, R. M.; Patil, P. S.; Bhosale, P. N. *Adv. Mat. Lett.* 2013, 4, 94-98.

[53] Cimino, A.; Porta, P.; Valig, M. *J. Am. Ceram. Soc.* 1966, 49, 152-156.

[54] Pandu, R.; Yadav, K. L.; Kumar, A.; Reddy, P. R.; Gupta, A. V. S. S. K. S. *Ind. J. Eng. Mater. Sci.* 2010, 17, 481-485.

[55] Stephen, Z. R.; Kievit, F. M.; Zhang, M. *Mater. Today* (Kidlington). 2011, 14, 330-338.

In: X-Ray Diffraction: Structure, Principles and Applications ISBN: 978-1-62808-591-4
Editor: Kaimin Shih

Chapter 8

CHARACTERIZATION OF MATERIALS OBTAINED BY AN INNOVATIVE INTEGRATED SYNTHESIS METHOD AIMED TO THE HYDROGEN TECHNOLOGY

Marcelo R. Esquivel *

Centro Atómico Bariloche - Comisión Nacional de Energía Atómica,
Bariloche, Río Negro, Argentina
Consejo Nacional de Investigaciones Cientificas y Técnicas (CONICET)
Centro Regional Universitario Bariloche, Universidad Nacional del Comahue,
Bariloche, Río Negro, Argentina

ABSTRACT

This chapter presents an innovative method consisting of the synthesis and further modification of materials aiming for the two stages of hydrogen technology. The selected materials are intermetallics (IM´s) with AB_5-type, and two synthesis steps are included in the method. The first one consists of the synthesis of the materials by mechanical alloying (MA) and mechanical milling (MM). The second one involves the adjustment of the material by changes of heating/annealing processes under different types of atmosphere, times, and temperatures. Materials obtained by annealing under inert atmospheres (Ar) are characterized for the application of thermal compression of hydrogen (TCH), while the materials obtained by heating under air are aimed for the use of gross purification of hydrogen (GPH).

LIST OF SYMBOLS

Latin

D, Crystallite Size (Å)

* Corresponding Author. Email: esquivel@cab.cnea.gov.ar.

M, Representative Composition of Lanthanides, $La_{0.25}Ce_{0.52}Nd_{0.17}Pr_{0.06}$
s, Strain (dimensionless)
t, Time (s,h)
T, Temperature (°C)
Greek
α, Degree of Reaction (Dimensionless)
ΔQ, Amount of Heat Evolved in a Calorimetric Thermal Event

Glossary

A, Annealing Treatment
GPH, Gross Purification of Hydrogen
H, Heating Treatment
IM´s, Intermetallic Compounds
MA, Mechanical Alloying
MM, Mechanical Milling
PCI, Pressure Composition Isotherm.
TCH, Thermal Compression of Hydrogen

1. Introduction

Observed as a successive set of pictures through time, materials science may be seen evolving as a tree of knowledge. As such, this tree consists of several branches, and, owing to a natural evolution of the research, the following ones become more important throughout the last decades:

- Research in new materials [1-2].
- Modification of traditional materials aiming to new applications [3-4].
- Investigation on the recovery of wastes [5-6].

The branches of this tree of knowledge are nourished by the daily effort of scientists working for both governmental agencies and private companies. Many of the research subjects are currently regulated by, in some way, the successful concept of "design of the material and the application the material is designed for". This concept can be summarized as a pair called *material-application*. It may be seen as a pair of puzzle pieces like the ones presented in Figure 1(a). The pair is successful because it ensures to focus the research work and corresponding resources to the selected goal as deduced from the perfect fit between the two pieces of Figure 1(b).

Although this concept is successfully applied in several cases, the other side of the coin is the unnecessary waste of resources, including the design of specific equipment and the specificity of the synthesis process. This specificity is related to the fact that once the material is obtained and used in the technological application, the whole process loses novelty and it becomes obsolete for using in another pair *material-application*. Then, the effort and resources used in the generation of the first successful pair *material-application* may not be used in the generation of other new pair. In most cases, the whole process should start all over again. Then, all effort done and resources used are probably successful but not necessarily optimized. As a bigger picture, this way of thinking might be seen as a waste of priceless resources.

Therefore, it is necessary to change the way of thinking on the design of materials and their further applications without losing the current successful correlation established by the specificity of the pair *material-application.* This new way of thinking should be made broader, therefore, a proposal for this goal is discussed in the chapter.

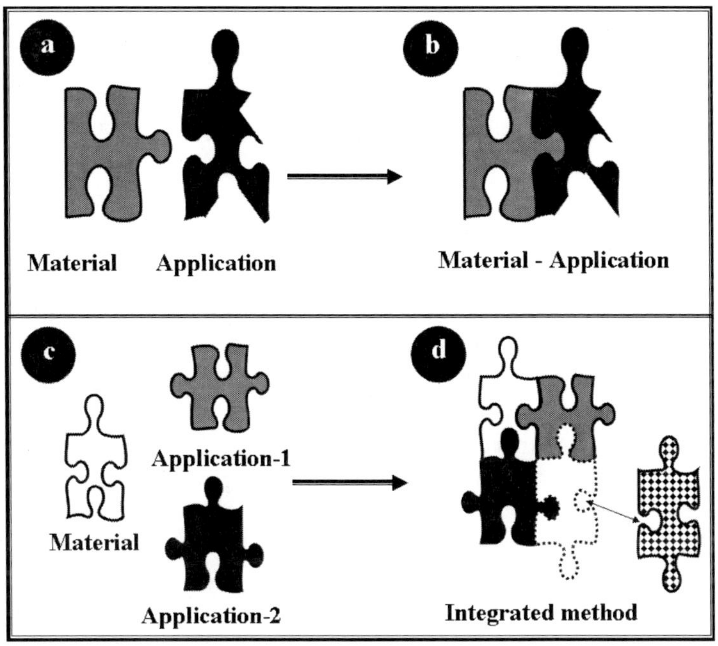

Figure 1. Successful *material-application* pair (a, b), a new way of design (c), and the generation of an integrated method (d).

2. Generation of the Idea

As discussed previously, a wider way of thinking is needed, and the first stage to be considered is the synthesis. This process should be specific enough to fulfill the requirements of the technological application but also leaving an *unfinished* product. This last characteristic implies that the product can be also used as intermediates or raw materials for the production of either a modified material or a new one. Strictly speaking, this new material is aimed to form other pair *material-application.* It may follow a design similar to that presented in Figure 1(c), and, to be successful, this concept should share the type and features of the manufacturing process. This way of production should also favor a decrement of the amount of resources and a unification of the wastes released in the whole production process. This concept will be successful only if integrated synthesis-modification of the material methods is used as implied in Figure 1(d). These methods should contain both the synthesis and the improvement of the materials. The concept of improvement in this case is related to further modifications of the same material. Of course, it has to be kept in mind that the material should be ready for a straightforward use in the desired application. The ultimate goal is the possibility to extend the method to a new material. In Figure 1(d), that goal is represented by the puzzle piece with the black and white pattern tiles designed to fit "the hollow place".

In this chapter, a case using the concept of integrated methods for material synthesis is discussed followed by further modification of properties. The synthesis is intended to obtain IM´s of the AB_5-type, and the further modification of properties should share almost all features of the manufacturing process. The materials are synthesized and characterized for two selected applications of the hydrogen technology: Gross Purification of Hydrogen (GPH) and Thermal Compression of Hydrogen (TCH).

3. Development of the First Stage: Synthesis Method

The synthesis method should be both economic and easy to scale up. As described previously, it should also leave an *unfinished* product. This last characteristic allows the production of different pairs of *material-application*. The material to synthesize is an AB_5-type intermetallic (IM) [7]. On these IM's, the A-constituent is a Lanthanide element or an alloy of them while the B-constituent is an element or alloy of elements from the groups 3 to 11 [8]. Both types make them in the typical $CaCu_5$ structure as presented in Figure 2.

The structure crystallizes in the *P6/mmm* space group. In this structure, lanthanide elements or their alloys are distributed in Wyckoff positions 1.a (0,0,0). The B-site elements are distributed between 2 c ($^1/_3$, $^2/_3$, 0) and ($^2/_3$, $^1/_3$, 0) and 3 g (½, 0, ½), (0, ½, ½,), ($^1/_2$, ½, ½) positions.

The most common composition studied is $LaNi_5$. It is due to the first discovery on the reversible absorption of hydrogen near room temperature for this particular AB_5 [9]. After that discovery, both the structure of $LaNi_5$-based IM's and hydrogen technology applications were studied together to reveal a very successful *pair* [10-13]. Studies ranged from the characterization of structural properties to their relation to hydriding properties [12-15], and the respective applications to the hydrogen technology [16-19]. This evolution was natural because the hydriding properties are strongly related to AB_5 structural features [20-21]. This evolution is a beautiful example of a successful *pair*. The synthesis of these IM´s is traditionally known but the main problems are related to the purity of the sample. It means that sample should consist only of AB_5-phase with no other site-structures. The other problem is the characterization of the composition in each representative site of the structure (A-B) [22]. The last characterization is not easy because elemental and structural techniques may only indicate the global composition of the phases present in the sample. Only an appropriate model of the structure can indicate the lanthanide composition and Ni-Ni substituents occupying the A and B sites [23].

The A-site may consist of pure lanthanide or an alloy of them. For economy reasons, this site is substituted by alloys of lanthanides. These alloys are a product of purification/separation processes and cheaper in comparison to pure lanthanides [24]. The generic name is Mischmetal (Mm).

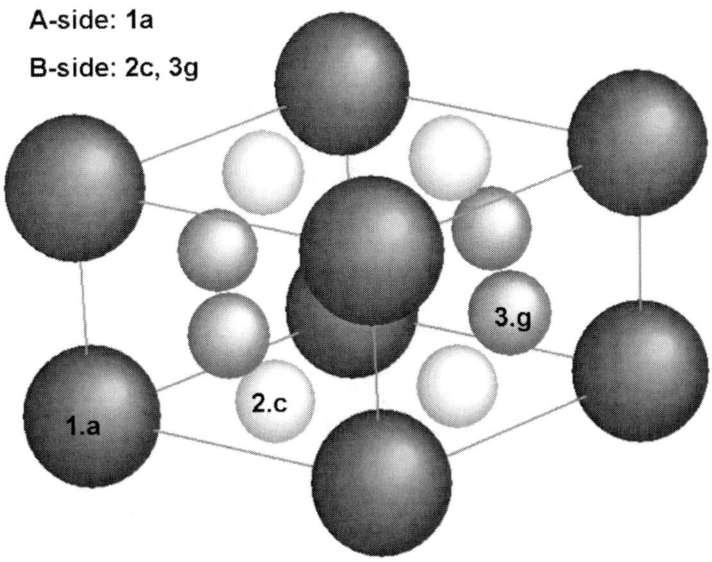

Figure. 2. The AB_5 ($CaCu_5$) structure. The Wyckoff positions of both elements of A and B sites are also presented.

This alloy is obtained from beneficiation of lanthanide-bearing minerals as a cheaper process with all lanthanides obtained together. Therefore, the Mm composition depends on the composition of the mineral and is obtained from namely xenotime, batnaesite or monazite. The most popular compositions include Ce, Nd, La and Pr. These alloys should present melting points similar to those of the pure components. It is expectable that most Mm melting temperatures are between Nd (1010 °C) and Ce (795 °C) [25].

Usually, most of the reported applications of the hydrogen technology use Ni and Ni-rich compositions on the B-site of the AB_5 [26]. Ni melting point is close to 1440 °C but many of Ni-replacements have large differences in melting points such as Al (660 °C), Sn (151 °C) and Zn (419,5 °C). Then, the synthesis of AB_5 from A and B constituents faces the following problems:

- The difference in melting points. As an example, over 600 ° C between Ce and Ni.
- The difference in vapor pressure between metals that leads differential vaporization.
- The AB_5 phase melts at temperatures higher than Ni (1440 °C). For example, $LaNi_5$ and $CeNi_5$ melting points are 1548 °C [27] and 1449 °C [28], respectively. Therefore, the synthesis of AB_5 from A and B constituents implies the use of high temperature methods.

These conditions define that the temperature of synthesis is the most relevant factor, and any of the three following methods was used to obtain these IM´s.

1. High temperature method (casting, melting) [29-32].
2. Chemical co-precipitation (CC) [33-34].
3. Mechanical alloying/milling (MA)/(MM) [23,35-45].

Before selection, pros and cons of the three methods are considered. Traditionally, these IM´s are obtained by high temperature synthesis methods. The know-how is now well known. It is also successful in most of the cases and the obtained IM presents a well crystallized microstructure. Provided that stoichiometry is controlled, especially when working with low melting temperature metals, such as Sn, the presence of secondary phases on the sample may be avoided [46]. Despite its advantages, the method requires high temperatures, at least over 800 °C, followed by annealing during time periods longer than 24 h [29-32]. These combined conditions of high temperature and long annealing times are a drawback for both technological and economic requirements [35]. These reasons are related to the energy consumption during the processing [35]. The IM´s obtained by high temperature methods favor the diffusion of the materials. Therefore, these IM's occur as chunks or large pieces. This presentation is not appropriate in the applications of hydrogen technology because the contact of the solid-hydrogen is maximized. Therefore, the IM´s obtained by this method need further treatments to decrease the average particle size, and this is the other drawback of the synthesis method.

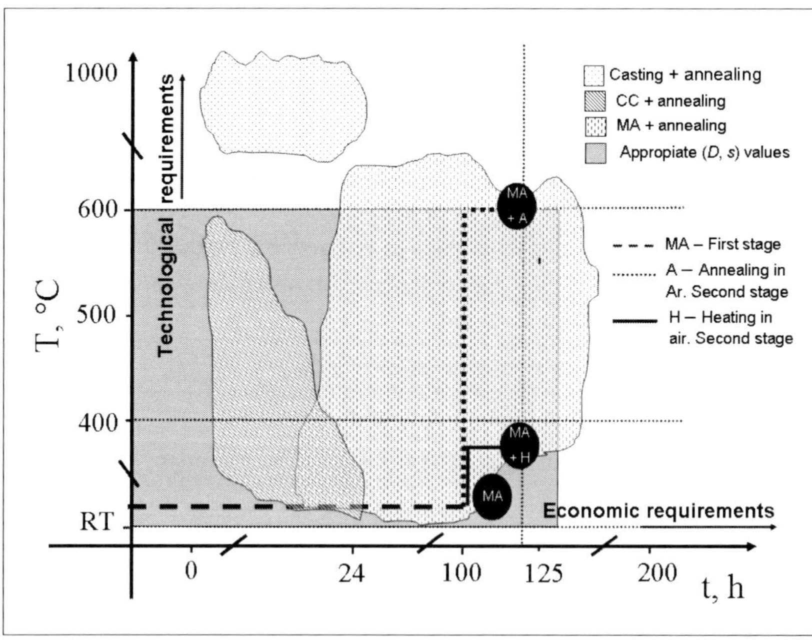

Figure 3. Comparison of the synthesis methods for IM´s.

Figure 3 presents a scheme describing technological and economic requirements for the synthesis and after-synthesis treatments for AB_5 type intermetallics. The variables selected are temperature (T) and time (t). Temperature is associated to the technological requirements while time is associated to the economical ones [35].

In Figure 3, the gray square represents a selection of moderate values of t and T to yield an appropriate sample microstructure after combined synthesis + annealing/heating. This moderate microstructure is the one needed for use in the desired application selected in this chapter [35]. In this schematic gray square, crystallite size distribution (*D*) is larger than 300 Å and strain (*s*) is smaller than 3 %. These parameters are representative of the microstructure of the sample. The operative upper and lower temperatures that limit the square are 600 °C and 400 °C. Temperatures higher than 600 °C demand the better technological conditions [35] and are associated with the non-optimized energy consumption. Temperatures lower than 400 °C hinder the diffusion by avoiding the grain growth [35].

Values of t and T for samples obtained by casting are represented by black dots in white background tile. The zone of synthesis (T > 800 °C, t > 12 h) followed by the annealing (T > 800 °C + t > 24 h) is out of the desired condition square.

The second method is chemical co-precipitation (CC). The control of stoichiometry and the temperature of synthesis are the main advantages of this method [36]. The AB_5 is also obtained with an appropriate final particle size distribution for the use on selected technological applications. Although *D* values might not be the desired, low temperature annealing at T < 600 °C solves the problem. The zone favored for this method is indicated by the diagonal black lines with the white background tile. As observed, the combined CC synthesis + annealing fell within the desired square. Nevertheless, the scaling-up and the purity of reactants are the main disadvantages of this method.

The third method is mechanical alloying (MA), and the synthesis of IM´s by MA is known for a pair of decades [47-49] with its high popularity in the last fifteen years [50]. In spite of the fact that the mechanical milling (MM), a technique similar to MA, is traditionally used in the ceramic industry and the know-how is well known [47], the method is economic and easy to scale up [23, 35, 40]. The material obtained presents a large particle size distribution. Microstructure issues can be corrected by low temperature annealing as observed on the short vertical line zone. The main drawback of this technique is the formation of secondary phases along the main AB_5 phase [35]. The advantages and disadvantages of these three techniques are summarized in Table 1.

Table 1. Comparison of the advantages and disadvantages of the synthesis methods. The techniques are compared to each other. Positive characteristics are indicated as (+). Negative characteristics are indicated as (-)

Technique / Features	Cast	Chemical Co-precipitation	Mechanical Alloying/Milling
Microstructure	-	++	+
Stoichiometry Control	+	++	-
Economics	+	++	++
Scaling up	+	-	++
Time of Processing	+	+	+
Necessity of Further Treatment	+	+	+

In this table, the microstructure is related to *D* and *s*. As explained previously, the synthesis method should be economic and easy to scale up. Economic methods are associated with the low energetic and technological requirements [35]. Samples obtained should have an adequate microstructure to be straightforwardly used in the selected technological applications. As observed in Table 1, all three techniques could be selected to synthesize the AB_5 intermetallics. However, when it comes to the selection of the lower temperature of synthesis, the melt casting option is usually eliminated. Both chemical co-precipitation and mechanical alloying can be selected. Despite the advantages of cast and chemical synthesis methods, mechanical alloying is selected. It is because this method falls within the technological and economical requirements and also assures the formation of an *unfinished product* suitable of further modifications [50].

4. Development of the Second Stage: Modification of Materials

This second stage should either change or modify the properties of the materials in order to fit to the selected application. This stage should share most of the unitary processes in order to decrease the costs of production and the waste of processing. This second stage is also limited by the technological and economic requirements of the synthesis method as discussed in the previous section. Many processing methods could be used, such as chemical, mechanical and surface treatments. Nevertheless, the synthesis and improvement should also fit with the desired application. By considering the limitations imposed in Figure 3, annealing under different conditions was selected [37, 39, 50]. IM´s obtained by MA annealed under Ar are intended to be used in TCH, and IM´s obtained under air are aimed to be used in GPH.

5. Characterization of Materials

5.1. Materials and Experimental Equipment

Low energy mechanical alloying was used to synthesize AB_5´s of different compositions using a UNI-Ball mill II mill (Australian Instruments) from raw materials. Integrated milling time was 100 h. The ratio of mass of balls to mass of sample (R) selected was 33/1. Powder mixtures and steel balls were set in stainless steel chamber. Samples handling was done under Ar in a glove box (Mbraum, Siemens Instruments) with the controls of oxygen and humidity. This processing time was enough to assure that the sample can reach the *final* stage [38, 40, 48]. Two AB_5 of compositions $La_{0.25}Ce_{0.52}Nd_{0.17}Pr_{0.06}Ni_5$ (AB_5-B) and $LaNi_{4.90}Fe_{0.10}$ (AB_5-A) were obtained. Raw materials were Ni (Sigma Aldrich, 99.99%) and La (Reacton, 99.99%) and mischmetal (Alpha Aesar, 99.7%) of nominal composition 52.0 mass% Ce, 25.6 mass% La, 16.9 mass% Pr, 5.5 mas% Nd. Mm composition was verified by neutron activation analysis (NAA) and energy dispersive spectroscopy (EDS).

Room temperature X-ray diffraction (XRD) measurements were done on a Philips PW 1710/01 Instrument with Cu K_α radiation (graphite monochromator). Diffraction patterns were analyzed by the Rietveld method using DBWS software [51]. Strain (*s*) and crystallite size (*D*) effects were estimated from diffraction peaks by empirically assuming a Gauss distribution and a Cauchy (Lorentz) component, respectively [52]. Particle size and morphology were observed by scanning electron microscopy (SEM). These microscopes were

FEG-SEM Nova NanoSEM 2030 and Philips SEM 515, both with energy dispersive X-ray spectroscopy (EDS) capability.

Differential scanning calorimetry (DSC) measurements were carried out using a DSC 2910 TA Instruments calorimeter under Ar and air. Non-isothermal measurements were carried out under heating rates varying from 5 °C min^{-1} to 25 °C min^{-1} at an Ar flow rate of 122 m min^{-1}. Isothermal measurements were done by heating the sample to the desired temperature under Ar at 25 °C min^{-1}. After that, the Ar flow was closed and isothermal measurement was conducted under the stagnant air. A Sieverts' type device was used to measure hydrogen sorption property.

5.2. Synthesis by MA

The diffractograms of the as-milled AB_5-A and AB_5-B samples are shown in Figure 4 (a) and (b), respectively. The comparison of the reference pattern of $LaNi_5$ is presented in Figure 4 (c) (PDF-0170126). MA consists of a continuous cycle of fracture and cold welding [40, 47-48]. As integrated milling time progresses (t_m), processing reduces crystallite size and induces deformation on the cell parameters of the sample [47-48]. As a result, the diffractograms of as-milled samples are poorly defined with broad peaks and a low peak to background ratio [35, 37-41]. This issue might complicate the identification of the phases obtained. Nevertheless, main peaks of the AB_5 structure are clearly observed. In the case of AB_5-B the identification is straightforward from reference peaks of Figure 4(c). The AB_5-A is isostructural to $LaNi_5$. Cell parameters for both AB_5 were estimated from the identified hkl peaks of Figures 4 (a) and (b). Results are summarized in Table 2. The samples are not single-phase and minor structures can also be observed. Unidentified peaks are indicated with question marks. Identification of these structures is not easy in as-milled samples due to the shape of the peaks. From the cell values calculated from peaks 101 and 110, it is deduced that both AB_5 have similar values on the *c*-parameter. It occurs because the change of lanthanide on the composition of the A-site produces changes on the basal parameters [42]. As described by a model developed previously [23], *c*-parameter changes are strongly related to changes on the composition of the B-site of AB_5. Then, these results agree with the main assumptions of that model [23].

To quantify the AB_5 microstructure, *s* and *D* values were obtained from the 101 and 110 hkl peaks. Results are summarized in Table 3, and the values obtained are associated with the microcrystalline microstructure according to sample withdrawn at *final-completion* milling stages [39-40].

The SEM image of the AB_5-A sample is presented in Figure 5. The image shows a conglomerate obtained after processing for 100 h. The conglomerate consists of a broad particle size distribution ranging from 5 to 100 μm. The larger particle shows traces of severe fracture and cold welding [38-40]. The fracture is evidenced by the clear cleavage observed in rectangle I while cold welding is identified by the layers of IM shown in rectangle II. This morphology is consistent with samples reaching the *final-completion* stages [40].

The composition of the sample was also analyzed by EDS. Examples of the analyses are shown with black circles, and the summary of results is presented in Table 4. It can be concluded that elemental composition is relatively homogeneous.

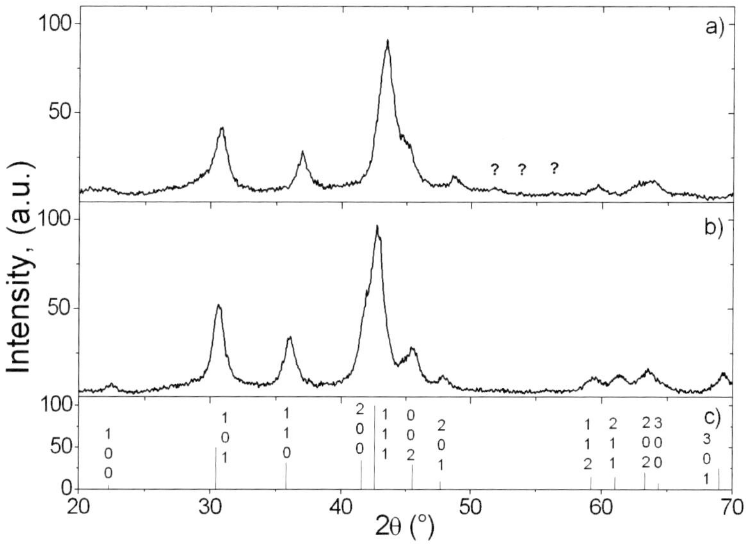

Figure 4. Experimental patterns of (a) AB5-A, (b) AB5-B, and reference pattern of (c) LaNi5 (PDF-0170126).

Table 2. Structural Parameters

Sample	Treatment	Cell Parameters				Composition Mass %		R_{wp} (%)
		AB_5			**Ni**	**AB_5**	**Ni**	
		***a* (Å)**	***c* (Å)**	**V_c (Å)^3**	***a* (Å)**			
AB_5-A	As-milled 100h	4.80	4	82.4	-------	----	-----	---------
	Annealed 600 °C, 24 h, Ar	4.907	3.977	82.93	3.530	93	6	12
AB_5-B	As-milled 100h	5	4	86	-------	----	----	------
	Annealed 600 °C, 24 h, Ar	5.011	3.977	86.53	-----------	----	----	------

Table 3. Microstructural parameters. M stands for $La_{0.25}Ce_{0.52}Nd_{0.17}Pr_{0.06}$

Sample	**Treatment**	**Phase**	**hkl**	**D ± 10 (Å)**	**s ± 0.05 %**
AB_5-A	As-milled (Fig. 4,7)	MNi_5	101	180	2
			110	180	2
	Annealed Ar (Fig. 7)		101	450	1
			110	550	1
AB_5-B	As-milled (Fig. 4)	$LaNi_{490}Fe_{0.10}$	101	190	2
			110	190	2
AB_5-A	Heated air (Fig. 12)	MO_2	311	>100	>1
		NiO	101	200	1
		Ni	200	349	1

The method of low energy mechanical alloying is done near room temperature [38]. Energy impingement is used to produce a continuous cycle of two complementary processes: fracture and cold welding. The cycle produces a solid-solid reaction between raw materials. Processing by MA usually occurs through the following four stages:

Initial: This stage is dominated by the fracture. Particle size is diminished in order to reach close sizes to favor the solid-solid reaction [38-40]. Since cold welding is negligible, compositional changes do not occur extensively. No compositional homogeneity is observed.

Intermediate: At this stage, cold welding influences the process. Real solid-solid reaction occurs and strong compositional changes occurs [38-40].

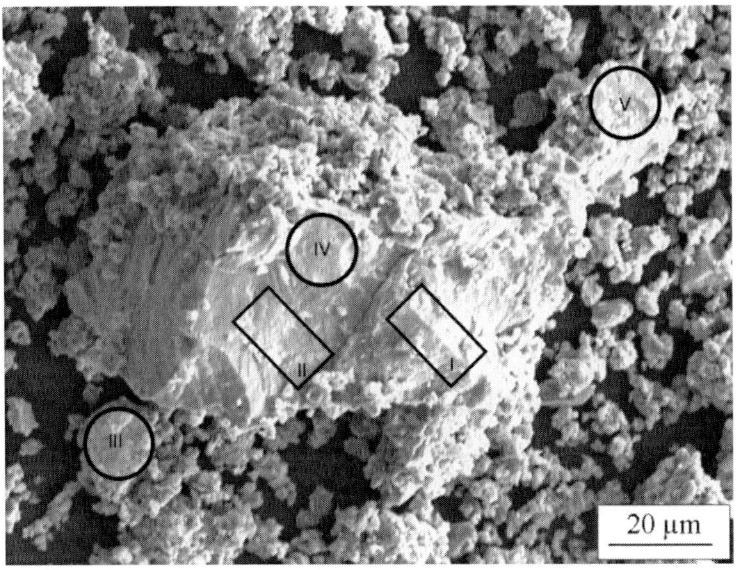

Figure 5. SEM image of the as-milled AB5-A.

Table 4. Results of EDS measurements. A-site elements are grouped as A in AB_5 structure. B-site elements are grouped as B in the AB_5 structure

Image/ Measurement		Sample	Composition (Atom%) ± 0.04					
			Ce	La	Nd	Pr	Ni	Fe
As-milled		AB_5-B	--------	100	----------	--------	0.97	0.03
			--------	100	--------	--------	0.97	0.03
			--------	100	-------	--------	0.98	0.02
			---------	100	----------	----------	0.97	0.03
Fig. 5	I	AB_5-A	51	26	16	7	100	---------
	II		52	25	16	7	100	--------
	III		52	26	15	7	100	--------
Fig. 13	I	AB_5-A	52	25	16	7	100	--------
	II		53	26	17	4	100	--------
	III		52	26	16	6	100	--------
	IV		53	25	17	5	100	--------

Final: At this stage, cold welding and fracture reaches equilibrium and share the control of the process. No further compositional changes should occur [38-40].

Completion: If material is stable, further milling will increase the energy of the surface of the particles. One option to decrease the surface reactivity is by cold welding of the material [38-40]. Being that the case, strong cold welding dominates the process and the particle size increases. The other option is the further destabilization of the main structure leading to the partial or total decomposition of the compound. Extended milling in AB_5 might induce equilibrium between these two processes. Strong cold welding was reported previously [40], and the destabilization of the main structure is a subject still under discussion.

The SEM image of Figure 5 shows agglomerate with strong fracture and cold welding. These processes dominate in both *final-completion* stages, and the image was compared to those of samples mechanically alloyed under similar conditions [39-40]. Comparison indicates that samples reached the same milling stage, and both EDS (Table 4) and XRD (Figure 4) indicate that a structure of the AB_5-type IM is formed. The MA process was used to synthesize an IM of the AB_5-type from raw materials. The process occurs near RT for 100 h. The obtained IM has a strong deformation of the surface due to the cycle of cold welding-fracture [39-40]. The microstructural parameters are indicated in Table 2, and the MA is represented by a straight black line on Figure 3. As observed, this first stage remains within the desired requirements zone, and the material is an *unfinished* product.

5.3. Improvement of the Material

As presented in Figure 1 (c-d), the *unfinished* product needs to be modified to fit the desired application. This material modification should be an improvement to the desired application for creating a perfect fit. This fit should be similar to the successful concept of *pair material-application*. In this case, the selected improvement consisted of an increase on the temperature. In one case, this increment is done under a protective atmosphere (Ar) to change the microstructure. In the other case, this increment is done to selectively to change the composition of the material.

5.3. a Annealing under Ar

The thermal heating under Ar of the as-milled sample is shown in Figure 6. The measurements were done at 5 $°C.min^{-1}$. The DSC curve shows no appreciable thermal event in all the measurement ranges. Since the $CaCu_5$-type structure melts congruently, no changes on composition should be expected. Taking the $LaNi_5$ as the most representative compound, no structure other than the hexagonal *P6/mmm* is found on the binary La-Ni phase diagram [27]. No thermal evolution related to thermal decomposition or structure change is observed in agreement with the thermodynamics of the lanthanide-nickel phase diagram [27, 28]. The annealing of as-milled samples might lead to the recrystallization of the main AB_5 phase. This occurs mostly in samples obtained by the high energy milling, because the impinged energy promotes the lack of intrinsic order or AB_5 amorphous phase [42].

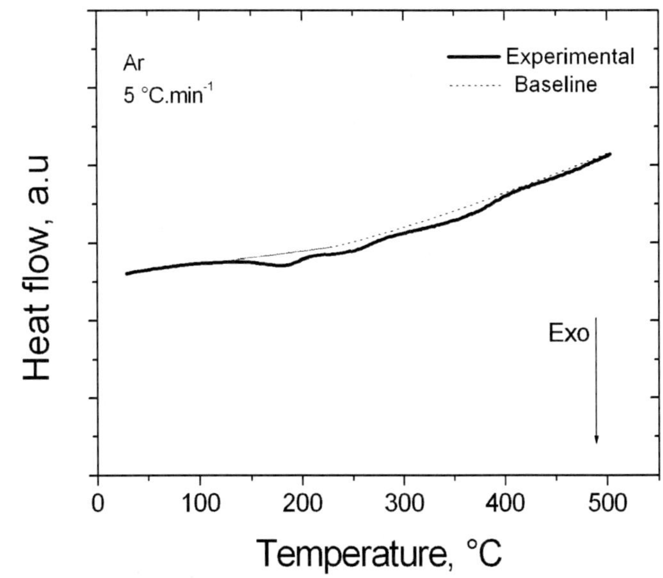

Figure 6. DSC measurement of as-milled AB_5-A sample.

This situation is not observed in low energy milling such as the one used in this work. It is because the amount of used energy only leads to micro-crystalline phases [23, 37-41]. Therefore, annealing under protective atmospheres only leads to the increment of crystallinity and the release of strain [37, 40-41, 53-58]. This effect is shown in Figure 7. The figure shows the as-milled and annealed samples of AB_5-A. Annealed samples were treated under Ar atmosphere at 600 °C for 24 h. As observed, this sample presents a well defined profile with sharp peaks and a large peak-to-background ratio.

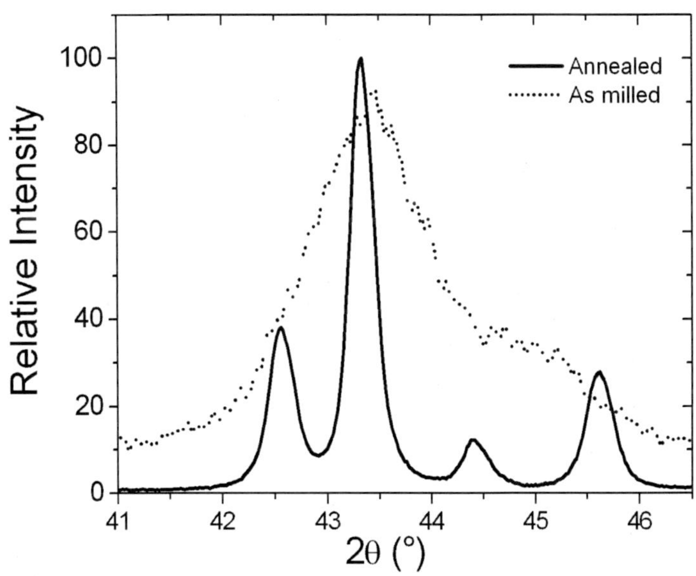

Figure 7. Diffractogram of the as-milled and annealed AB_5-A samples.

The severe change in microstructure is associated with the resolution of the peaks corresponding to the annealed sample. The as-milled sample presents a broad peak with shoulder. Instead, the annealed sample shows four peaks in the same 2θ range. The changes on microstructure are quantified by calculating the *D* and *s* values for the (101) and (110) peaks. The summary of calculated values is shown in Table 3. The *D* values of annealed sample are twice larger than the as-milled sample. On the contrary, *s* values are half of the as-milled sample. *D* increment is a process related to the grain growth [37-38] while *s* decrement is related to the release of stress. The calculation result is in agreement with the evolution of the same peaks in the annealed sample. The refined sample of AB_5-A is presented in Figure 8, and the sample presents Ni as a minor phase. A summary of the refined results is presented in Table 2, and the formation of minor phases is not a clear process. Since the sample was milled up to *final-completion* stage, no further modification regarding to the starting reactants should be expected [39-40]. However, the release of strain might be done by partial decomposition of the main phase. Therefore, these side-phases are formed during milling process due to the deviations of stoichiometry of starting reactants [41]. Another method is the partial decomposition after milling. This situation occurs because the final material state can be highly activated. One way to release the stored energy is through the welding particle surface process which leading to the increment of size. Another way is commonly known as the cold welding method by decomposing the bulk to stable phases. The representation of this stage is shown in Figure 3 by dot line. Samples following the synthesis approach by MA and annealing in Ar are labeled as MA + A in Figure 3. Significant changes on microstructure after such treatment are observed for all AB_5 samples [50-55].

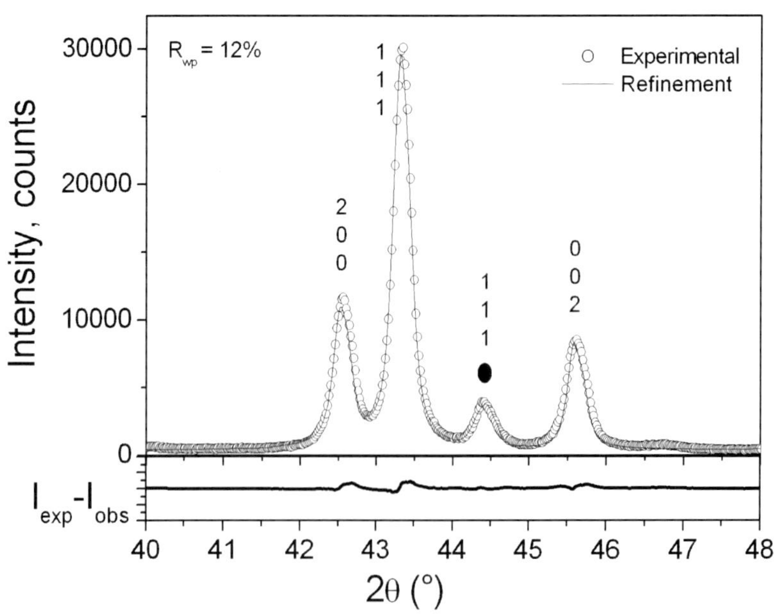

Figure 8. Details of the Rietveld refinement result of AB_5-A sample. Ni peaks are indicated with the black circle.

5.3. b Heating under Air

Thermal decomposition of AB_5 samples, together with the reaction products in air at the target temperatures, have been studied previously [50]. From that work, it was observed that the composition of AB_5 samples is similar to AB_5-A with a minimum destabilization temperature of 120 °C. The formations of $La_{0.25}Ce_{0.52}Nd_{0.17}Pr_{0.06}O_2$ and NiO were found at above 350 °C. In the case of similar composition for AB_5 samples, isothermal measurements show that the maximum temperature of destabilization is 200 °C [54].

Figure 9 (a) shows the isothermal DSC measurement corresponding to AB_5-A sample at 250 °C under air atmosphere. The isotherm shows the heat flow as a function of time. In presence of O_2, IM is thermally decomposed to:

$$La_{0.25}Ce_{0.52}Nd_{0.17}Pr_{0.06}Ni_5 \rightarrow La_{0.25}Ce_{0.52}Nd_{0.17}Pr_{0.06} + 5Ni \quad (1)$$

$$La_{0.25}Ce_{0.52}Nd_{0.17}Pr_{0.06} + O_2 \rightarrow La_{0.25}Ce_{0.52}Nd_{0.17}Pr_{0.06}O_2 \quad (2)$$

$$Ni + O_2 \rightarrow 2NiO \quad (3)$$

These three reactions are not reversible. Reaction (3) is reported to occur at above 400 °C for NiO production [56]. Therefore, only reactions (1) and (2) are considered at 250 °C. The corresponding reaction temperature of Mm is a function of time, and it is shown in Figure 9 (b). In the curve, the $\alpha(t)$ can be expressed as follows:

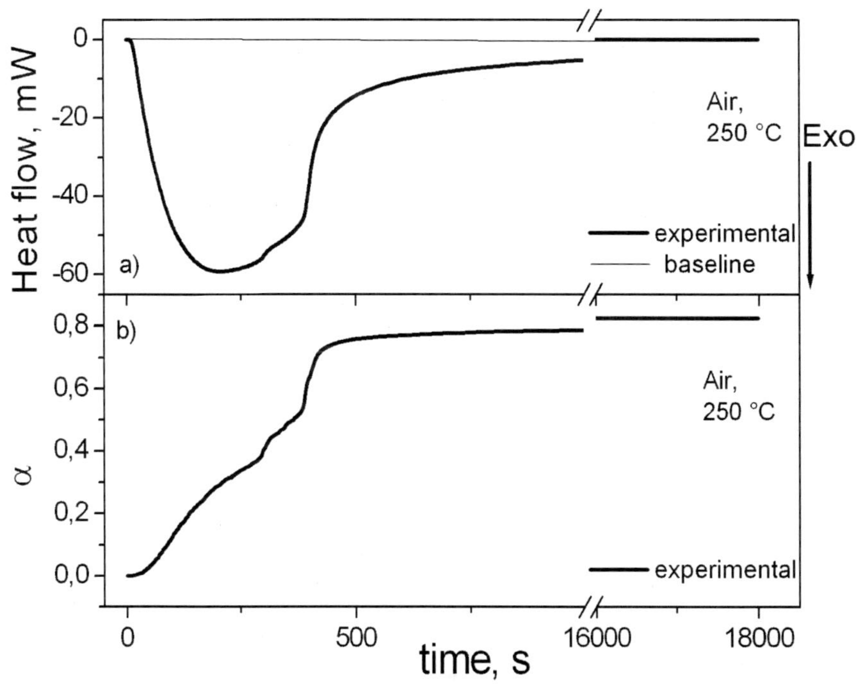

Figure 9. a) Thermal evolution of AB_5-A in air at 250 °C. b) The evolution of the same process expressed as α vs. t.

$$\alpha(t) = \frac{\Delta Q \exp}{\Delta Qt} \tag{4}$$

where: ΔQt considers the heat of AB_5 compound decomposition and the heat of MO_2 compound formation (ΔH^f), which represents a lanthanide-dioxide of composition $La_{0.25}Ce_{0.52}Nd_{0.17}Pr_{0.06}O_2$. These values are -25 kJ/mol and 872 kJ/mol, respectively. The ΔQexp corresponds to the heat integration derive from the experiment. From observation, MO_2 is not completely formed at this temperature and is reaching a maximum α value of 0.84.

To further investigate the conditions for the formation of MO_2-Ni composite, XRD measurements are carried out at 275 and 300 °C. Results are shown in Figure 10. At 275 °C, MO_2 is detected after treatment for 2 h. The comparison between the as-milled sample and the heated sample at 275 °C can be observed with a wide peak from the diffractogram at 2θ of 45 - 49 °. This peak is corresponding to the (220) peak of MO_2. The shoulder along the (111) peak of the as-milled sample is corresponding to the (111) peak of Ni. The sample heated in air at 300 °C for 800 s is also shown. Although the crystallinity is poor, it is clearly observed a larger crystalline evolution of (111) peak for the Ni in this sample. No significant changes in MO_2 crystallinity are observed. Longer treatment time would lead to the formation of NiO at 300 °C.

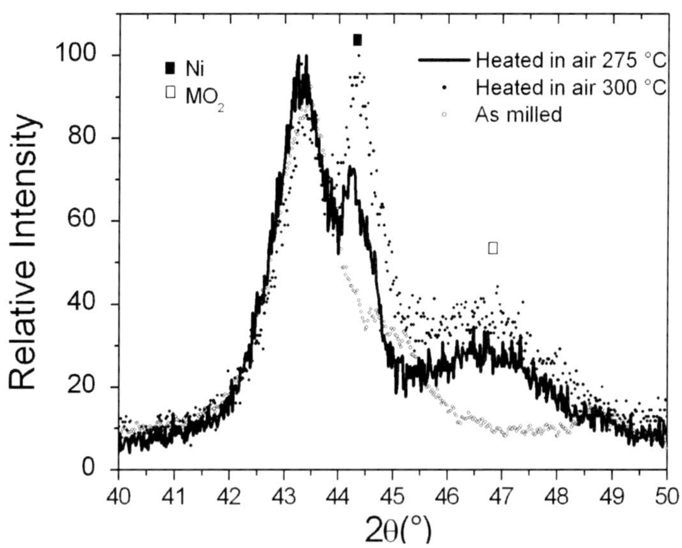

Figure 10. Diffractogram of samples treated at 275 and 300 °C. Main MO_2 and Ni peaks are indicated.

This temperature represents an important limit. If allowed, the formation of MO_2 is completed at 300 °C. The value of α is presented as a function of time in Figure 11, and the maximum amount of α is calculated by considering the formation heat of MO_2 (Reaction 2) and the decomposition heat of AB_5 (Reaction 1). The formation of NiO might also occur in parallel to Reaction 2. Nevertheless, the kinetics of the production of NiO is three times slower than the one of MO_2. It means that, at 300 °C, only with reaction times larger than 24 h can observe the reaction kinetics using this technique.

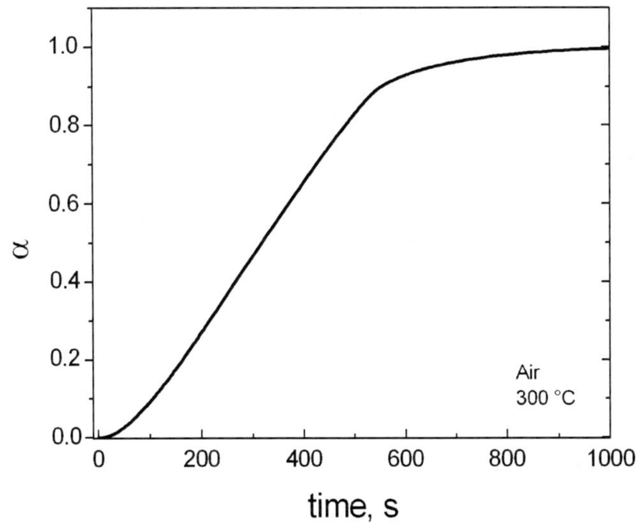

Figure 11. Evolution of the thermal decomposition of AB_5 at 300 °C. The value of α is expressed as a function of the heats generated from Reactions (1) and (2).

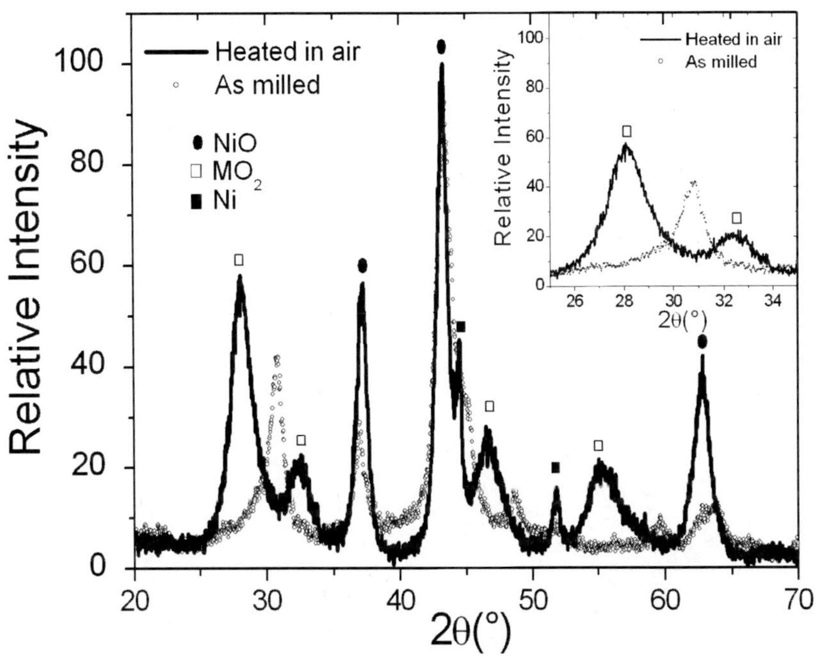

Figure 12. Diffractogram of the as-milled and the heated samples. Inset: details of 2θ from 25 to35 °.

To obtain MO_2-Ni composite, the heating temperatures in air should be below 300 °C but as close as possible to this temperature in order to accelerate the kinetics of the formation of oxide products. Once 300 °C is reached, NiO is also formed but with a rate three times slower than that of MO_2. Figure 12 shows the XRD data of AB_5-A sample heated at 350 °C in air for

24 h (black line). The diffractogram of the AB_5-A is also presented in Figure 12 (hollow dots line).

As observed, the unstable IM leads to the formation of $La_{0.25}Ce_{0.52}Nd_{0.17}Pr_{0.06}O_2$ and partial formation of NiO from Reaction (3). The microstructure of the oxides is characterized by calculating the *s* and *D* values. To perform this calculation, isolated peaks were selected. The results are summarized in Table 3. As observed, the microstructure is different for each compound presented in the sample. The obtained values show that Ni crystallite size is larger than those of either NiO or MO_2. These differences are related to nucleation and growth processes which make the composite appropriate for some applications [50]. The treatment consists of the formation of composites MO_2-NiO/MO_2-Ni shown in Figure 3 by the black line. From Figure 10 to Figure 12, it can be seen that temperatures below 300 °C should be selected to obtain MO_2-Ni composite. Heating time in air should be selected according to both economics restrictions and the kinetics of oxide formation. When the temperature is close to 300 °C, it should represent the best compromise between these two factors. This type of composite is aimed to be used in gross purification of hydrogen [50]. For this application, the metal presents catalytic properties, while the MO_2 structure acts as a filter [57]. As a composite, the most important aspect is the intimate contact between the metals and the oxides. A backscattered electron image of an AB_5-A sample heated in air at 300 °C is shown in Figure 13. No difference in bright between lanthanides and Ni was observed, and thus an intimate mixture is achieved. The white dots show the areas measured by EDS. As observed in Table 4, the EDS results show the homogeneous A-5B composition that is well-agreed with the backscattered electron image. It also indicates that the composite presents no segregation of the metals and the oxides.

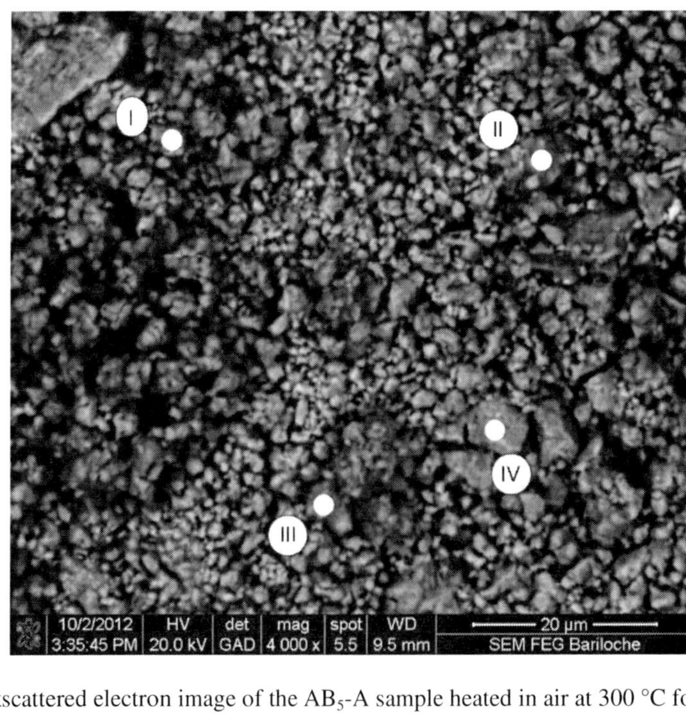

Figure 13. Backscattered electron image of the AB_5-A sample heated in air at 300 °C for 24 h.

6. Application to TCH

The synthesis by MA (white-piece of puzzle in Figure 1 (c)) and the improvement by thermal annealing of IM for AB_5-type have been reported [39-40]. The related application is TCH (the grey-piece of puzzle in Figure 1 (c)) [37]. The application of AB_5-B to the one-stage TCH is presented in Figure 14. The material is synthesized by MA and annealed 24 h in Ar according to operation conditions in Figure 3. The graph presents the pressure-composition isotherms (PCI) at 25 and 90 °C. This kind of measurement describes the interaction of the IM with H_2 in near equilibrium conditions.

The material-application pair function in an operation scheme is shown as follows. The fresh hydrogen enters the hydride-forming container through valve A at point I. At this temperature, the IM forms a hydride according to the following reaction:

$$AB_5 + 3\,H_2 = AB_5H_6 \qquad (5)$$

The composition-pressure evolution is shown in Figure 14. After hydriding, the system reaches condition II, and the hydride is formed with hydrogen safely stored as a solid. At this point, the valve A is closed to isolate the system, and the device is heated up to 90 °C. Since the system is with a closed volume, the pressure is increased. The system reaches equilibrium according to Reaction (5) under the new pressure conditions at point III. Valve B is then opened and hydrogen is released from the system. The IM desorbs hydrogen according to the reverse of Reaction (5) reaching condition IV. At this point, valve B is closed and the system is isolated. The device is cooled to reach point I.

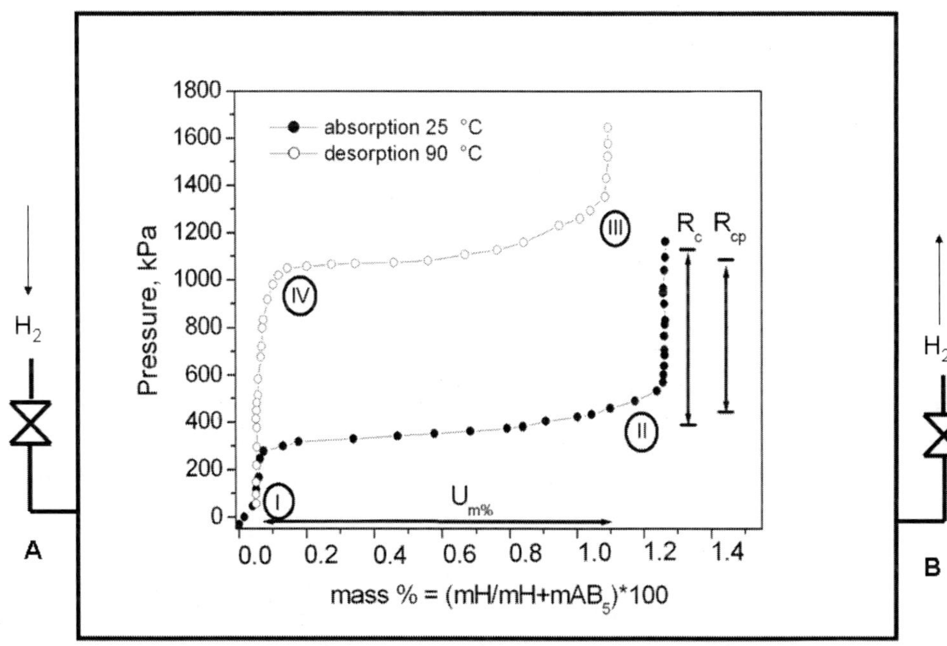

Figure 14. A scheme of TCH process using AB_5-B obtained by integrated synthesis method.

The cycle is completed with a useful hydrogen mass percentage ($U_{m\%}$) of 1.05. This value represents the difference between the absorbed mass percentage of hydrogen at the low temperature and the desorbed mass percentage of hydrogen at high temperature. This cycle also has a standard compression ratio (Rc) of 2.93 and a practical compression ratio (Rcp) of 2.04. The first value represents the ratio of middle plateau pressure at high and low temperature. The second value represents the ratio between the lowest plateau pressures of the high temperature PCI and the highest pressure of the low temperature PCI.

The TCH is an open-cycle application [37, 50, 55, 58]. It means that in every cycle of TCH a new fresh load of hydrogen is supplied. This condition is with the disadvantage of introducing air/oxygen to the system. The selected IM´s resist the contact with air at temperatures higher than 90 °C [50], and samples with similar compositions resist the contact with air at temperatures as high as 200 °C [50].

7. Composites Aimed to Application to GPH

MO_2-Ni composites can be prepared by two successive steps as described in Figure 3. The synthesized process is MA. Under the target conditions, the final stages of milling are reached after 100 h [39-40]. At these MA stages, the unfinished AB_5 product is obtained. At this condition, the IM represents the white puzzle pieces in Figure 1 (c). The AB_5 is decomposed when heated in air. The one evaluated in this chapter leads to the formation of MO_2/Ni and MO_2/NiO composites. Temperatures below 300 °C lead to the formation of MO_2/Ni composites. The difference in the crystallinity of the composite is one of the most important features of this improvement. The Ni crystallinity is twice as large as the one of MO_2 presented in Table 3. In addition, the separation of AB_5 due to the selective oxidation of A-site leads to the formation of an intimate mixture as observed in FEG-SEM image in Figure 11. The formation of the oxide leads to a porous structure [57]. Then, the porous oxide (MO_2)-crystalline metal (Ni) can be synthesized according to the integrated synthesis method in Figure 3. This composite is the link to the black puzzle pieces in Figure 1 (c). The experimental measurements related to the pair AB_5-GPH is currently under way.

Conclusion

The evolution of the pair material-application reveals a successful strategy for the production and the material use in a variety of devices. The IM´s of the AB_5 type in the hydrogen technology application is an example of this model [10-13]. Although the close relation among them has produced a significant growth on the structure of the material and the quality of the applications, an unsolved issue is that the developed production method only serves for some specific purposes. Other pair material-applications might not be produced under such a scheme and their designed machinery may be useless for the other purposes. This chapter focuses on the change of design process, and discusses the production of a first unfinished product that is modified to produce two new different products. Each of them is aimed to be applied to the two applications of hydrogen technology. The characteristics of the production process are limited due to the economic and technological reasons [35]. The AB_5-intermetallics are synthesized by MA [37-38, 50], and the process can be implemented near room temperature and the technique is well known to favor the scale-up plan [47-48]. The improvement for the as-milled material is done according to the same technological and economic reasons [35]. These materials were aimed to be applied in TCH and GPH technologies, and these two applications provide an example for developing the proposed material synthesis method.

ACKNOWLEDGMENTS

The author wishes to thank Consejo Nacional de Investigaciones Científicas y Técnicas of Argentina (Project N PIP 0109), Agencia Nacional de Promoción Científica y Tecnológica of Argentina (Project N PICT 0092), and Universidad Nacional del Comahue of Argentina (Project N B183 for partial financial support.

The author would like to particularly appreciate Dr. María del Carmen Ruiz[1] and Dr. Analía L. Soldati[2] for their help of reviewing this chapter prior to the submission.)

REFERENCES

[1] Shi, S.; Hwang, J.Y. *Int. J. Hydrogen Energy* 2007, 32, 224-228.

[2] Zones, S.I.; *Micropor. Mesopor. Mat.* 2011, 144, 1-8.

[3] Martinez, C.; Corma, A. *Coord. Chem. Rev.* 2011, 255, 1558-1580.

[4] Chan, G.; Mooney, J. *Trends. Biotechnol.* 2008, 26, 382-392.

[5] Consonni, S.; Giugliano, M.; Massarutto, A.; Ragazzi, M.; Saccani, C. *Waste Manage.* 2011, 31, 2057-2065.

[6] Font Vivanco, D.; Puig Ventosa, I.; Gabarell Durany, X. *Waste Manage.* 2012, *32*, 2496-2510.

[7] Villars, P.; Calvert, L.D. In *Pearson´s Handbook of Crystallographic Data for Intermetallic Phases;* Editor, 2 nd Ed.: *Pearson´s Handbook of Crystallographic Data for Intermetallic Phases* ISBN: 0871704161; American Society for Metals; Materials Park, OH, 1991, vol 1, pp 207-219.

[8] Nassau, K.; Cherry, L.V.; Wallace, W.E. *J. Phys. Chem. Solids* 1960, 16, 123-130.

[9] Van Vucht, J.H.N.; Kuijpers, F.A.; Bruning H.C.A.M. *Philips Res. Rept.* 1970, 25, 133-135.

[10] Van Mal, H.H.; Buschow, K.H.J.; Miedema, A.R. *J. Less Comm. Met.* 1974, *35*, 65-76.

[11] Mendelsohn, M.H.; Gruen, D.M. *Mat. Res. Bull.* 1978, 13, 1221-1224.

[12] Mendelsoh, M.H.; Gruen, D.M., Dwight, E. *J. Less Comm. Met.* 1979, 63, 193-207.

[13] Van Mal, H.H.; Buschow, K.H.J., Miedema, A.R. *J. Less Comm. Met.* 1976, 49, 473-475.

[1] Dr. María del Carmen Ruiz (Instituto de Investigaciones en Tecnología Química, INTEQUI, CONICET, INTEQUI, Chacabuco y Pedernera 5700, San Luis, Argentina; Facultad de Química, Bioquímica y Farmacia, Universidad Nacional de San Luis, INTEQUI, Chacabuco y Pedernera 5700, San Luis, Argentina)

[2] Dr. Analía L. Soldati (Centro Atómico Bariloche, CNEA–CONICET. Av. Bustillo 9500, S.C. de Bariloche (R8402AGP), Río Negro, Argentina)

[14] Miedema, A.R.; Buschow, K.H.J.; Van Mal, A.R. J. *Less Comm. Met.* 1976, 49, 463-472.
[15] Lundin, C.E.; Lynch, F.E.; Magee, C.B. J. *Less Comm. Met.* 1977, 56, 19-37.
[16] Gruen, D.M.; Mendelsohn, M.H.; Sheft, I. *Solar Energy* 1978, 21, 153-156.
[17] Suda, S. *J. Less Comm. Met.* 1984, 104, 211-222.
[18] Huston, E.L.; Sandrock, G.D. *J. Less Comm. Met.* 1980, 74, 435-443.
[19] Bernauer, O.; Halene, C. *J. Less Comm. Met.* 1987, 131, 213-224.
[20] Yoshikawa, A.; Matsumoto, T.; Yagisawa, K. *J. Less Comm. Met.* 1982, 88, 73-79.
[21] Valøen, L.O.; Zaluska, A.; Zaluski, L., Tanaka, H., Kuriyama, N.; Ström-Olsen, J.O.; Tunold, R., *J. Alloys Compd.* 2000, 306, 235-244.
[22] Percheron-Guégan, A. ; Lartigue, C. ; Achard, J.C. *J. Less Comm. Met.* 1985, 109, 287-309.
[23] Obregón, S.; Andrade Gamboa, J.J ; Esquivel, M.R. *Procedia Materials Science,* 2012, 1, 156-163.
[24] Powell, J.E. ; In Non Metallic Compounds Elsevier, Gschneidner, Jr., K.A., Eyring, L; 1st Ed., *Handbook of the Physics and Chemistry of Rare Earths 21,* Amsterdam, 1979, vol. 3, pp. 81-109.
[25] Palmer, P.E.; Burkholder, H.R.; Beaudry, B.J., Gschneidner Jr., K.A. *J. Less Comm. Met.* 1982, 87, 135-148.
[26] Sandrock, G.; Bowman, Jr., R.C. *J. Alloys Compd.* 2003, 356-357, 794-799.
[27] Dischinger, J.; Schaller, H-J. *J. Alloys Compd.* 2000, 312, 201-210.
[28] Palumbo, M.; Borzone, G.; Delsante, S.; Parodi, N.: Cacciamani, G.; Ferro, R.; Battezzatti, L.; Baricco, M. *Intermetallics* 2004, 12, 1367-1372.
[29] Sakai, T.; Matsuoka, M.; Iwakura, C. In Handbook of the Physics and Chemistry of the Rare Earths, Gschneidner Jr., K.A.; Eyring, L.; 1st Ed.; *Handbook of the Physics and Chemistry of the Rare Earths,* Elsevier, Amsterdam, 1995, vol. 21, pp. 133-178.
[30] Buschow, K.H.J.; Van Mal, H.H. *J. Less Comm. Met.* 1972, 29, 203-210.
[31] Schlapbach, L.; Anderson, I.; Burger, J. P. *Mat. Sc. Tech.* 1993, 3.A., 340-341.
[32] Srivastava, S.; Srivastava, O.N. *J. Alloys Compd.* 1998, 267, 240-245.
[33] Genshi, W.; Xianglong, W.; Deying, S.; Panwen, S. *Int. J. Hydrogen Energy* 1989, 14, 45-47.
[34] Shahgaldi, S.; Yaakob, Z.; Mat Jalil, N.; Tasirin S.M. *J. Alloys Compd.* 2012, 541, 335-337.
[35] Obregón, S.A.; Andrade Gamboa, J.J.; Esquivel, M.R. *Int. J. Hydrogen Energy* 2012, 37, 14972-14977.
[36] Lenain, C.; Aymard, L.; Salver-Disma, F.; Leriche, J.B.; Chabre, Y.; Tarascon, J.M. *Solid State Ionics,* 1997, 104, 237-248.
[37] Esquivel, M.R.; Rodriguez, M.G., *Energy Materials: Materials Science & Engineering for Energy Systems* 2009, 4, 145-149.
[38] Cerón-Hurtado, N.M.; Esquivel, M.R. *Int. J. Hydrogen Energy* 2010, 495, 541-544.
[39] Esquivel, M.R.; Meyer, G. *Mater. Sci. Forum* 2008, 570, 72-77.
[40] Esquivel, M.R.; Meyer, G. *J. Alloys Compd.* 2007, 446-447, 553-562.
[41] Esquivel, M.R.; Gennari, F.C.; Andrade Gamboa, J.J.; Meyer, G. In *Sohn International Symposium on Advanced Processing of Metals and Materials;* F. Kongoli, 1st, TMS, San Diego CA, Vol. 4, 517-526.
[42] Blanco, V.; Esquivel, M.R. *Adv. Powder Technol.* 2013, 24, 86-92.

[43] Liang, G.; Huot, J.; Schultz, R.; *J. Alloys Compd.* 2011, 320, 133-139.

[44] Jurczyk, M.; Smardz, K.; Rajewski, W.; Smard, L. *Mat. Sc. Eng.* A 2001, 303, 70-76.

[45] Chen, Z.; Su, Y.; Lü, M.; Zhou, D; Huang, P. *Mat. Res. Bull.* 1998, 33, 1449-1455.

[46] Higashiyama, N.; Matsuura, Y.; Nakamura, H.; Kimoto, M.; Nogami, M.; Yonezu, I.; Nishio, K. *J. Alloys Compd.* 1997, 253-254, 648-651.

[47] Suryanarayana, C. *Prog. Mat. Sc.* 2001, 46, 1-184.

[48] Lu, L; Lai, M.O. *Mechanical alloying;* Kluver Academic Publishers: Boston, MA, 1998; pp. 140-145.

[49] Koch, C.C.; Whittenberger, J.D. *Intermetallics* 1996, 4, 339-355.

[50] Esquivel, M.R.; Zelaya, E.; Andrade Gamboa, J.J.; Obregón, S. *Procedia Materials Science* 2012, 1, 172-179.

[51] Young, R.A; Sakthivel, A.;Moss, T.S.; Paiva-Santos, C.O. *J. Appl. Cryst.* 1995, 28, 366-367.

[52] Langford, J.I:, Delhez, R.; De Keijser, T.H.H.; Mittemeijer, E.J., *Aust. J. Phys.* 1998, 41, 173-187.

[53] Talagañis, B.A.; Esquivel, M.R.; Meyer, G. *Int. J. Hydrogen Energy* 2009, 34, 2062-2068.

[54] Talagañis, B.A.; Esquivel, M.R.; Meyer, G. *J. Alloys Compd.* 2010, 495, 541-544.

[55] Talagañis, B.A.; Esquivel, M.R.; Meyer, G. *Int. J. Hydrogen Energy* 2011, 36, 11961-11968.

[56] Syrukri, T.; Ban, Y:; Ohya, Y.; Takahashi, Y. *Mat. Chem. Phys. 2003*, 78, 645-649.

[57] Esquivel, M.R.; Zelaya, E. *Adv. Appl. Ceram.* 2011, 110, 219-224.

[58] Cerón-Hurtado, N.M.; Esquivel, M.R. *Int. J. Hydrogen Energy* 2012, 37, 10376-10379.

In: X-Ray Diffraction: Structure, Principles and Applications ISBN: 978-1-62808-591-4
Editor: Kaimin Shih

Chapter 9

DEPTH-RESOLVED MEASUREMENT OF THE 3D RESIDUAL STRESS STATE IN SURFACE ENGINEERED ALUMINIUM BY SYNCHROTRON DIFFRACTION

***D. T. Asquith*[1], *D. J. Hughes*[2], *D. G. Hattingh*[3], *M. N. James*,[4,3], *J. R. Yates*[5] and *A. Yerokhin**[6]**

[1]Sheffield-Hallam University, UK
[2]ILL, Grenoble, France
[3]Nelson Mandela Metropolitan University, SA
[4]University of Plymouth, UK
[5]University of Manchester, UK
[6]University of Sheffield, UK

ABSTRACT

Evaluation of the complex three dimensional residual stress states in manufactured components is an important part of assessing the durability of safety critical components. Many experimental techniques have been used in attempts to quantify residual stresses. Few are capable of resolving the full stress tensor in engineering components. In this work a method for measuring the three dimensional residual stress state has been proposed and demonstrated experimentally. Strains have been measured in machined and shot-peened aerospace aluminium alloy specimens by X-ray diffraction in the ESRF Synchrotron Facility. Data from four orientations have been collected so that both the magnitude and direction of the principal stresses can be determined. The problems of gauge volume size and specimen orientation relative to the beam were addressed and resolved. Effects of the specimen location relative to the beam, translation hardware repeatability, beam slits and spectrum acquisition time are discussed. The differences observed between a 2D and 3D evaluation of the same data suggest that measurement techniques using only in-plane data can be misleading so care should be taken when using such data to guide simulations.

* Corresponding author: Tel: +44-11422-25510; Fax: +44-11422-25943; E-mail: A.Yerokhin@sheffield.ac.uk

1. Introduction

Almost all manufacturing processes, from the casting of a billet to the final surface finishing of a component, result in residual stresses being locked into the product. These residual stresses are likely to vary in magnitude and orientation throughout the structure. They will interact with intrinsic defects in the material, with the geometric features of the component and with the stresses generated by the service loads. Beneficial compressive residual stresses, such as those introduced by shot-peening will enhance the integrity of the structure, but tensile stresses can contribute to premature failure by fracture or fatigue.

The measurement of residual stresses has been the subject of extensive research over a period of many years. Several reviews and reports on the sources and measurement of residual stresses have been published, including those by Withers and Bhadesia [1,2] and the NPL [3]. Reimers, et al. [4], Pyzalla [5] and James, et al. [6] have described the use and advantages of hard X-ray synchrotron radiation in measuring the residual stress state of engineering features, such as friction stir welds. Frequently, when collecting diffraction data one assumes prior knowledge of the principal stress directions [7, 8]. Often this is taken to be the parallel and perpendicular to the rolling direction of a plate. This may well be a reasonable assumption in some circumstances, but it cannot be a universal truth. The difficulty in testing this assumption is in acquiring data from a sufficient number of directions to resolve both the magnitude and direction of the strains.

It is possible with the use of high power x-ray (synchrotron) radiation to penetrate much greater distances in many structural materials than with conventional lab sources; approximately 10mm in aluminium. This enables the measurement of strains at a number of points throughout the surface and bulk of a specimen without any destructive procedures. Consequently this allows the analysis of a number of directions of strain to be made and retain the sample for further work. By careful orientation of the illuminated gauge volume employed in the strain measurements, it is possible to achieve similar spatial resolution for all directions of strain measured, thus providing a set of data which can be reasonably combined into the three principal strains. This work describes one approach for measuring sufficient directions of strain to calculate a three dimensional stress state. As an example data from an experiment involving shot-peened aluminium using measurements made at the ESRF, Grenoble, France is presented. Furthermore it shows that the 3D profile obtained provides significantly more data with some subtle differences to that typically obtained through a 2D profile, in plane evaluation. Although demonstrated with cold-worked aluminium the technique is equally applicable to any metallic material within the limitations of the x-ray source attenuation.

Experimental Technique

2.1. θ-2θ Strain Measurements

While several approaches to evaluating strain using diffraction exist, this method utilises the phenomenon of Bragg peak shift with applied elastic strain, which is evaluated in a θ-2θ

configuration by comparison with a strain free lattice parameter. This technique is briefly described here. When the diffraction geometry satisfies the Braggs' condition whereby:

$$n\lambda = 2d \sin \theta$$

Constructive interference occurs producing an intense peak. The position of this peak in terms of the diffraction angle, θ, varies with a change in the crystallographic lattice spacing, d, and the diffracting wavelength, λ. Evaluating a particular hkl peak with a fixed wavelength of x-rays provides information of the particular lattice spacing, d. If an elastic strain is applied to the material, its d value changes accordingly. Two peaks for a strained and strain free material are shown in Figure 1.

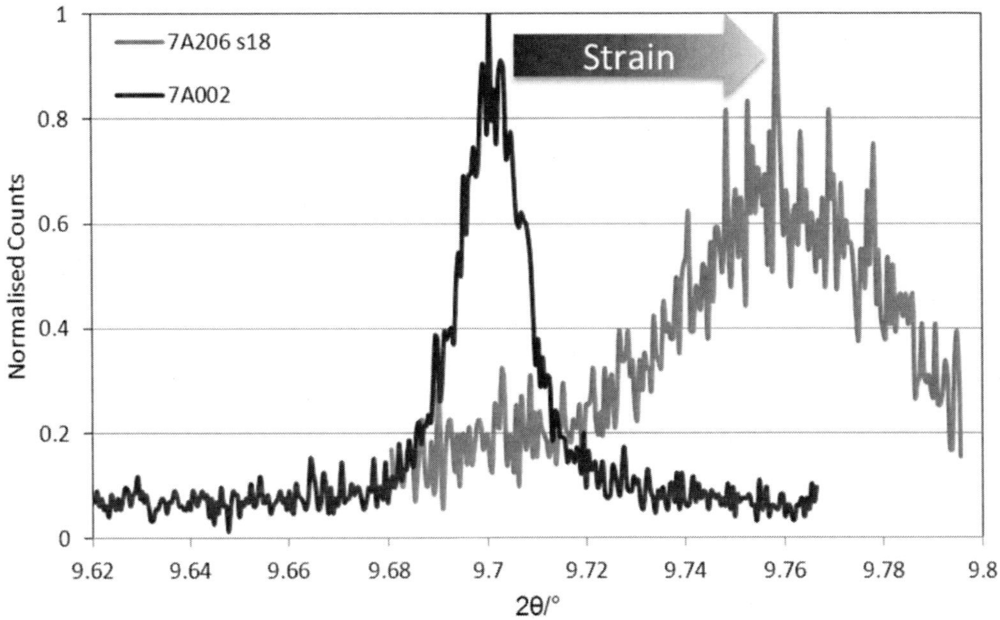

Figure 1. Peak shift due to elastic strain in an aluminium alloy.

Using a source of highly collimated x-rays such as those produced in a synchrotron allows the definition of a small diffracted gauge volume, as the detected beam is effectively coherent due to the acceptance crystals we can accurately define both the location and direction of measured strain as a function of the experimental parameters. Due to the polycrystalline nature of metallic materials it is possible to make measurements on the same *hkl* plane in several physical orientations of the specimen which is necessary to calculate a full strain tensor.

Due to the dependence of the crystallographic spacing on materials properties, such as those modified by alloying and processing, it is necessary to make an evaluation of a representative unstrained lattice parameter. One effective approach for doing this is described by Hughes, et al. [9] and involves relaxing the strain without extensively modifying the microstructure by creating small comb teeth in a representative specimen and averaging the *d* measurements across several teeth.

2.2. Specimen Orientations for Individual Strain Measurements

Firstly, it must be noted that residual stresses are evaluated from measurements of residual strain. In general, measurements in at least five different directions would be required, if the full three dimensional stress state were to be evaluated without *a priori* knowledge, or assumption, of the principal directions. This can be reduced to four individual directions provided three of these can be considered to lie in a plane defined by two of the principal directions. This allows us to assume that the third principal direction is perpendicular to the defined plane and can be measured individually as a fourth direction. In the experiment presented here the process of shot-peening the specimen surface creates conditions which satisfy the above assumption thereby simplifying the experimental work.

The four required strain directions are identified here as ε_a, ε_b, ε_c and ε_d, they are demonstrated in Figure 2 relative to the specimen shown geometry. Having obtained a suitable measure of strains for a given point in the material we must then evaluate the principal directions of strain, this can be done simply by applying Mohr's circle.

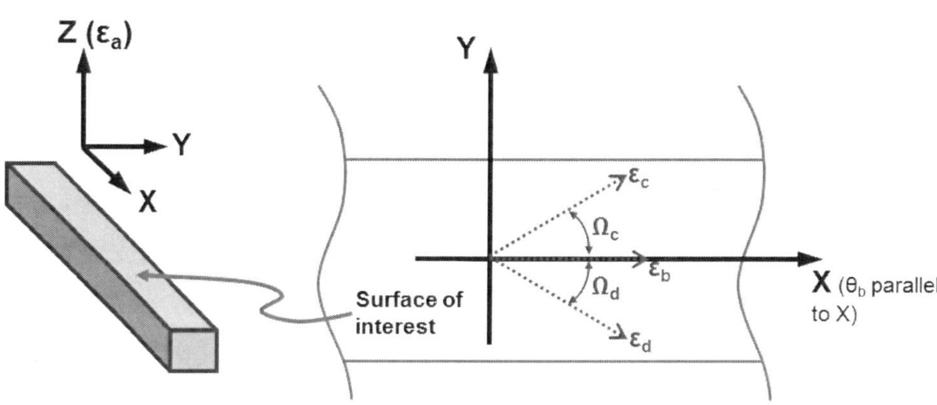

Figure 2. Measured strain directions relative to specimen, the remaining direction of strain (ε_a) is coordinate with Z in the left-hand diagram, and it corresponds to one of the principal directions.

The experimental challenge is to orientate the specimen with respect to the beam in such a way that sufficient sets of data are obtained. Two characteristics of the beam make this difficult. The first is the attenuation of the beam as it passes though the aluminium. This becomes critical when exploring the sub-surface layers, and also when the there is a shallow angle of incidence between the beam and the specimen. As the diffracted portion of the beam approaches the axis of the longer specimen dimension, the length of attenuating material the beam must pass through increases thus reducing flux at the detector, and increasing the necessary count time.

Having defined a plane in the specimen surface in which three of the required directions lie also allows one setup to be used to measure these. It is necessary to rotate the specimens about the experimental θ axis, normal to the plane, but this facility is provided for as a basic diffractometer requirement. This setup utilises gauge volume $G_{b,c,d}$ and is, in this case, a transmission arrangement.

The first measured direction, ε_b, is measured nominally in the rolling direction (X) and requires the specimen to be orientated at the diffraction angle θ to the beam. The measured directions ε_b and ε_c are obtained by rotating the specimen toward and away from the beam relative to ε_b. In this case an angle of 30° was used as a compromise between attenuation and error in resolved strain. Figure 3 shows this relative to the specimen geometry and beam with an image of the actual setup for clarity. The fourth direction, (Z) perpendicular to this defined plane, requires the specimens to be rotated about the axis of the beam and so necessitates a different experimental setup using gauge volume G_a in reflection. These measurements are inherently slower since the gauge volume is slightly smaller and as such longer count times are required. The setup used for obtaining ε_a is less easily represented with a 2D schematic, Figure 4 shows a 3D visualisation with a photograph for comparison with Figure 3.

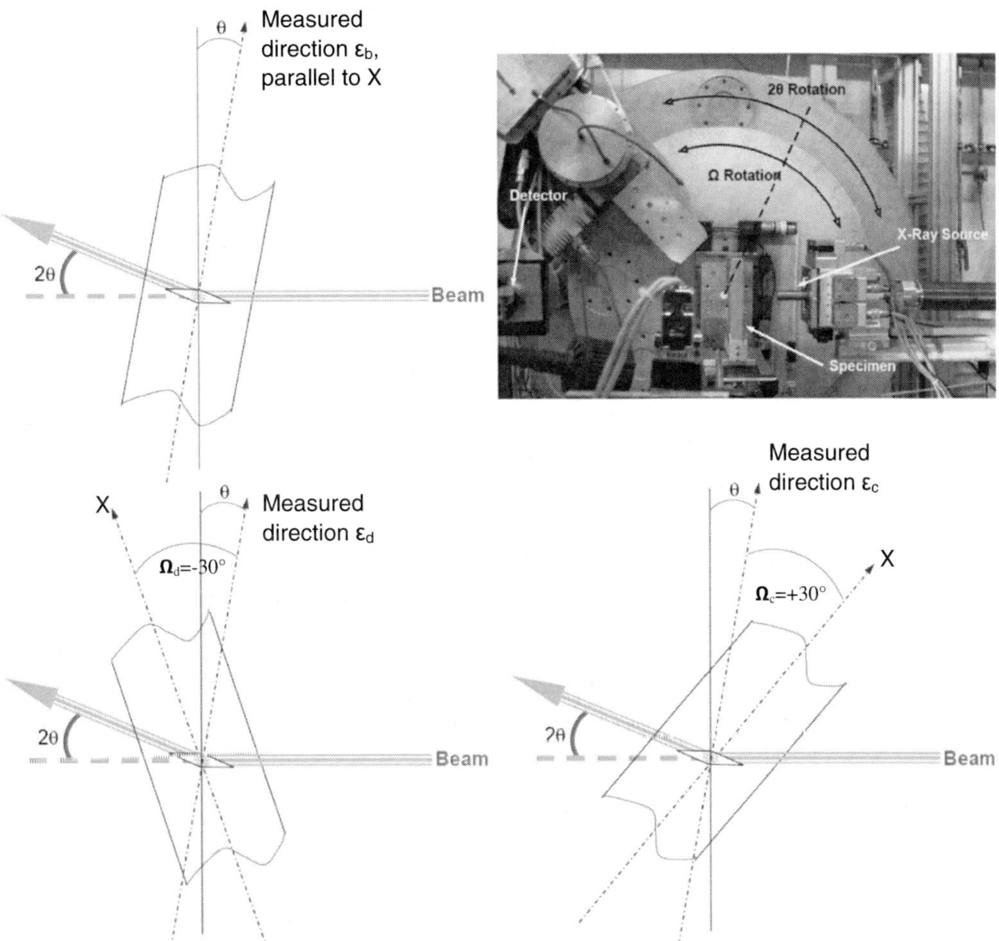

Figure 3. Specimen orientations for in-plane strain measurements (transmission) and actual setup on experimental beamline ID31.

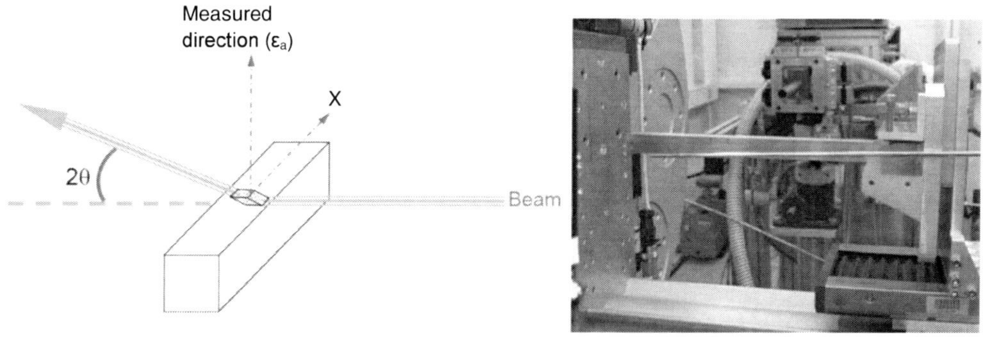

Figure 4. Specimen orientation for out-of-plane strain measurement (reflection) and setup on experimental beamline. Note that the diagram observes from a different angle to the photograph, in which the arrow denotes the beam direction.

2.3. Gauge Volumes for in-Plane and Out-of-Plane Measurements

The second important consideration is the 'gauge volume', that is, the size and shape of the illuminated volume of material causing the diffraction of the beam. The spatial resolution of the strain gradient with respect to depth depends, amongst other factors, on the size and shape of the gauge volume. Changing the gauge volume according to the desired resolution and orientation is the key to be able to collect sufficient data to resolve the three dimensionality of the stress state. By choosing gauge volume dimensions that are small in the stepping direction spatial resolution can be maximised. To provide a good average of the strain state present it is necessary to maximise other dimensions, thereby including a sufficient number of individual grains in the diffracting orientation. This also reduces the count time required for each data point. Since strains are to be measured in different directions and using both transmission and reflection, it is unreasonable to assume a single shape of gauge volume will suffice for all measurements. In this work two different arrangements were used to optimise the spatial resolution for different measurement directions. By designing the gauge volumes to provide uniform resolution in the step direction for each direction of strain, it is possible to correlate individual data points allowing the combination of individual measurements to evaluate principal directions and then subsequently the stresses.

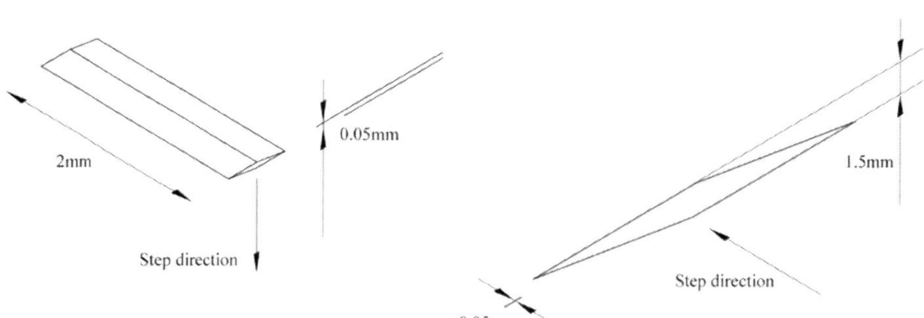

Figure 5. Gauge volume shapes (not to same scale), volume G_a is on the left, $G_{b,c,d}$ the right.

Figure 5 shows the shapes of both gauge volumes used in the following case study. In practice the gauge volumes are defined by limiting the horizontal and vertical dimension of the beam, it is absolutely critical to the success of such experiments that this is both accurate and repeatable with well-engineered beam slits. Having established a limited gauge volume from which to diffract, the specimen is then orientated relative to this such that the desired direction of strain can be evaluated; it can then be stepped relative to the gauge volume to obtain strain variation with respect to the step direction. In all cases the step direction is parallel to the shortest dimension, thus providing highest resolution.

3. Case Study

For this chapter a case study is taken from research presented by Asquith [10], and specimens were produced from the structural aluminium alloy 7050-T7451. Plates of 20mm thickness were sectioned and machined into specimens measuring 20×20×50mm as shown in Figure 6 Specimens were shot-peened at the University of Sheffield using a Tealgate Precifeed system, the peened surface is identified in Figure 5 as a hatched surface. Parameters were taken from prior work on the optimisation of the shot-peening process undertaken at the University of Sheffield [11].

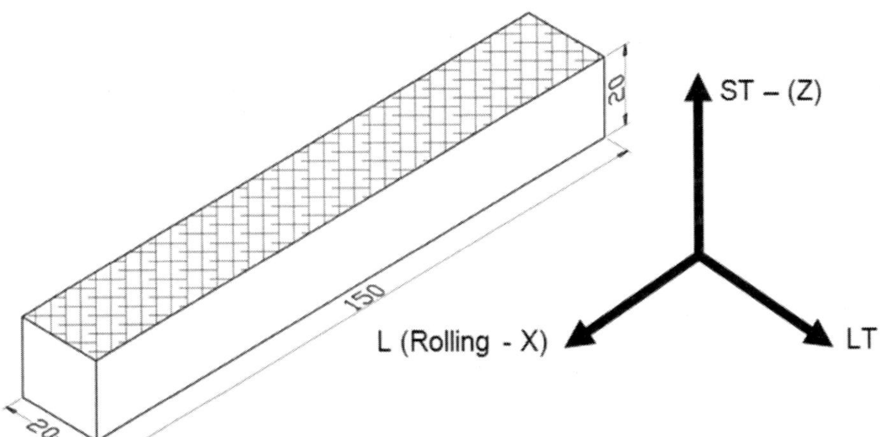

Figure 6. Tested aluminium alloy plate with dimensions in mm (surface of interest hatched).

Residual stresses were evaluated by high resolution synchrotron diffraction at the ESRF, Grenoble (beam-line ID31). The strains were measured using a highly collimated monochromatic hard X-ray beam (E=60keV, λ=0.206nm). The measurements were taken starting from the surface with an increment size of 20μm to a depth of 0.5mm and then in 50μm increments to 1mm. For each depth increment, the strain was measured in four different directions (three in the plane of, and one normal to, the specimen surface) allowing principal strains to be resolved using Mohr's circle and subsequently evaluated the three principal residual stresses using Hooke's law, as discussed in section 2. The two principal directions in the surface plane were assumed to be orthogonal. The changes in residual elastic strain were evaluated using the Braggs' equation:

$$\varepsilon = \frac{\theta - \theta_0}{\tan\theta_0} = \frac{d - d_0}{d_0}$$

where θ_0 and d_0 = 0.1213346±0.000264nm correspond to the unstrained Al (311) peak centre and lattice spacing respectively, obtained using a sample that had been machined into a comb profile with ≈1mm^2 teeth in order to relieve the residual stress.

4. RESULTS

4.1. Measured Strains

Figure 7 below shows strain measured in four directions with respect to depth in the specimen surface. The spatial resolution required throughout the depth is not uniform, and more detail is needed in the near surface where the strain profile changes more rapidly. Toward the centre of the specimen the profile is less altered by the shot-peening. Therefore, the gradient is lower and less frequent measurements are required to characterise the strain field. As may be expected, the in-plane measurements show similar behaviour and are typical of the residual strain profile commonly presented for shot-peened aluminium. Of particular interest though is the out of plane strain. This is considered in some cases to be zero, particularly if a state of plane stress is assumed, but can clearly be seen here to vary significantly with depth and exhibits substantial tension in the near surface.

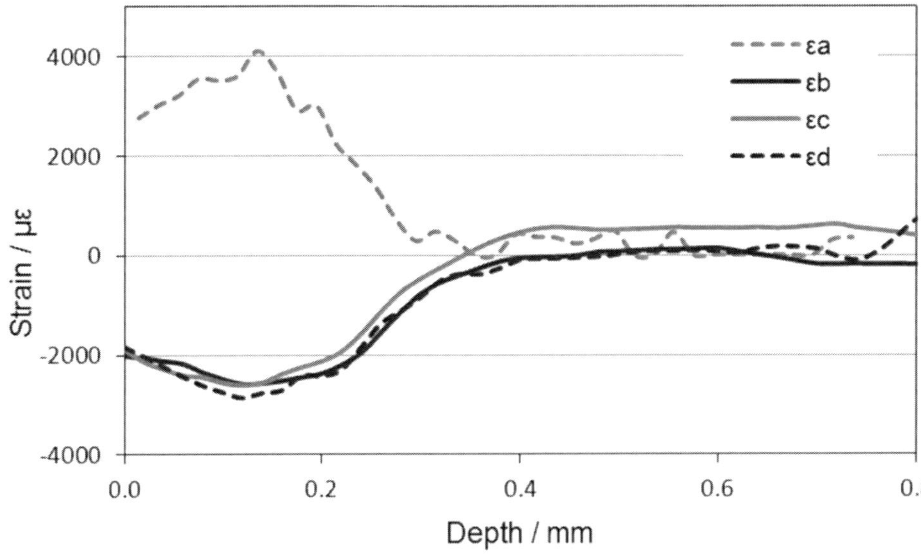

Figure 7. Strains measured in shot-peened AA7050 in different directions with increasing depth in the specimen.

4.1. Evaluating Principal Strains

Assuming that two of the three principal strains lie in the plane of the surface, we can define the normal direction as a third principal. It then remains to translate the three measured in-plane directions to the principal directions. This process requires two stages; firstly, the three measured in-plane strains must be translated to two arbitrary but orthogonal axes. Secondly the principal strains and their direction relative to the specimen geometry can be obtained from the orthogonal strains by applying Mohr's circle. Translating three strains to two orthogonal axes can be done by applying the general translational formulae, *Equations 1-3*, and commonly these are applied in strain gauge measurements.

$$\gamma_{xy} = \frac{\varepsilon_d - \varepsilon_c}{\cos\theta_d \sin\theta_d - \cos\theta_c \sin\theta_c}$$

Equation 1

$$\varepsilon_y = \varepsilon_c - \varepsilon_b \cos^2\theta_c - \frac{\gamma_{xy} \cos\theta_c \sin\theta_c}{\sin^2\theta_c}$$

Equation 2

$$\varepsilon_x = \varepsilon_b - \varepsilon_y \sin^2\theta_b - \gamma_{xy} \sin\theta_b \cos\theta_b$$

Equation 3

Using the values of ε_x, ε_y and γ_{xy}, Mohr's circle can be constructed and principal values of ε_1, ε_2 and θ evaluated as shown in Figure 8.

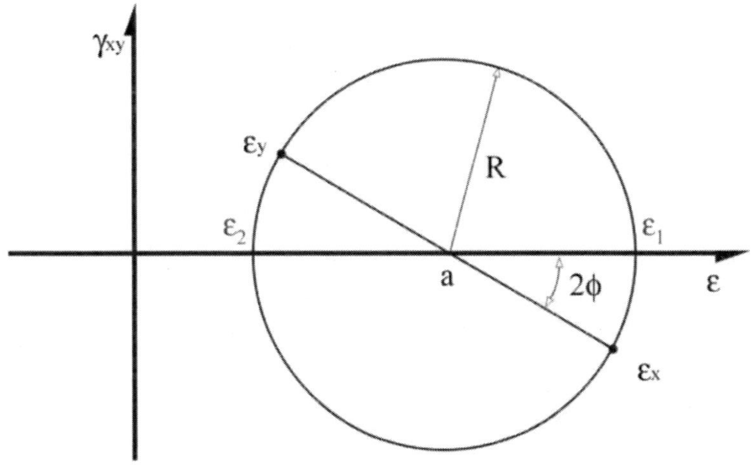

Figure 8. Mohr's circle of strain.

The centre point, a, and radius, R, of Mohr's circle can be evaluated as:

$$a = \frac{\varepsilon_x + \varepsilon_y}{2}$$

$$R = \sqrt{\left(\frac{\varepsilon_x - \varepsilon_y}{2}\right)^2 + \left(\frac{\gamma_{xy}}{2}\right)^2}$$

Having evaluated the geometric properties of Mohr's circle, it is relatively simple to determine the principal stresses, ε_1 and ε_2, and the angle between ε_1 and ε_x, ϕ:

$$\varepsilon_1 = a + R$$

$$\varepsilon_2 = a - R$$

$$\phi = \frac{ArcCos\left(\varepsilon_x - a \big/ R\right)}{2}$$

By measuring strains at the same point as $\varepsilon_{b,c,d}$ but perpendicular to the plane they define, i.e. the third principal direction we can complete a full analysis of the strains in three dimensions, these are shown in Figure 9.

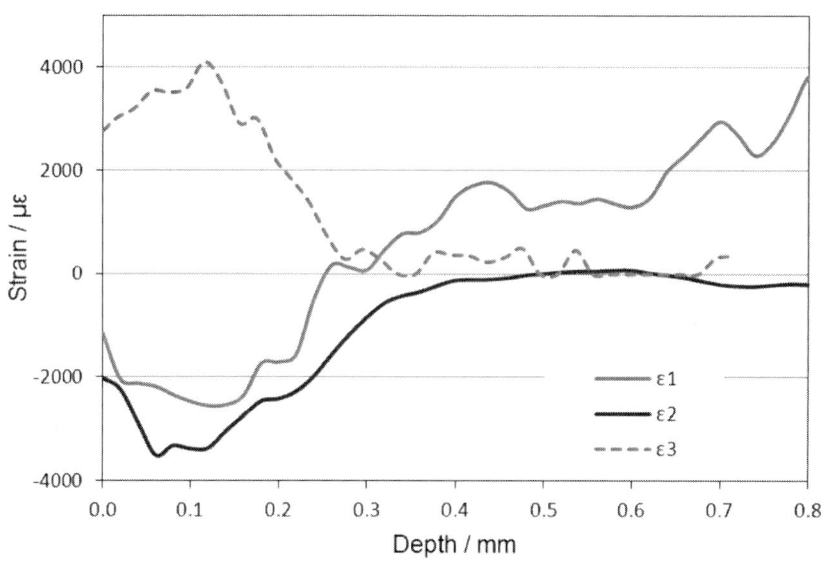

Figure 9. Principal strains evaluated from the measured data in Figure 8.

5. Triaxial Stress Distribution

Having obtained the strain in three principal directions it is a straightforward process to evaluate the residual stresses in three dimensions. This is achieved by applying three independent solutions of the general form of Hooke's law, *Equations 4-6*.

$$\sigma_{xx} = \frac{E}{(1+\nu)(1-2\nu)}((1-\nu)\varepsilon_x + \nu(\varepsilon_y + \varepsilon_z))$$

Equation 4

$$\sigma_{yy} = \frac{E}{(1+\nu)(1-2\nu)}((1-\nu)\varepsilon_y + \nu(\varepsilon_z + \varepsilon_x))$$

Equation 5

$$\sigma_{zz} = \frac{E}{(1+\nu)(1-2\nu)}((1-\nu)\varepsilon_z + \nu(\varepsilon_x + \varepsilon_y))$$

Equation 6

where E = 73 GPa_is the modulus of elasticity and v = 0.33 is Poisson's ratio.

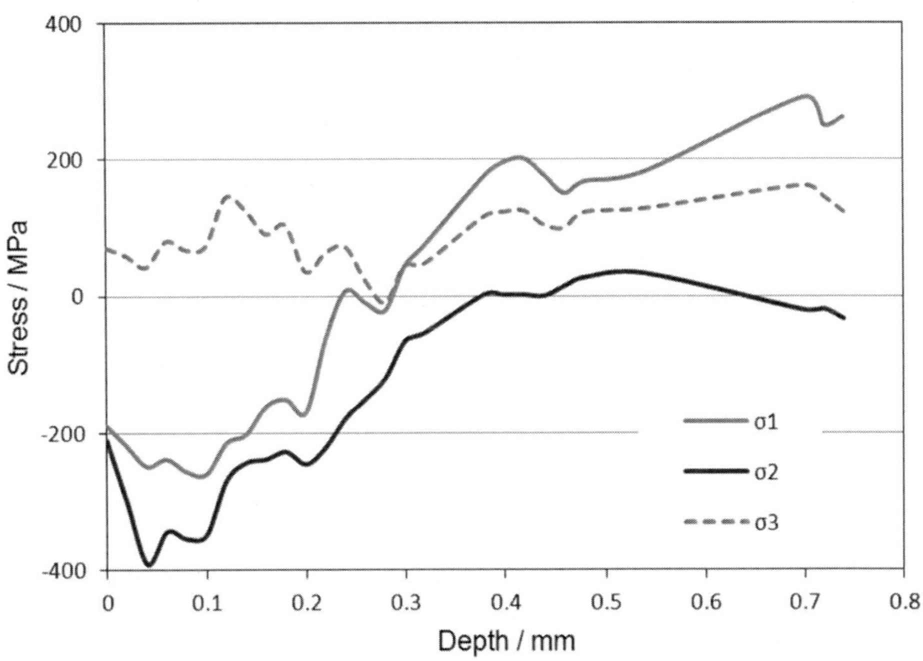

Figure 10. 3D residual stress state evaluated from strains presented in Figure 9.

The data obtained from this type of analysis will be useful in the context of predictive modelling and numerical simulation of materials. The three dimensional residual stress state has been obtained for the strains presented in Figure 9 and is shown in Figure 10.

5.1. Comparison with 2D Stress Distribution

Figure 11 shows the 2D stresses produced from the same strain data shown in Figure 7 but assuming a state of plane stress, i.e. omitting the out of plane strains and thereby only considering strains from the planar surface of the specimen. The peak compressive stresses occur at ~350MPa at a depth of 100μm. The 3D stress plot shows a peak value at ~400MPa, significantly higher than that shown in 2D. Differences in stress levels can be seen at all depths by comparing the two figures, particularly near the bulk in the 3D plot where σ_1 and σ_2 differ by ~400MPa, whereas in the 2D plot the stresses are only 80MPa apart. Differences on this scale can have a major bearing on predictive models, and whether beneficial or detrimental cannot be overlooked.

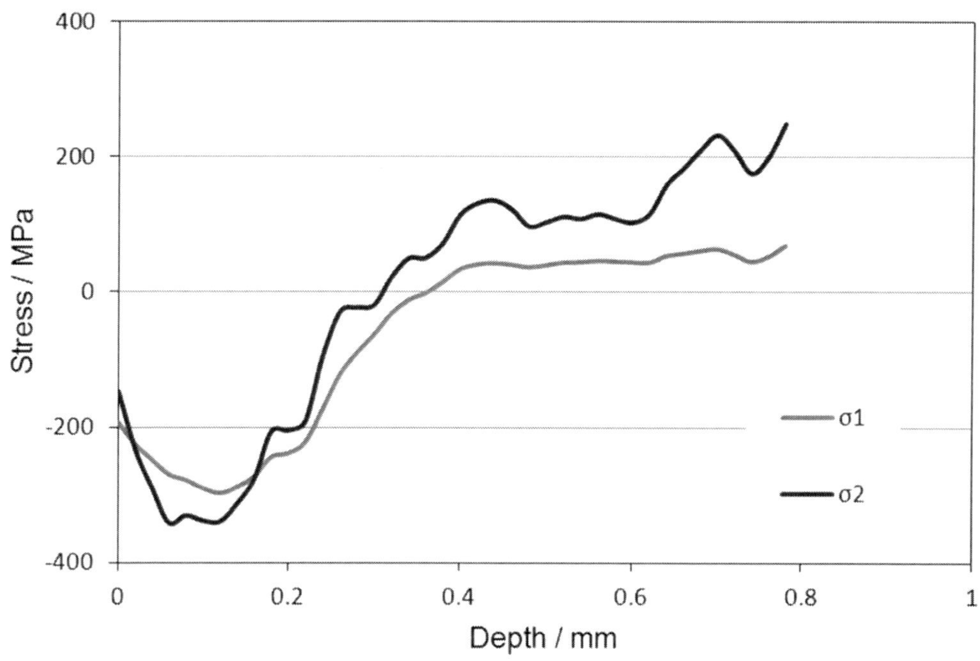

Figure 11. 2D (in-plane) stress distribution produced from the in-plane measured strains.

The curves for 2D principal stresses cross over at 160μm, and this corresponds to an angular shift in the dominant strain direction. By contrast, in Figure 10 the first two principal stress curves (σ_1 and σ_2) do not cross at all suggesting that omitting the third principal strain leads to inaccuracies in angular data, and therefore provides misleading data for analysis and modelling, as subtle changes in the orientations of principal directions can affect the output of numerical simulations. The desire to increase validity of and, therefore, confidence in simulations requires that experimental data is of the highest possible detail.

The third principal stress (σ_3) in Figure 10 is often assumed to be negligible in magnitude thereby justifying its omission. It can clearly be seen from this plot that a stress profile exists in this direction and furthermore this stress is tensile. It is widely accepted that tensile stresses are detrimental with regard to the fatigue life. Although loading in the direction of this stress is not typical in most situations, it is probable that occasions will occur where this is not desirable.

5.2. Error Analysis

There are several sources of experimental errors in synchrotron diffraction, and some of which can be quantified and incorporated in data presentation. Firstly, it is necessary to consider the influence of user error; problems with mechanical alignment and repeatability are best overcome by rigorous experimental technique and careful use of equipment. Secondly, mechanical errors in the sample translation hardware can result in errors in the diffracted depth, this can be quantified based on manufacturer's specifications for the hardware and incorporated in the depth evaluation.

A more quantifiable amount of uncertainty arises during the fitting process is associated with the confidence in the fitted curve to the raw data obtained. The software used to evaluated spectra in this experiment provides a figure for uncertainty in peak position, as do many peak fitting packages. This can be related directly to an uncertainty in strain and subsequently stress using the relationship between θ and ε by substituting the value for strained peak (θ) centre for its error in peak centre.

$$\epsilon = \frac{\theta - \theta_0}{\tan \theta_0}$$

A further approach to error analysis has been approximated by Chen [12] based on the peak height and is expressed as:

$$U_\varepsilon = U_\theta cot\theta = \frac{U_C cot\theta}{2} \approx \frac{W cot\theta}{3\sqrt[3]{C_T}}$$

where:

U_θ= Error in diffraction angle θ (radians)
U_C= Error in peak centre (radians)
W = Peak width at FWHM
C_T= Total number of counts in peak

Errors calculated in such experiments are generally small, of the order of a few percent, and as such difficult to observe on plots unless exaggerated. The most significant but difficult to quantify error is that associated with the evaluation of an unstrained lattice parameter. It is difficult to be absolute with such a figure, since it varies with the local chemical composition and is therefore heavily dependent on the particular alloy used and also its process history. As such the composition and therefore the crystal lattice vary with position in the material and can provide a more significant source of uncertainty.

CONCLUSION

A method for measuring the three-dimensional residual stress state has been proposed and demonstrated experimentally. The problems of gauge volume size and specimen orientation relative to the beam were addressed and resolved. Some key considerations arising from this work are that:

- Location of the specimen relative to the beam is critical
- Repeatability in the translation hardware is also important
- Flexibility in the beam slits is important to define different gauge volumes
- Time required to obtain full-field data is large so require a carefully designed experiment.

The differences observed between a 2D and 3D evaluation of the same data suggest that measurement techniques using only in-plane data can be misleading, so care should be taken when using such data to guide simulations.

ACKNOWLEDGMENTS

Financial support provided for AY by the UK Engineering and Physical Sciences Research Council (EPSRC) is acknowledged with thanks.

REFERENCES

[1] Withers, P. J.; Bhadeshia, H. *Materials Science and Technology.* 2001 17 355–365.
[2] Withers, P. J.; Bhadeshia H. *Materials Science And Technology.* 2001 17 366–375.
[3] Lord, J. D.; Grant, P. V.; Fry, A. T.; Kandil, F. A. *Materials Science Forum.* 2002 404-4 567–572.
[4] Reimers, W.; Pyzalla, A.; Broda, M.; Brusch, G.; Dantz, D.; Schmackers, T.; Liss, K. D.'; Tschentscher, T. *J Mater Sci Lett.* 1999 18 581–583.
[5] Pyzalla, A. *J Nondestruct Eval.* 2000 19 21–31.
[6] James, M. N.; Hattingh, D. G.; Hughes, D. J.; Wei, L. W.; Patterson, E. A.; Da Fonseca, J. Q. *Fatigue Fract Eng Mater Struct.* 2004 27 609–622.
[7] Webster, P. J.; Hughes, D. J.; Mills, G.; Vaughan, G. B. M. *Materials Science Forum.* 2002 404-4 767–772.
[8] James, M. N.; Hughes, D. J.; Chen, Z.; Lombard, H.; Hattingh, D. G.; Asquith, D.; Yates, J. R.; Webster, P. *J. Eng Fail Anal* 2007 14 384–395.

[9] Hughes D J, James M N, Hattingh D G and Webster P J 2003 The Use of Combs for Evaluation of Strain-free References for Residual Strain Measurements by Neutron and Synchrotron X-ray Diffraction *Journal of Neutron Research* 11 289–293.

[10] Asquith, D. T. *Measurement of 3D residual stresses in shot-peened aluminium alloys;* MPhil Thesis. University of Sheffield: Sheffield, UK 2004.

[11] Rodopoulos, C. A.; Curtis, S. A.; de los Rios, E. R.; Solis Romero, *J. Int J Fatigue.* 2004 26 849–856.

[12] Chen, Z. *Determination of residual stresses using synchrotron x-ray techniques;* MSc Thesis. Salford University: Salford, UK 2001.

In: X-Ray Diffraction: Structure, Principles and Applications ISBN: 978-1-62808-591-4
Editor: Kaimin Shih

Chapter 10

IN SITU NANOFOCUSED X-RAY DIFFRACTION COMBINED WITH SCANNING PROBE MICROSCOPY

***T. W. Cornelius*[1*], *F. Mastropietro*[1], *O. Thomas*[1] *and T. U. Schülli*[2]**

[1]Aix-Marseille Université, CNRS, IM2NP UMR,
Faculté des Sciences et Techniques, Campus de Saint-Jérôme, Marseille Cedex, France
[2]European Synchrotron Radiation Facility (ESRF), Grenoble Cedex, France

ABSTRACT

In the recent past, low-dimensional materials have attracted enormous attention because of the spatial confinement which modifies properties, such as the electronic band structure, or the vibrational band structure as well as the mechanical properties. In addition, strain is a very important aspect concerning the physical properties of nanostructures.

While for transmission electron microscopy studies electron transparent specimens are needed, X-ray techniques are non-invasive and allowing for high-resolution strain analysis. Within the last decade, great advances were made in the development of various focusing optics – Kirkpatrick-Baez mirrors (reflective), compound refractive lenses (refractive), and Fresnel zone plates (diffractive) – for synchrotron radiation. Nowadays, synchrotron X-ray beams are routinely focused down to the range of 100 nm, facilitating the investigation of single nanostructures such as nanowires and nano-islands.

Recently, scanning probe microscopes (SPMs) have been developed for *in situ* studies on synchrotron beamlines. Scanning probe microscopes provide information on surface morphology and they allow for a local characterization of a wide number of physical properties with a resolution going down to the atomic scale. Moreover, specific SPMs, such as atomic force microscopes (AFMs), allow for manipulating and modifying the matter at the nanoscale. The *in situ* combination of SPM and X-ray techniques opens the path to novel imaging capabilities, such as photocurrent imaging. When X-rays impinge on the sample photoelectrons are ejected from the surface. This photocurrent can be detected by means of the SPM tip. The X-ray absorption and, hence, the yield of excited photoelectrons is increased at the electronic edges of the various elements. Thus,

* Corresponding author: Tel :+33-491288013; Fax: +33-491282793 Email adresse : thomas.cornelius@im2np.fr.

photocurrent images taken at different element-specific electronic edges give access to the chemical composition of the sample with a resolution in the range of few 10 nm. Moreover, X-ray diffraction studies with nanofocused X-ray beams combined with the *in situ* manipulation of single nanostructures using a contact atomic force microscope allows the study of the elastic and the plastic behavior of individual nano-objects. Such an *in situ* AFM was developed at the European Synchrotron Radiation Facility (ESRF) in Grenoble (France).

INTRODUCTION

Basics of X-ray Diffraction

X-rays cover a spectrum of electromagnetic radiation with a wavelength ranging from 0.01 to 10 nm corresponding to energies in the range of 100 eV to 100 keV. Typically, X-rays with energies of up to 4 keV are regarded as "soft" X-rays and from about 4 to 100 keV as "hard" X-rays due to their penetrating abilities. X-rays were first discovered by Wilhelm Conrad Roentgen in the year 1895 during gas discharges. In the year 1901 Roentgen received the very first Nobel Prize for physics for his discovery. Later on, in the year 1912 Max von Laue performed, together with Walter Friedrich and Paul Kipping, the first X-ray diffraction studies on crystals and received the Nobel Prize two years later in 1914. The diffraction patterns are an example of X-ray wave interferences and they are a direct evidence for the periodic atomic structure of crystals postulated for several centuries. After von Laue's pioneering work, the English physicist Sir W.H. Bragg and his son Sir W.L. Bragg developed a relationship (Bragg's law) which connects the observed scattering with reflections from evenly spaced planes of distance a within a crystal (see Figure 1) [1]. A diffraction signal is recorded when the diffracted X-rays interfere constructively, i.e. the difference in path length $2\Delta s$ of the diffracted X-rays is an integer multiple of their wavelength λ.

$$2\Delta s = n\lambda = 2a\sin\theta \tag{1}$$

The whole set of lattice planes are measured by rotating the sample by an angle θ and simultaneously rotating the detector by twice this angle (for symmetric reflections). This approach is schematically displayed in Figure 2. The incident wave possesses the wave vector $\vec{k}_{in} = 2\pi/\lambda$. For an elastic scattering process, the values of the wave vectors moduli for the diffracted X-rays and for the incident X-rays are the same: $|\vec{k}_{in}| = |\vec{k}_{out}|$.

The construction of the Ewald sphere (see Figure 3) allows the determination which lattice planes – represented by the reciprocal lattice points – result in a diffraction signal for a given wavelength λ of incident radiation [2]. The radius of the Ewald sphere is given by the modulus of the incident wave vector $|\vec{k}_{in}|$. The difference between the wave vectors of the incident and the diffracted waves is defined as $\vec{q} = \vec{k}_{out} - \vec{k}_{in}$. Since both the incident and the diffracted wave vectors have the same length $|\vec{k}_{in}| = |\vec{k}_{out}|$, the reciprocal lattice points which may result in a diffraction signal must lie on the Ewald sphere.

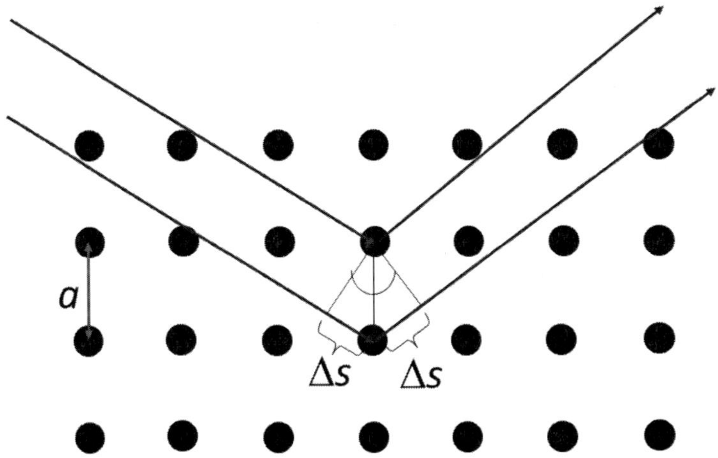

Figure 1. Schematic of X-ray diffraction on a crystal with a lattice parameter *a*. In order to record a Bragg reflection, the difference in path length $2\Delta s$ of the diffracted X-rays has to be a multiple integer of the X-ray wavelength.

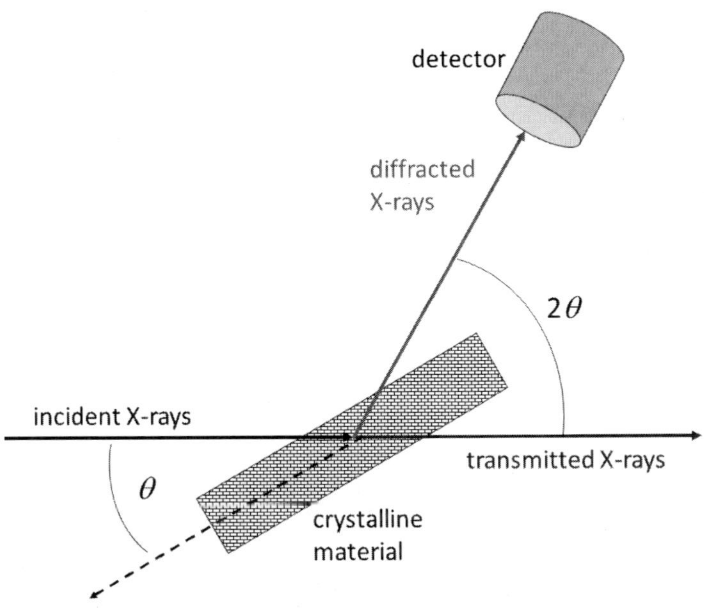

Figure 2. Schematic view of a θ-2θ-scan configuration.

Mathematically, the amplitude of x-ray diffraction in the far field regime and within kinematic approximation is a Fourier transform of real space which transforms into reciprocal space with $\rho(r)$ being the electron density.

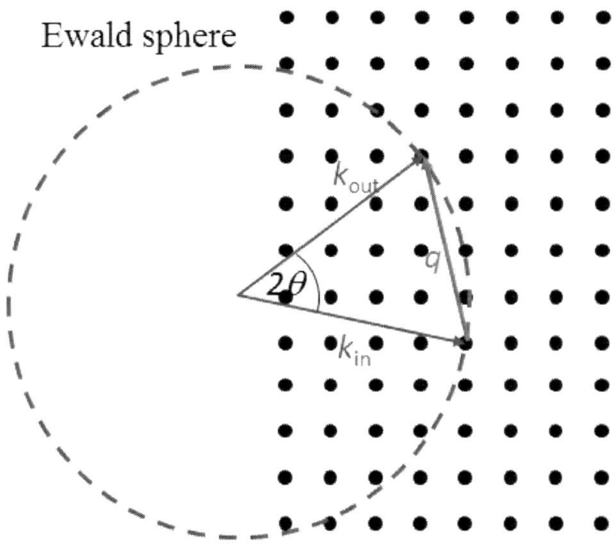

Figure 3. Schematic of the Ewald sphere whose radius is given by the length of the incident wave vector $\left|\vec{k}_{in}\right|$. The diffraction points which may result in a diffraction signal must lie on the Ewald sphere.

$$F(\vec{q}) = \int \rho(\vec{r}) \cdot e^{-i\vec{q}\vec{r}} d\vec{r} \tag{2}$$

The Fourier transform $F(q)$ is generally a complex function and, thus, has a magnitude $|F(q)|$ and a phase $\phi(q)$.

$$F(\vec{q}) = |F(\vec{q})| e^{i\varphi(\vec{q})} \tag{3}$$

The measured quantity in an experiment is the scattered intensity $I \propto F(\vec{q}) \cdot F(\vec{q})^*$, the information carried by the phase is lost.

As a general rule, one may note that extended objects in real space are small in reciprocal space while small objects in real space result in extended diffraction signals in reciprocal space. Therefore, an infinite three-dimensional crystal of periodically arranged atoms results in periodically arranged sharp Bragg reflections. A surface can be regarded as an infinite two dimensional plane with a finite thickness of a few atomic layers in the third dimension. This confined thickness in real space leads to an extended reflection in reciprocal space – the crystal truncation rod (CTR). In the mathematical description, the Fourier transformation of an infinite two-dimensional plane results in a one-dimensional line standing perpendicular on the plane – the CTR. This crystal truncation rod connects the Bragg reflections and it is the weakest in the center between two reciprocal lattice points. Studies of the CTR intensity give access to the surface morphology [3, 4].

If the size of the investigated object – a single grain in case of polycrystalline material or a nanosized monocrystal – is smaller than the X-ray beam with sufficient angular resolution, the shape of the structures influences the diffraction pattern. Each surface creates a crystal truncation rod. Thus, for instance, for a hexagonal object the shape factor results in a six-fold symmetrical diffraction pattern (see Figure 5). This shape factor is independent of the scattering vector and, hence, it appears at *every* Bragg reflection.

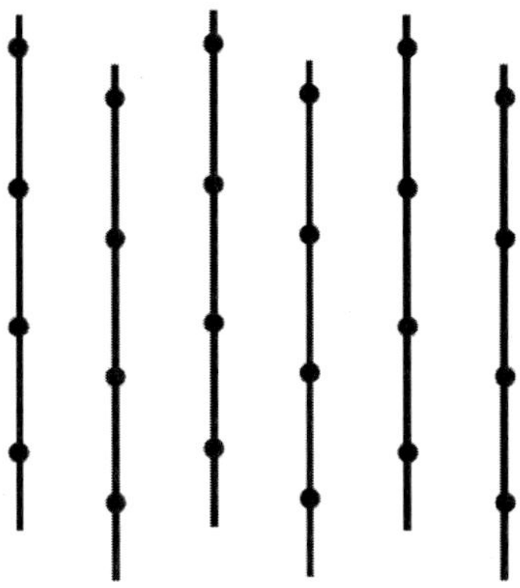

Figure 4. Schematic of crystal truncation rods connecting the Bragg reflections.

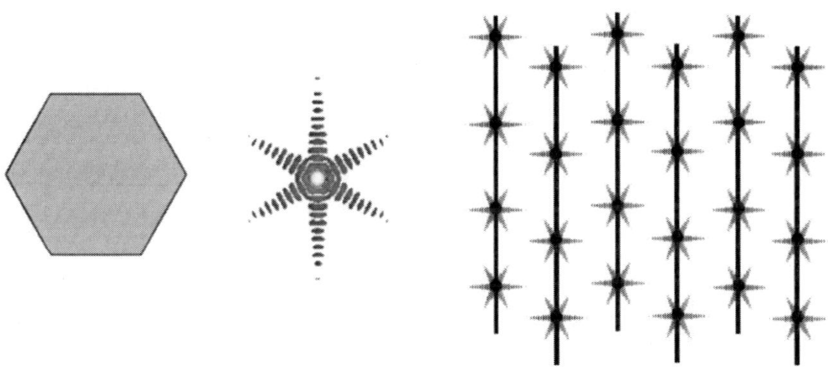

Figure 5. Schematic of a hexagonal object (left) whose shape factor results in a six-fold symmetrical Bragg reflection (centre) appearing at every reciprocal lattice point (right).

X-ray diffraction is a very precise tool for measuring the lattice parameter and detecting its variations. It allows for recording strains with a resolution of $\Delta a/a \sim 10^{-4}$ and it can yield the six elements of the strain tensor by measuring three independent Bragg peaks. A perfect crystal with a lattice parameter a_0 exhibits a Bragg reflection at a defined angle with its Darwin width. When atoms are displaced by a vector *u*, meaning the crystal is strained, the angular position and the width of Bragg reflections are affected. Compared to an unstrained crystal with a lattice parameter, a_0, a uniform tensile (compressive) strain leads to a larger (smaller) lattice parameter, a_1. This change in lattice plane distance results in a simple shift of the Bragg reflection to a lower (higher) diffraction angle, which may easily be measured with a high precision. In case of non-uniform strains, the lattice plane distance is reduced for one part of the sample while for the other part it is increased. Therefore, the Bragg peak width is increased due to the larger variation of lattice plane distances.

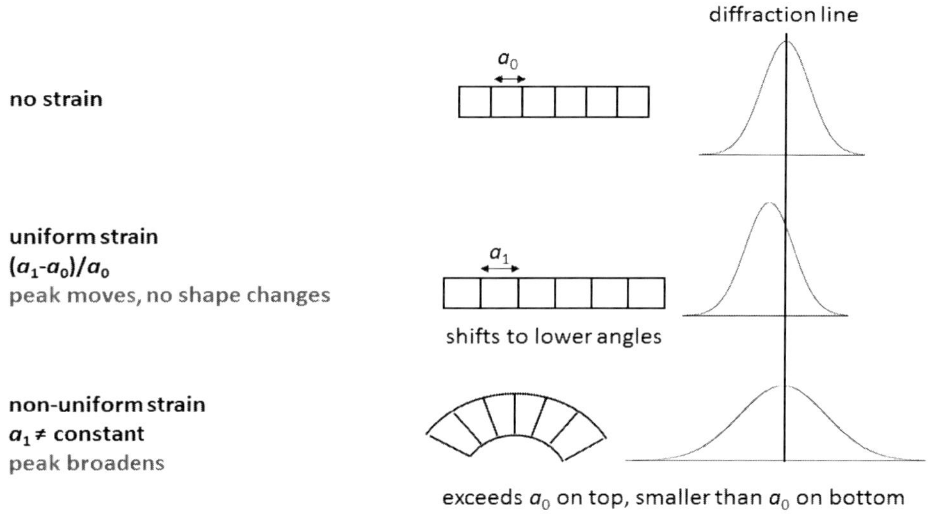

Figure 6. Schematic of strain and its effect on a diffraction signal. A uniform tensile (compressive) strain leads to a simple shift of the Bragg reflection to a lower (higher) diffraction angle. Non-uniform strains result in an increase of the width of the Bragg reflection which remains at its diffraction angle.

On the one hand, strain may already be present in nanostructures in their as-grown state. As an example, during epitaxial growth of a layer on a substrate, the strain generated in the growing layer is caused by the misfit between the two lattice parameters. The strain in fully coherent epitaxial heterostuctures (without extended defects) can be theoretically described through the linear elastic theory. The layer lattice parameter shrinks to accommodate coherently the one of the substrate. In a good approximation, the substrate is considered much thicker than the layer and all the lattice accommodation mainly occurs in the thin film. The stored elastic energy increases with the film thickness and beyond a critical thickness plastic or elastic relaxation may occur. Plastic relaxation occurs via the introduction of misfit dislocations at the interface. Elastic relaxation is the result of a transition from 2D to 3D growth (Stranski-Krastanov transition) with additional free surfaces. While for many years it was aimed on reducing or even avoiding strain in functional materials, it has been shown for some systems that stress may not be detrimental to the physical properties (optical and electronic). The device performance may sometimes even be improved for strained structures with respect to unstrained crystals, as shown for the mobility in transistor heterostructures [5].

On the other hand, strain may be applied mechanically. Recently, the development of atomic force microscopy paved the way to the investigation of elasticity and plasticity at the nanometer scale. Several works demonstrated, as in the case of one-dimensional nano-objects, that under tension, bending, or compression they exhibit mechanical properties quite different from the corresponding bulk materials. In addition, at variance with what is predicted by continuum elastic theory, the elastic properties are also size-dependent. This effect has been already demonstrated via experimental approach or calculations in the case of GaN [6], Au [7], and Ag [8] nanowires.

3D X-Ray Diffraction

A nanostructure is expected to exhibit an extended three-dimensional diffraction pattern in reciprocal space due to its small size and to an eventually inhomogeneous strain. In this section, we will shortly review different techniques to record the full three-dimensional intensity distribution in reciprocal space. For example, we will show the results for SiGe islands which were grown by liquid phase epitaxy on Si (001). The islands have truncated pyramidal shapes (with four (111)-oriented side facets) with square bases (see Figure 7). The island size is pre-determined by the Ge content which also results in a lattice expansion compared to pure Si [9]. The corresponding lattice expansion can be inferred from Vegard's law [10]. Here, the Ge content amounts to 12 % resulting in a SiGe island with a base width of 1 μm, a height of 500 nm, and a top facet with a side length of about 300 nm.

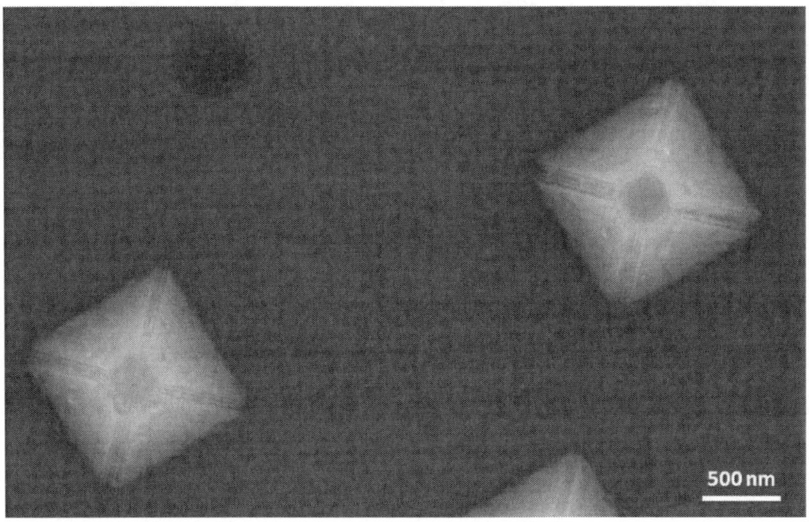

Figure 7. Scanning electron microscopy image of SiGe islands epitaxially grown on Si (001).

The calculated three-dimensional reciprocal space map (3D-RSM) for a SiGe island modeled by finite-element method simulations and fast-Fourier transformations is displayed in Figure 8 (a). The diffraction signal includes the Si (004) Bragg peak, the crystal truncation rod (CTR) of the Si substrate, and the signal of the $Si_{88}Ge_{12}$ island with the CTRs originating from the (111) side facets of the truncated pyramid. The semi-transparent plane represents the detector plane cutting through the reciprocal space. Figures 8 (b) and (c) display the simulated intensity in this plane and the corresponding experimentally recorded diffraction pattern, respectively. Both images show the same features, namely the diffuse scattering close to the Si (004) Bragg reflection, the substrate CTR, and the island induced diffuse signal at lower q_z values.

In Bragg geometry, three-dimensional intensity distributions can be obtained by three different techniques which are schematically depicted in Figure 9 (a) - (c). The dashed and dotted lines indicate the variation of the wave vector for the incident k_{in} and the diffracted beam k_{out}, while the solid black lines illustrate the movement of the 2D detector in reciprocal space when recording the 3D intensity distribution. In the first approach, the rocking scan, the angle of incidence of the X-rays is varied while the detector remains at its angular position. Thus, in reciprocal space the detector moves along the Ewald sphere mapping out the reciprocal lattice point. Hence, the Bragg signal stays at the same position on the detector while the CTR signal moves. In the second method, the θ-2θ scan or radial scan, the sample and the detector are simultaneously rocked where the latter is moved by twice the angle of the former. These angular changes in real space results in a movement of the detector plane along the q_z vector in reciprocal space leading to an immobile CTR signal while the Bragg signals move on the detector. The third technique does not contain any movements of diffractometer motors but varies the X-ray energy. In reciprocal space, a scan in energy results in the same behavior of the detector as for a radial scan.

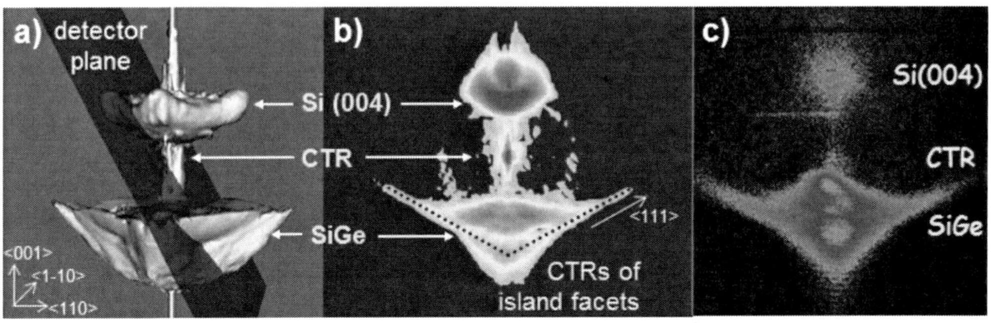

Figure 8. a) Simulated three-dimensional reciprocal space map of a SiGe island on Si (001) showing the diffuse scattering close to the Si (004) Bragg peak, the substrate crystal truncation rod, and the signal of the SiGe island including the CTRs of the (111) side facets of the pyramid. The semi-transparent plane represents the projection of the 2D detector in the reciprocal space at the measured Bragg angle. b) Simulated and c) experimentally measured 2D diffraction maps showing similar features.

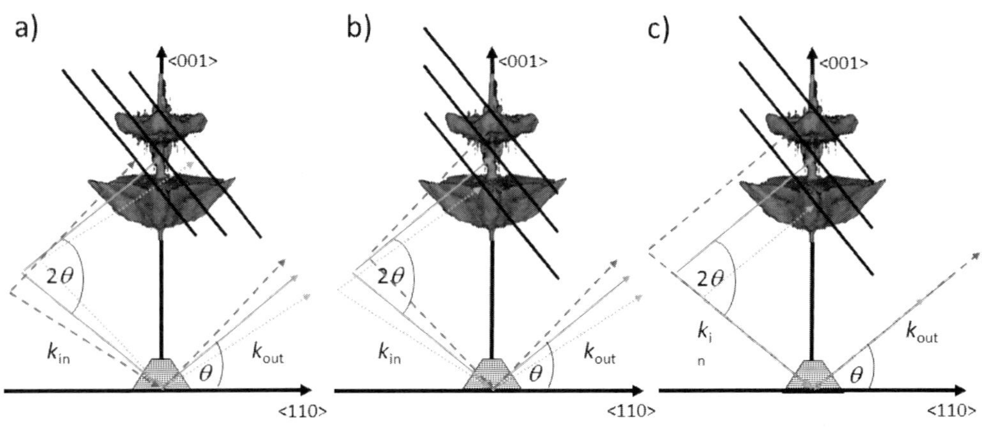

Figure 9. Schematics of the detector movement in reciprocal space for a) a rocking curve scan, b) a θ-2θ scan (radial scan), and c) an energy scan. The dashed and dotted lines indicate the variation of the incident and diffracted beams.

In case of the energy scan, care has to be taken in maintaining the intensity of the incident X-rays, since the undulator curve is rather narrow as shown in Figure 10. The intensity drops by ~80% when changing the X-ray energy by 100 eV without adjusting the undulator gap while it is fairly constant when aligning the undulator for each energy value. The intensity fluctuations for the energy scan with undulator gap adjustment probably originate from the positioning accuracy of the undulator gap.

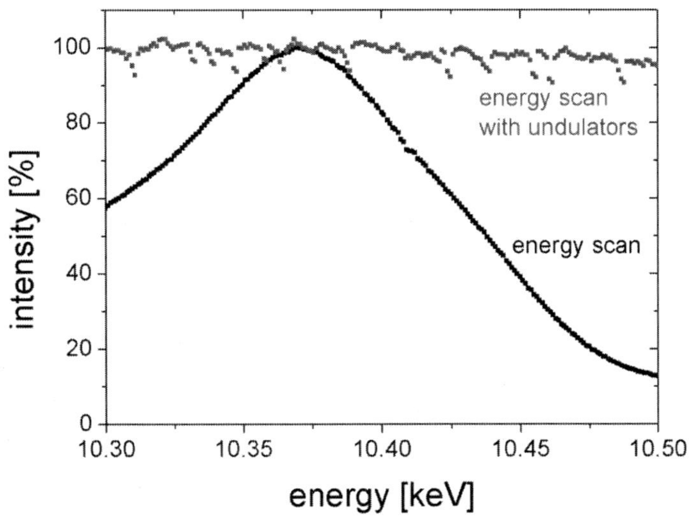

Figure 10. Normalized incident intensity as a function of energy with and without adjusting the undulator gap. The intensity fluctuations for the energy scan with undulator gap adjustment originate from the positioning accuracy of the undulator gap. Reprinted with permission from [11].

Sequences of X-ray diffraction images recorded while rocking the sample in an angular range of $\Delta\theta = \pm 0.5\,°$ ($\Delta Q = \pm 0.82$ nm^{-1}) and tuning the energy from 10.5 to 10.3keV ($\Delta Q = \pm$ 0.475 nm^{-1}) are presented in Figure 11 (a) and (b), respectively [11]. The third technique (θ-2θ scan) is omitted here since the behaviour of the diffraction signals as a function of the scanned angle is the same as for the energy scan. In both sequences, the CTR signal is encircled to highlight its position in the 2D cuts through the 2 types of RSM scans. The presented sequences show exactly the features described above: For the rocking scan the CTR moves through the detector plane and the Bragg reflections remain at their positions while for the energy scan the CTR is immobile and the Bragg peaks move through the detector plane.

These kinds of scans allow for obtaining the complete 3D information of one Bragg reflection of nanostructures. The 3D intensity distributions in Q space obtained from the rocking curve scan and the energy tuning scan – from image sequences like the one displayed in the figure above – are presented in Figure 12(a), and (b), respectively. Both distributions show the same features and they are in agreement with the simulated 3D intensity distribution (see Figure 8(a)). The intensity in the plane cutting through the Si (004) Bragg peak is caused by air scattering. A part of the X-rays diffusely scattered in front of the sample exhibits the correct incident angle on the substrate and fulfils the Bragg conditions. This leads to a region of parasitic intensity in reciprocal space. The amount of air scattering and, hence, the parasitic intensity plane can be reduced by inserting an aperture close to the sample position (not done here).

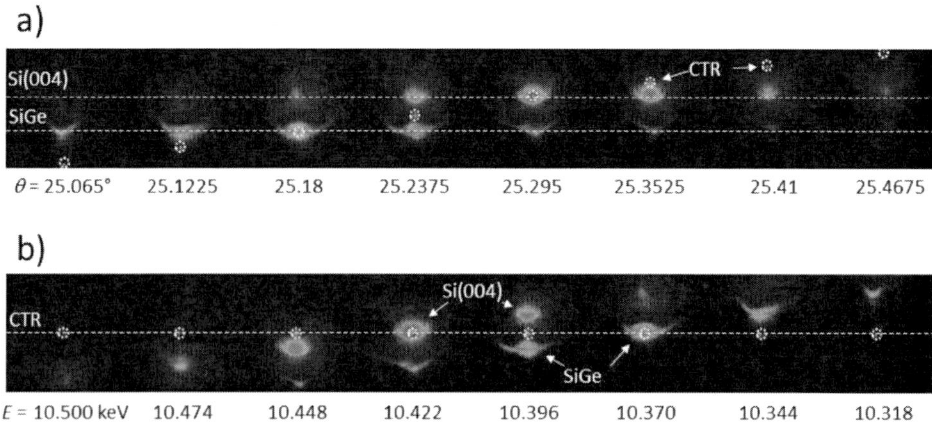

Figure 11. Sequences of XRD 2D patterns recorded during a) a rocking curve and b) an energy scan. To highlight the respective movement, the position of the CTR is marked by circles for both cases. Reprinted with permission from [11].

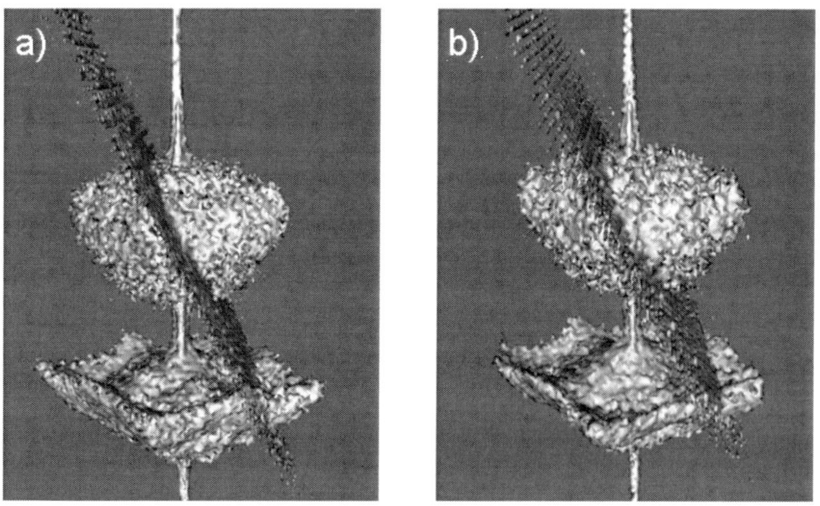

Figure 12. Three-dimensional reciprocal space maps in Q space obtained from a) a rocking curve scan and b) an energy scan. Reprinted with permission from [11].

FOCUSING OPTICS

In standard laboratory diffractometers, the size of the X-ray beam is typically in the range of few millimeters. Such broad X-ray beams allow for studying an ensemble of nanostructures but they do not give access to the properties of one single nano-object. In order to examine an individual nanostructure the size of the X-ray beam has to be in the same order as the size of the object, i.e. the X-rays have to be focused down to the 100-nm regime.

The refractive index for electromagnetic waves is described by the following equation

$$n = 1 - \delta - i\beta \tag{4}$$

where the real part 1-δ and the imaginary part β define the critical angle of reflection and the absorption, respectively. The real part of the refractive index scales quadratic with the atomic number ($1-\delta \propto Z^2$), while the absorption rises with Z^4. For visible light the refractive index is larger than unity, e.g. it amounts to 1.33 in water. In the case of hard X-rays, typical values for δ and β are in the range of 10^{-6} and 10^{-9}, respectively. Thus, the refractive index for hard X-rays is very close to unity and, in addition, $n < 1$. This fact complicated the development of focusing optics for hard X-rays. Recent progresses in fabrication of diffractive, refractive, and also reflective X-ray optics applied at 3rd generation synchrotron radiation sources gave rise to tremendous advances of micro- and nanofocusing optics.

Note that the synchrotron X-ray source, meaning the electron beam, is not a point source but it is extended. Therefore, focusing optics only allow for a demagnification of the source size which is typically in the range of several ten micrometers. Nowadays, the length of a typical synchrotron undulator beamline is around 50 m and the distance of the focusing optics from the sample varies between few centimeters up to the meter range. Thus, demagnification ratios ranging from several tens to one thousand are obtained resulting in focal sizes in the range of few hundreds of nanometers. While focusing optics installed very close to the sample position offer a very small focus, they limit the possibility of installing special sample environments such as heating stages, cryostats etc. Currently, many synchrotron beamlines are extended in length opening the possibility to further reduce the focal size, or for keeping the same focus as before for optics with larger working distances and, thus, for allowing the installation of sample environments. The smallest focal spot obtained at the 1 km long beamline BL29XUL of SPRING-8 amounted to 7 nm in vertical direction [12]. Besides the advantages concerning the focal size, the elongation of beamlines also increases the lateral coherence length of the X-ray beam which is important for diffraction studies using coherent beams [13, 14, 15, 16].

Reflective Optics

Conventional mirrors, as used for visible light, at normal incidence cannot be used for hard X-rays, because the reflectivity is too low. A high reflectivity is obtained for grazing incidence when the incident angle is smaller than the critical angle of reflection which is defined by the real part of the refractive index

$$1-\delta = \cos\theta_c \tag{5a}$$

$$\theta_c \approx \sqrt{2\delta} \tag{5b}$$

Taking into account $\delta \sim 10^{-6}$, the critical angle θ_c is typically of the order of few milliradians and it increases with increasing Z. To reduce the astigmatism Kirkpatrick and Baez proposed the use of two spherical and cylindrical mirrors in a crossed configuration [17]. Figure 13 shows such geometrical arrangement which became known as Kirkpatrick-Baez (KB) mirrors.

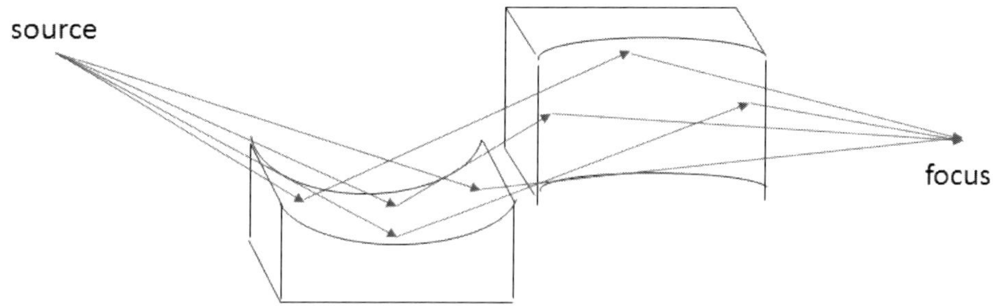

Figure 13. Schematic view of the two mirror Kirkpatrick-Baez system.

A big advantage of mirrors as focusing optics is their achromaticity, i.e., the focal length is independent of the energy and respective wavelength of the incident radiation. Thus, they are used for focusing monochromatic beams as well as white beams. In addition, they allow for varying the energy of the incident X-ray beam which is mandatory for absorption and spectroscopy studies.

Refractive Optics

Refractive lenses made of glass are among the most widely used optical components for visible light. For a long time refractive lenses were considered unfeasible for X-rays due to the weak refraction and strong absorption. However, in the 90s of the last century Snigirev et al. demonstrated that hard X-rays can be focused by concave lenses [18, 19]. The concave shape results from the fact that the refractive index of hard X-rays is smaller than unity. The focal length *F* which is defined by the equation below is only positive for lenses exhibiting a negative radius *R*, i.e., having a concave shape.

$$\frac{1}{F} = \frac{2(n-1)}{R} \tag{6}$$

Thus, for a biconcave lens with a parabolic profile $x^2 = 2Ry$ as depicted in Figure 14, the focal length is given by the following equation.

$$F = R/2\delta \tag{7}$$

For a typical radius of curvature of 200 μm and considering the value for δ being in the range of 10^{-6}, the focal length is in the range of several tens to one hundred meters. In order to achieve a focal length of the order of 1 m, several lenses have to be stacked behind each other. This arrangement is known as compound refractive lens (CRL). To minimize absorption and due to the fact that the absorption rises with Z^4, the lenses have to be prepared out of light materials such as Be, B, C, and Al. The aperture for a lens with a thickness $2y_0 + d$ amounts to $2R_0 = 2\sqrt{2Ry_0}$.

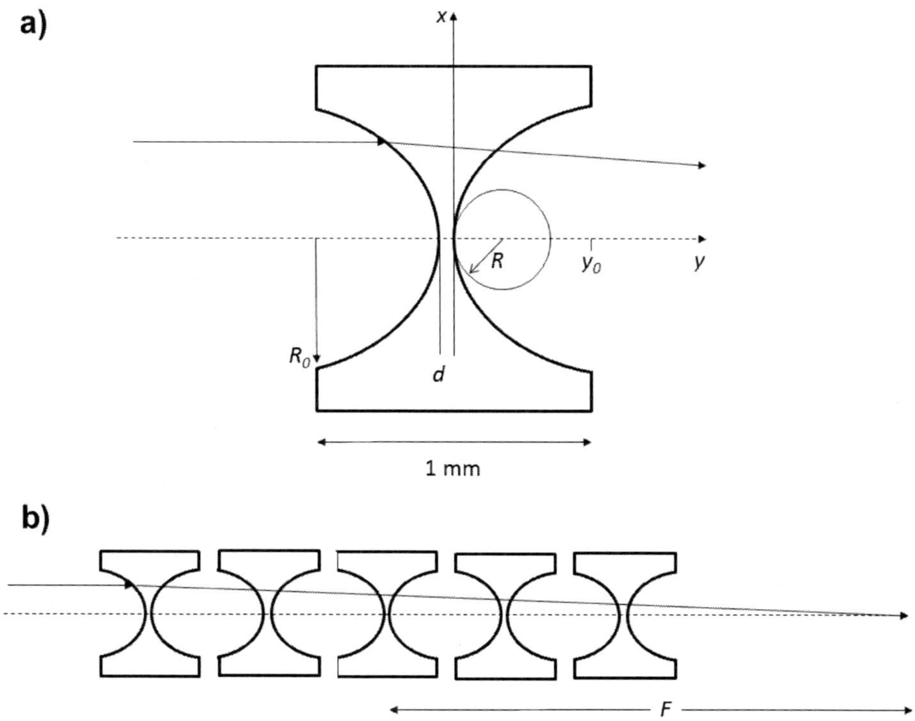

Figure 14. Schematics a) of a biconcave lens with radius *R* and b) the arrangement of several lenses as compound refractive lens with focal length *F*.

Diffractive Optics

Diffraction refers to phenomena which occur when waves encounter obstacles. When light passes through a double slit, a circular wave emanates from each slit which interfere with each other. This experiment was first performed by Thomas Young in the year 1803. For a grating with an arbitrary number of slits the interference fringes are described by the following equation

$$m\lambda = d \sin \theta_m \tag{8}$$

where *m* is the order of diffraction, *d* is the inter-slit distance, and θ_m is the diffraction angle of the diffraction order *m*.

Diffraction gratings do not have to consist necessarily of straight slits. For instance, a Fresnel zone plate is a circular diffraction grating. Its focusing properties were first discussed in the late 19th century, and Baez originally suggested its use as X-ray optical element. A FZP consists of a series of *N* concentric rings with a radius of $r_N^2 = N\lambda F$ (see Figure 15 [20]).

The width dr_n of the rings decreases with increasing radius. The focusing capability is based on constructive interference of the wavefront modified by the passage through the zone plate. Such FZPs are conventionally fabricated by electron beam lithography or by holographic techniques [21, 22].

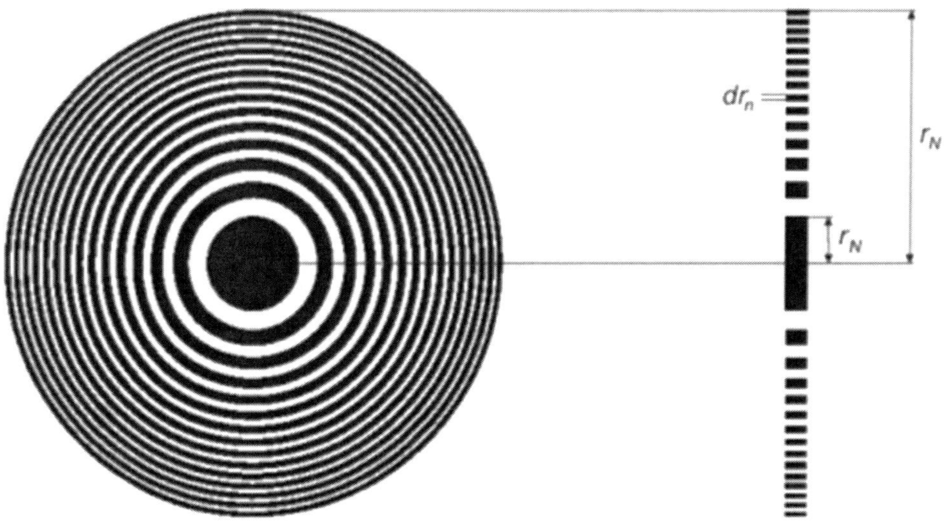

Figure 15. Schematic of a Fresnel zone plate [taken from 20].

Besides the first order of diffraction higher diffraction orders *m* exist as well as the 0th order. These additional diffraction orders deteriorate the first order beam focus and, hence, they have to be eliminated. For this purpose, a pinhole – the order sorting aperture (OSA) – is introduced as close as possible to the focal point of the first order. However, the OSA still allows the 0th order to penetrate and, thus a central beam stop (CBS) which has at least the same size as the OSA, is installed right in front of the Fresnel zone plate. Figure 16 displays a schematic of this complete setup for a Fresnel zone plate.

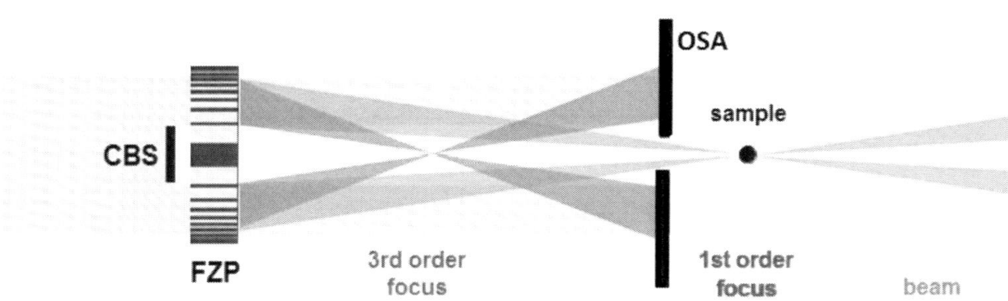

Figure 16. Schematic of a Fresnel zone plate (FZP) setup which consists of the FZP itself, an order sorting aperture (OSA) for selecting one particular diffraction order of the FZP (here the 1st order of diffraction), and a central beam stop to absorb the 0th diffraction order.

Just as the Rayleigh resolution of a light microscope lens is determined by its numerical aperture (NA) the Rayleigh resolution of a zone plate is determined by its maximum diffraction angle $NA = \lambda/2dr_n$, so that the Rayleigh resolution is 1.22 dr_n/m. Typical values for diffraction limited by an X-ray energy of about 10 keV and with a focal distance of 10 cm by lithographically produced FZPs are below 100 nm. The Fresnel zone plate is a chromatic focusing device and, thus, the focal length $f = D \cdot dr_n/(m\lambda)$ as well as the depth of focus $DOF = \pm 4dr_n^2/(m\lambda)$ depends on the wavelength (or energy) of the incident X-rays. At an energy of 10 keV, the focal length and the *DOF* of the FZP (dr_n = 70 nm, d = 200 μm) used at beamline ID01 at the European Synchrotron Radiation Facility (ESRF) in Grenoble amount to 112.9 mm and ±160 μm, respectively.

Advantages/Disadvantages of Different Focusing Optics

The three different types of focusing optics have their respective advantages and disadvantages. Mirrors are achromatic devices and, thus, they allow focusing monochromatic radiation as well as white beams. Hence, they facilitate energy changes necessary for X-ray absorption spectroscopy or 3D X-ray diffraction by the energy tuning approach. In contrast, diffractive and refractive optics are chromatic with a bandwidth of about $\Delta E/E \sim 10^{-3}$. The chromaticity, i.e., the dependence of the focal distance as a function of the energy is more pronounced for refractive lenses than for diffractive gratings. The focal length is directly proportional to the energy for Fresnel zone plates while it rises with E^2 for refractive lenses. Concerning the preservation of the coherence of the beam, diffractive gratings are very well suited since they do not alter the wavefront of the incident beam. The asperity of mirror surfaces and refractive lenses disturb the wavefront. Great advances have been made within the last years concerning the polishing of mirror surfaces rendering possible the fabrication of surfaces with an rms well below 1 Å over a mirror length of 1 m and, thus, minimizing the wavefront perturbation.

Besides the influence of the focusing optics on the beam properties they may also limit the space available to install specific sample environments for *in situ* studies. While the distance for all focusing optics from the sample position is the same in order to obtain the same focal spot size, the order sorting aperture used in combination with Fresnel zone plates limits the most the experimental setup. Typical distances of the OSA from the sample position are in the range of few centimeters for hard X-rays.

Micro- and Nanofocused X-ray Diffraction

Scanning X-ray Diffraction Mapping

It is rather straightforward to perform X-ray diffraction experiments using a broad X-ray beam with several hundred micrometers in size. However, it is not that easy when individual micro- or nanostructures shall be investigated employing highly focused X-ray radiation. For this purpose, the structures have to be located and the X-ray beam has to be focused upon them. In the following we will shortly review the scanning X-ray diffraction mapping technique which allows for locating and selecting individual micro- and sub-micrometer structures [23]. Optical microscopy images of adjacent regions of a Si substrate with SiGe islands having a base width of 3.2 μm and a height of 1 μm are presented in Figure 17(a). This sample was raster scanned by means of a microfocused X-ray beam. The beam was focused down to a size of 3 and 5 μm in vertical and horizontal direction, respectively, using Be compound refractive lenses. Generally, in order to obtain an X-ray diffraction map of the sample, the only requirement is that different areas on a sample produce sufficiently different scattering patterns in reciprocal space. Thus, by keeping the reciprocal space position fixed, an intensity modulation is observed while the sample is raster scanned in real space. Here, the intensity contrast is given by the fact that the lattice strain in the SiGe islands caused by the composition and the relaxation of the island leads to a separation of the SiGe Bragg reflection from the Bragg peak of the Si substrate (as shown in Figure 8). While raster scanning the sample in the X-ray beam, the diffraction intensity originating from the islands is recorded. The scanning X-ray diffraction map (SXDM) presented in Figure 17(b) shows the distribution of islands on the sample. The high intensities (orange and red colors) represent island positions, whereas low intensities (yellow to blue colors) represent the areas in between islands. For ease of comparison, the optical microscopy image (red squares in Figure 17(b)) is superimposed with the SXDM revealing an excellent match. Thus, by means of this technique particular islands can be identified and it is straightforward to select and investigate one specific structure.

The two circles in the SXDM indicate two different islands, namely a standard truncated pyramidal shaped island (full circle) and a flat island (dashed circle). Two-dimensional reciprocal space maps taken around the Si (004) Bragg reflection for the island ensemble, the truncated pyramidal island, and the flat island are displayed in Figure 18 (a), (b), and (c), respectively. Scanning electron micrographs of the two islands are presented as insets in the central and the right reciprocal space maps. The RSMs show the Si (004) Bragg peak, the substrate CTR, and the signal of the SiGe islands including the CTRs originating from the island side facets. Two additional streaks are visible where one originates from the monochromator and the second one is caused by a saturation of the detector while measuring the highly intense Si Bragg peak. The reciprocal space maps for the island ensemble and the standard truncated pyramidal island show exactly the same features while for the RSM of the flat island the island signal is located at lower q_z, indicating a higher Ge concentration for this island compared to the standard one. By means of finite-element method simulations combined with microfocused X-ray diffraction, it was found that the SiGe islands consist of two parts. The lower third of the island possesses a larger Ge content than the upper two thirds. The flat island is exactly this lower third of a standard SiGe island. It is not yet understood why the island growth stops after the first third of growth in about 1% of all the islands. Not until the emergence of the microfocused X-ray beam, the different island growth stages could be investigated by X-ray diffraction techniques.

A high-resolution scanning X-ray diffraction map of a $Si_{88}Ge_{12}$ island with a base width of 1 μm and a height of 500 nm is displayed in Figure 19 showing the X-ray diffraction patterns as a function of the position of the X-ray beam with respect to the SiGe island. The X-ray beam was focused down to 200 x 300 nm^2 using a Fresnel zone plate and it was scanned in steps of 100 nm across the island. When the beam spot hits the center of the island top facet, a symmetric diffraction pattern is obtained including the Si (004) Bragg reflection, the substrate CTR, and the complete signal of the SiGe islands. Since the beam focus is smaller than the island size, the CTRs originating from the island side facets are not strongly pronounced. However, when the X-ray beam is focused on one of the side facets of the island the diffraction signal contains mainly the facet CTR. Thus, this high-resolution scanning X-ray diffraction map demonstrates the feasibility to study the spatial origin of certain diffraction signals and to focus the X-ray beam on a particular feature of the sample for further investigations.

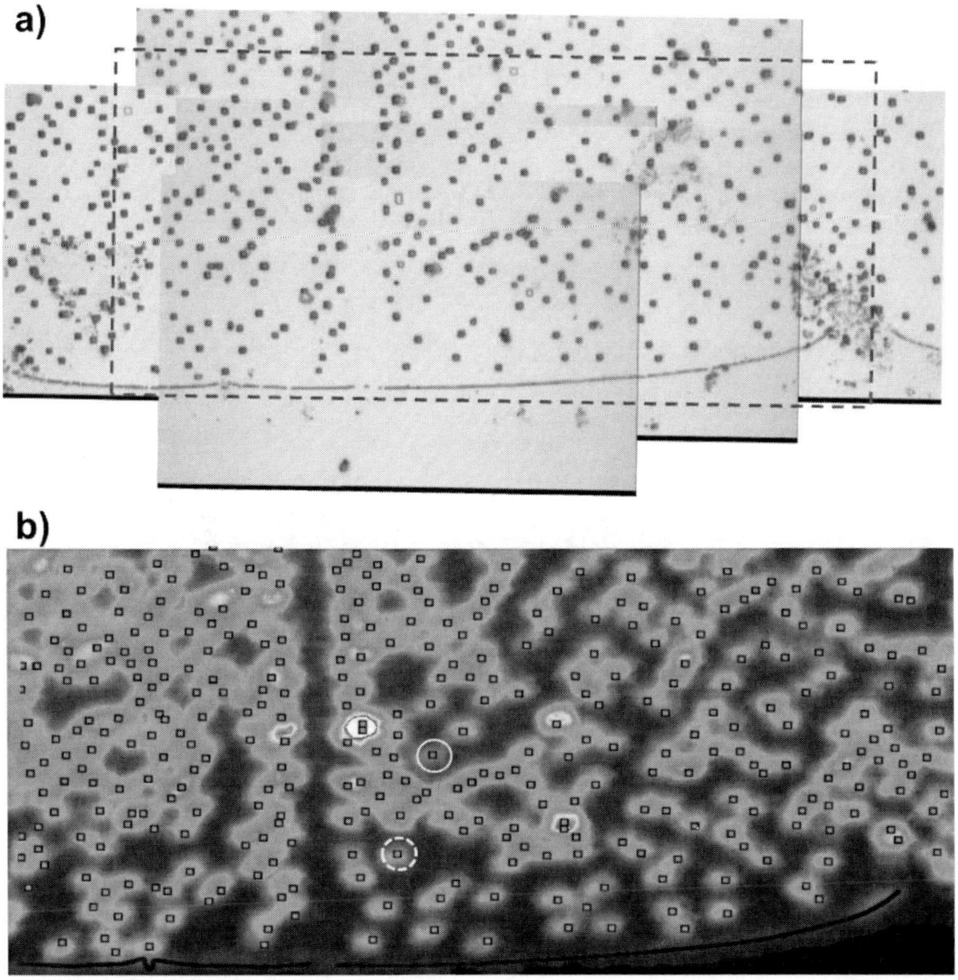

Figure 17. a) Optical microscopy images of adjacent regions of SiGe islands with a base width of 3.2 μm and a height of 1 μm epitaxially grown on Si (100). b) Superposition of a scanning X-ray diffraction map of the area marked in part (a) and the optical microscopy image in part (a). The two circles in the SXDM indicate two different islands, namely a standard truncated pyramidal shaped island (full circle) and a flat island (dashed circle). (Images from ref. [23]).

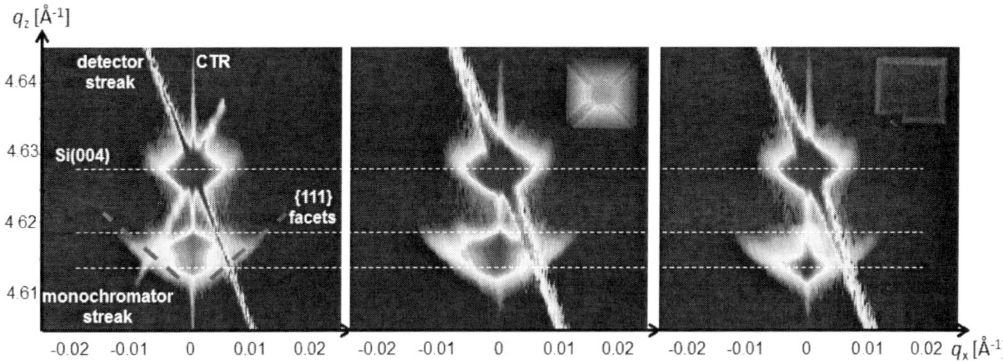

Figure 18. Two-dimensional reciprocal space maps taken around the Si (004) Bragg reflection for a) the island ensemble, b) the truncated pyramidal island, c) and the flat island. The insets in part (b) and (c) are scanning electron micrographs of the two different islands.

Figure 19. High-resolution scanning X-ray diffraction map of a $Si_{88}Ge_{12}$ island with a base width of 1 μm and a height of 500 nm taken in steps of 100 nm using a X-ray beam focused down to 200 and 300 nm in horizontal and vertical direction, respectively. The X-ray diffraction patterns as a function of the beam position are overlaid with a schematic of a SiGe island.

3D X-ray Diffraction with Nanofocused Beam

Generally, nanostructures exhibit an extended intensity distribution in reciprocal space. As demonstrated above, three different techniques exist for recording 3D reciprocal space maps – rocking scan, θ-2θ scan, and energy scan. While these techniques can be used without any restrictions using broad X-ray beams on nanostructure ensembles, challenges arise for studies on individual nano-objects using nanofocused beams.

One main issue concerning the collection of 3D diffraction data using one of the first two techniques (rocking scan and radial scan) is the large sphere of confusion of the existing diffractometers (typically few tens of micrometers over a 360° rotation) as compared to the size of nanostructures and nanofocused beams (a few hundreds of nanometers). Even though the angular variation of the angle of incidence is typically 1° only, the investigated nanostructure may move out of the beam. In the special case of coherent diffraction even a movement of the sample within the beam focus may be detrimental since different parts of the wavefronts are diffracted. The acquisition of 3D patterns requires thus the re-alignment of the structure in the nano-beam for every rocking angle, resulting in an increase of the measuring time. Furthermore, the presence of complex sample environments for *in situ* XRD measurements may demand a limitation of sample movement to reduce or avoid any vibrations induced by the diffractometer movement. This is achieved by scanning the X-ray energy. For the energy tuning approach the use of *achromatic* focusing optics such as KB mirrors is an obvious choice since the optics-sample distance is invariant with the X-ray energy, while in the case of *chromatic* optics (such as CRLs and FZPs) the change in focal length $F(E)$ has to be considered for each energy step.

On the other hand, one may adjust the distance between the focusing optic and the sample according to $F(E)$ [11] which scales linearly with E in the case of a Fresnel zone plate and with E^2 for compound refractive lenses [24]. In Figure 20, the measured horizontal focal size for 37 Be CRLs positioned 90 cm in front of the sample position at different energies is presented, where the CRL-sample distance was adjusted for each energy level. Note that the parallelism of the lenses with respect to the X-ray beam was aligned after each translation. Both the size and the position of the focal spot at different energies were determined by "knife edge scans". For this purpose, a 250 μm thick tantalum wire was scanned through the X-ray beam while the absorption was recorded. Figure 20 (a) displays three horizontal knife-edge scans at 10.3, 10.4, and 10.5 keV where the focal distance of the CRLs was adjusted for each energy level. The variation of both the focal size and the position of the focal spot is negligible in comparison with both the sample and the beam size (Figure 20 (b)). The latter was determined by the slope of the curve obtained by the knife-edge scan. The measured focal spot size was ~ 3.5 and ~ 2.5 μm in the horizontal and vertical direction, respectively.

On the other hand, note that the beam focus possesses an extension allowing for a variation in energy without losing the focal spot on the sample. In addition, the depth of focus can be extended by reducing the numerical aperture, i.e., by reducing the illuminated area of the focusing optics. For instance, the focal depth of the fully (200 x 200 μm^2) illuminated Fresnel zone plate at the ID01 beamline at ESRF amounts to about 0.3 mm. This focal depth is increased when reducing the illuminated area of the FZP [25] and, thus, the energy of the incident X-ray beam may be changed within a wider range without losing the focal spot on the structure under investigation. For a fully illuminated FZP the X-ray energy may be varied only by ~ 30 eV which is not sufficient of a complete 3D XRD. The reduction of the illumination area to 50 x 50 μm^2 results in a focal depth of about 1.5 mm and, hence, the energy may be varied within 150 eV without losing the focal spot on the sample.

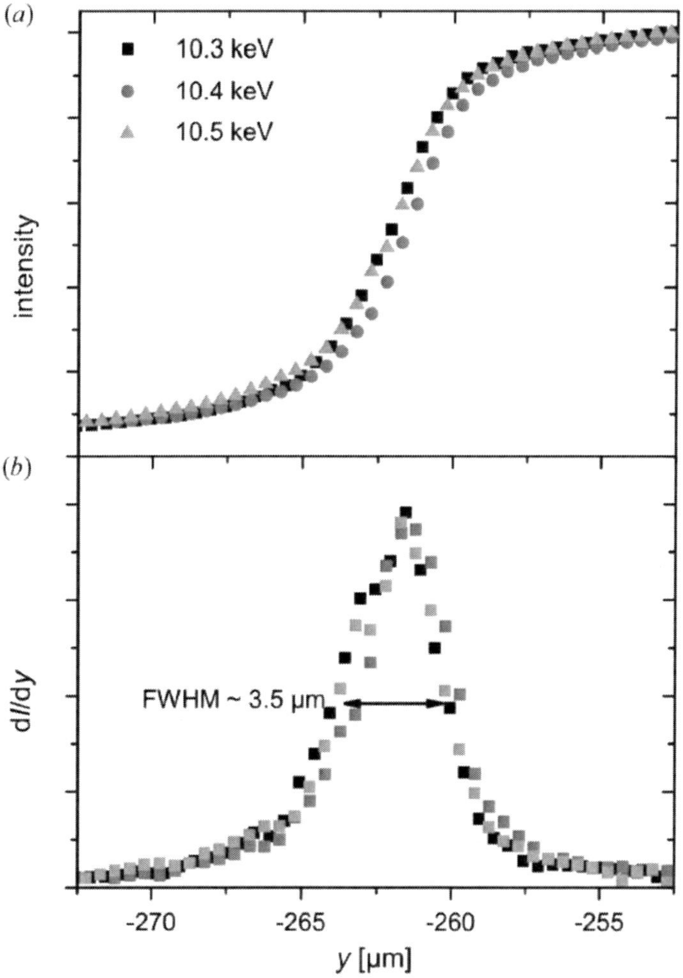

Figure 20. Beam focusing with compound refractive Be lenses: a) Normalized intensity during the "knife edge scans" of a 250 μm thick tantalum wire at 10.3, 10.4, and 10.5 keV. b) Derivatives of the intensity during the knife edge scans shown in part (a). The full width at half maximum (FWHM) indicates the horizontal focal size of ~3.5 μm. Reprinted with permission from [11].

IN-SITU APPLICATIONS

In this section, we will review various possible *in situ* applications of X-ray methods combined with scanning probe microscopes (SPMs). Scanning probe microscopes exhibit an extremely high lateral resolution. They provide information on surface morphology and allow for a local characterization of a wide number of physical properties with a resolution going down to the atomic scale. Moreover, specific SPMs allow manipulating and modifying the matter at the nanoscale. For instance, atomic force microscopes allow for applying and detecting forces in the nanonewton range and, thus, the mechanical behavior of nanostructures can be investigated. X-ray diffraction gives access to the strain as well as to defects. Hence, complementary information such as the elastic strain, defect creation and defect evolution can be obtained by *in situ* nanofocused diffraction during mechanical testing using an atomic force microscope. In addition, while SPMs suffer the lack in chemical sensitivity, X-ray absorption techniques give access to the chemical composition of the sample in this study. When X-rays impinge on a surface and the energy of the incident X-rays surpasses the ionization threshold, two effects occur: (i) photoelectrons are knocked out of an atom and (ii) optical luminescent light is emitted due to recombination processes. Optical fibers at scanning near-field optical microscopes may be used as detector for the X-ray Excited Optical Luminescent (XEOL) light [26, 27, 28] or the SPM tip can be used as a detector for the photoelectrons [29, 30]. A schematic view of a SPM-tip as photoelectron detector is presented in Figure 21. The atomic scattering factor for the forward scattering direction is $f(0) = f_1 + if_2$, where the imaginary part is derived from the atomic photoabsorption and the real part is related to the imaginary part by the Kramers-Kronig dispersion. Both parts strongly vary at the electronic edges of an element resulting in a stronger absorption of the incident X-rays and, hence, a variation of the XEOL and the photoelectron yield. Thus, the yield as a function of the X-ray energy gives access to the chemical information in the sample.

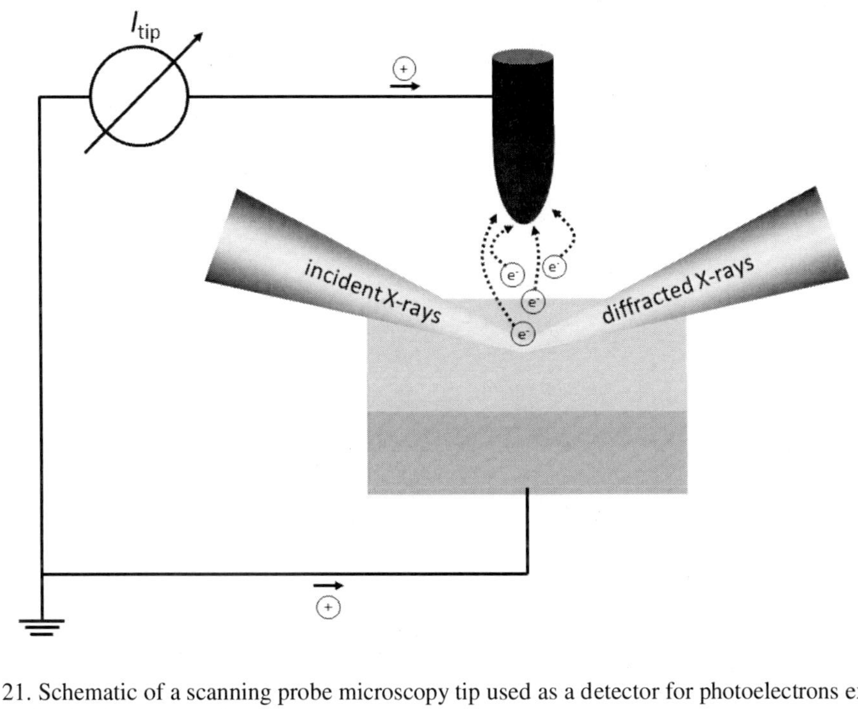

Figure 21. Schematic of a scanning probe microscopy tip used as a detector for photoelectrons excited by X-rays impinging on a surface.

The typical size of the X-ray beams is in the range of few hundred nanometers up to several micrometers. Hence, the spatial resolution is very limited. In contrast, the spatial resolution of SPMs is in the nanometer or sub-nanometer range. The SPM tip may be used as detector for the optical luminescent light or the photoelectrons and, thus, the combination of the two techniques - SPM and X-ray absorption - give access to high chemical resolution maps allowing for the investigation of single nanostructures. Note that the ejected electrons are collected by the complete SPM tip including lateral surfaces. In order to improve the lateral resolution, the area of the active SPM tip has to be minimized by covering it with a sufficiently thick insulating layer. These "smart tips" where only the very end of the SPM tip is active and serves as electron detector allow for X-ray absorption spectroscopy studies with a resolution of about 10 nm [31, 32].

Besides the direct detection of the photoelectrons or the X-ray excited luminescence the change in capacitance or variations in a tunneling current can be recorded by a scanning probe [33, 34, 35]. For instance, the photoionization of defects by X-ray absorption changes the capacitance resulting in site-selective X-ray absorption spectra of the defects. These new experimental techniques pave the way to novel investigations of various physical properties of individual nanostructures. As described above several different combinations of X-ray absorption spectroscopy techniques and scanning probe microscopy have been developed in the recent past. Concerning the combination of SPM and X-ray diffraction, much less applications exist so far. For *in situ* spectroscopy methods, broad beams can be employed since the SPM tip serves as the detector and, thus the resolution is defined by the SPM-tip and not by the beam size. In contrast, for X-ray diffraction with high spatial resolution, the highly focused X-ray beams are needed. A schematic view of the experimental setup for *in situ* nanofocused X-ray diffraction in combination with scanning probe microscopy is displayed in Figure 22. The incident X-ray beam is monochromatized and then focused onto the sample which is mounted in the SPM. The X-rays diffracted by the sample are recorded by a 2D detector. The mutual alignment of the SPM-tip and the nanofocused X-ray beam is a challenging task which is additionally complicated by the alignment of a low-dimensional structure with respect to both the SPM-tip and the X-ray beam.

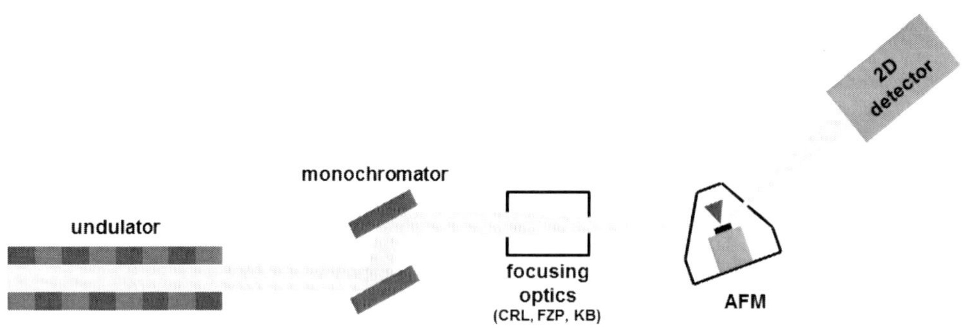

Figure 22. Schematic of the experimental setup including monochromator, focusing optics, the *in situ* SPM, and a 2D detector for recording the diffracted X-rays.

While scanning probe microscopes image the topography of the sample surface, nanofocused X-ray diffraction gives access to the crystalline orientation as well as the strain state of individual grains or nanostructures. In addition, SPMs allow for contacting structures electrically, for manipulating and deforming nanostructures affecting their strain state, their defect density, etc. Thus, the combination of these two techniques allows for following the changes of stress and strain in single nanostructures *in situ* as a function of an applied force or an applied electrical potential. Note that X-rays impinging on a surface may lead to a polarization and, thus, to an electrostatic charging of the material affecting the interaction forces between the SPM tip and the sample. Thus, X-rays may influence scanning probe microscopy imaging of the surface. On the other hand, the SPM tip may influence the XRD measurement by partly shadowing the sample or by additional scattering of the X-ray beam at the tip.

Photocurrent Imaging (PI) Combined with Nanofocused X-ray Diffraction

A specially adapted atomic force microscope was developed at the European Synchrotron Radiation Facility in Grenoble which can be installed on a diffractometer at synchrotron beamlines [36]. It allows for both illuminating the sample installed therein and for recording the diffracted X-rays in a wide angular range. Thus, many different Bragg reflections – symmetric as well as asymmetric – are accessible. In the following, recent results obtained with this AFM in combination with microfocused X-rays are presented. The experiments were performed at the ID01 beamline at ESRF where the X-ray beam was monochromatized to 9 keV using the double-bounce channel-cut Si111 monochromator and focused employing Be compound refractive lenses. The diffracted X-rays were recorded by a MAXIPIX 2D pixel detector with a pixel size of 55 x 55 μm^2. The force sensor used in this AFM is a quartz tuning fork (see Figure 23) with a resonance frequency of 2^{15} Hz = 32768 Hz. Any stress in the tuning fork is converted into charge thanks to the piezoelectric properties of quartz. Thus, resonating the tuning fork induces an alternating stress field which causes an alternating current that can be read-out and allows for the determination of the resonance frequency of the tuning fork. For this purpose, electrical conductive leads are deposited alongside of the tuning fork prongs (red and blue lines in Figure 23). In contrast to an ordinary AFM where the bulky detection system prevents the access to the sample with other tools, the tuning fork with a size of 5 mm allows for the illumination of the sample with X-rays in a wide angular range. The AFM-tip is an electrochemically etched tungsten wire which is glued on one of the two tuning fork prongs. The additional mass of the tungsten tip results in a diminution of the resonance frequency of the tuning fork. Typical resonance frequencies of the system tuning fork – AFM-tip are in the range of 30 to 31 kHz.

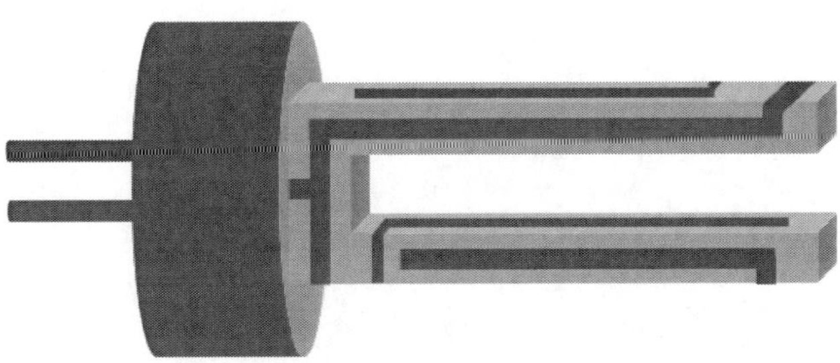

Figure 23. Schematic of the quartz tuning fork used as force sensor in the *in situ* AFM. The lines on the tuning fork represent conductive leads.

In Fig 24 (a), a scanning electron micrograph of bismuth islands deposited on silicon is presented. A bismuth film with a thickness of 500 nm was sputter deposited on Si (100) and, subsequently, the film was structured by UV lithography. The Bi islands are polycrystalline and they consist of grains with a size of about 500 nm. This sample was scanned in the X-ray beam by moving the complete AFM laterally and the photoelectrons emitted from the sample were collected by the AFM-tip. When the X-ray beam hit the tip itself, it also ejects photoelectrons which are re-collected by the AFM-tip. Figure 24 (b) displays the resulting photocurrent image (PI) of the scanned area marked by the dashed rectangle in part (a). The semi-metallic Bi islands are clearly visible with respect to the semiconducting Si substrate. In addition, the AFM-tip is also observed in the upper central part of the photocurrent image.

Simultaneously to the photocurrent image the diffraction signal at the Bi (112) Bragg reflection was recorded. As can be seen in part (c) only very few grains are oriented along this direction. The two images are superposed in part (d) showing the exact location of the grains oriented. For example, the (112) planes are parallel to the surface within the islands. Thus, for polycrystalline samples the exact determination of the position of certain grains with respect to specific structures is not possible by scanning X-ray diffraction alone. Only the combination with photocurrent imaging or other *in situ* imaging techniques allows for the unambiguous determination of the position of specific grains. In addition, this combination allows for the positioning of the AFM-tip with respect to the focused X-ray beam as well as the selection of specific grains in polycrystalline microstructures for further studies.

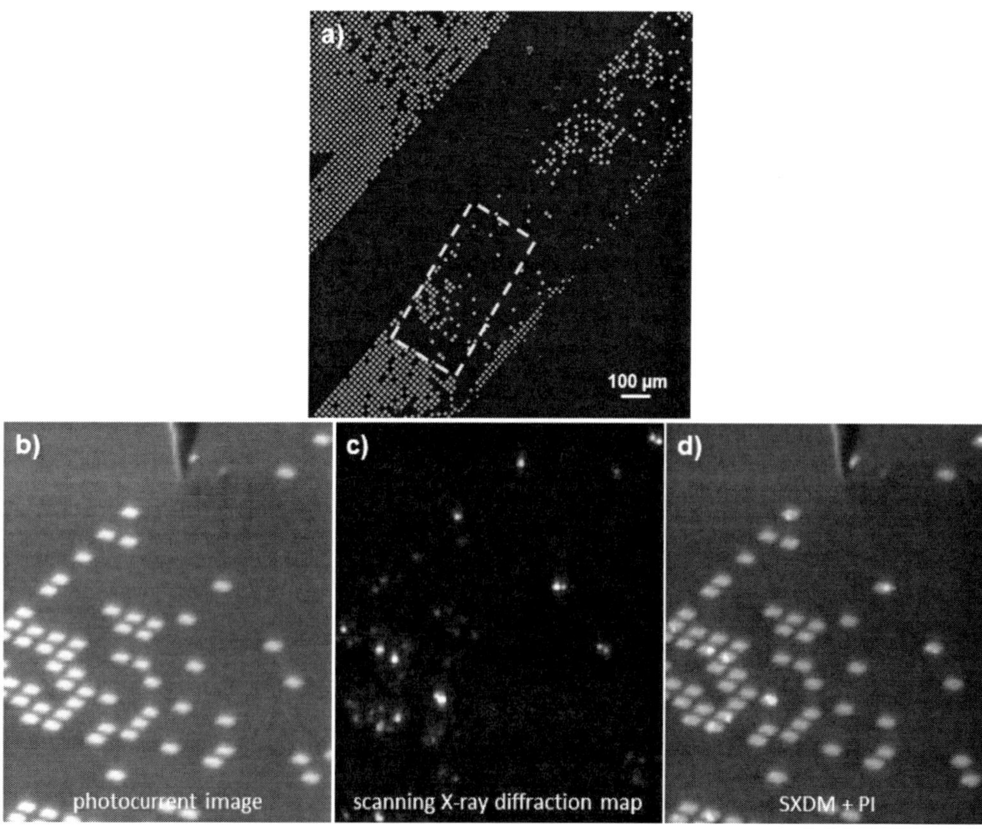

Figure 24. a) Scanning electron microscopy image of polycrystalline Bi islands sputter deposited on Si (001). b) Photocurrent image (PI) and c) scanning X-ray diffraction map (SXDM) of the area marked by the dashed rectangle in part (a). d) Superposition of the two images shown in parts (b) and (c).

Figure 25(a) presents a 2D XRD pattern at the Bi(110) Bragg reflection clearly showing a part of the Debye-Scherrer ring revealing the polycrystallinity of the Bi islands. The scattering in the horizontal direction on the detector is directly related to the tilting of the investigated structure. For the analysis, the detector was divided into four sections, thus, giving access to the tilting of various Bi grains (as shown in Figure 25 (a)). A scanning X-ray diffraction map was recorded reading out the four parts of the detector separately. Part (b) displays the SXDM where the different colors represent the different detector areas and, hence, differently oriented grains. Simultaneous to the diffracted X-rays the photocurrent was measured. Figure 25(c) displays the superposition of the scanning X-ray diffraction map and the photocurrent image demonstrating the excellent alignment capabilities of the AFM-tip by this method.

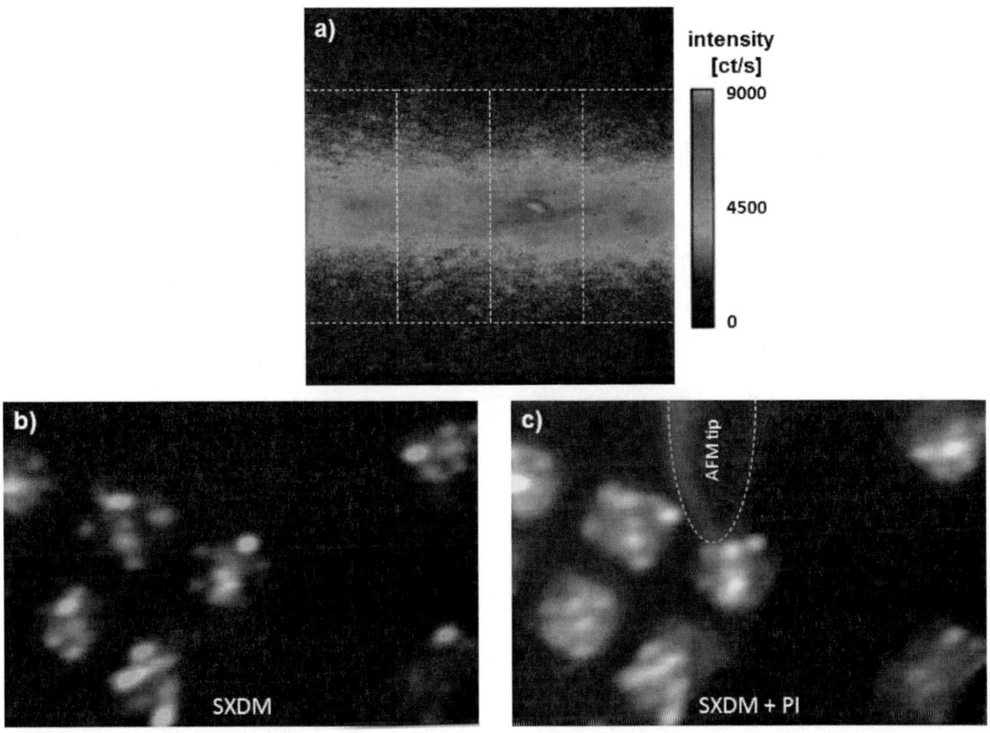

Figure 25. a) 2D X-ray diffraction pattern of a Bi island at the Bi112 reflection. b) Scanning X-ray diffraction map of Bi islands. The color code represents grains diffracting into the four different sections of the detector marked as dashed rectangles in part (a) and, thus represent different tilting angles of the grains. c) Superposition of a photocurrent image and the SXDM shown in part (b).

In Situ Combination of Atomic Force Microscopy and Nanofocused X-ray Diffraction

Atomic force microscopes do not only allow for imaging the topography of a sample but also allow for manipulating and modifying structures on the nanoscale. For instance, well defined forces in the range of nanonewtons can be applied on single nanostructures. Thus, the mechanical behavior of low-dimensional materials can be investigated by *in situ* mechanical testing of nanostructures using an atomic force microscope and simultaneous nanofocused X-ray diffraction giving access to the elastic strain as well as the formation of defects and their evolution.

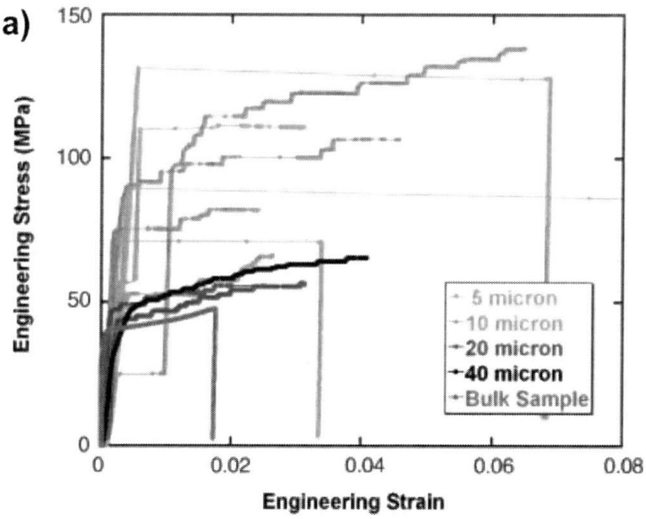

(a)

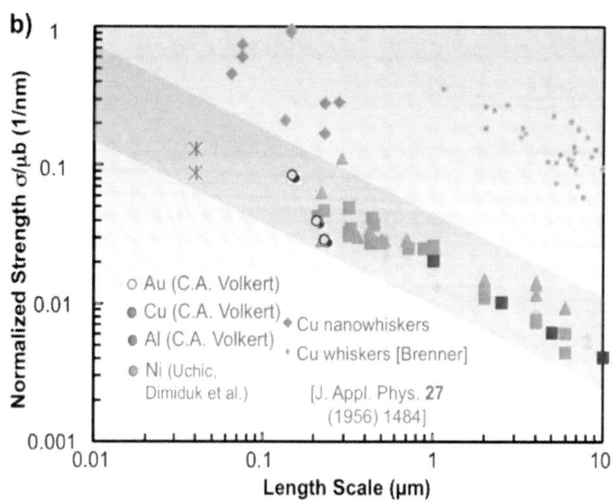

(b)

Figure 26. a) Stress-strain curve of focused ion beam milled Ni micro-pillars. Reprinted with permission from [37]. b) Strength versus size for micro- and nano-sized specimen. Reprinted with permission from [38].

With the advent of the focused ion beam (FIB), Uchic et al. prepared metal micro-pillars by FIB milling and studied their mechanical properties by compression tests [37]. Figures

26(a) and (b) present stress-strain curves for Ni micropillars and the normalized strength as a function of the pillar diameter for fcc materials [38], respectively. As can be seen, the yield stress, which is the stress needed to overcome the elastic deformation and enter the plastic regime, increases with decreasing pillar diameter. This behaviour became known in literature as "smaller is stronger" [37, 39, 40]. A lot of speculation has been involved in the interpretation of the non-continuous stress-strain curves and the yield stress increase that are observed when deforming micro-pillars. The smaller the pillar is, the less dislocations are available, and thus the larger is the yield stress. Qualitatively, larger pillars (> 10 μm) should contain many dislocations and behave like bulk material: dislocation sources can be found easily and hardening takes place by dislocation-dislocation interaction. As the diameter decreases few dislocations can be activated as sources (possibly spiral sources) and give rise to strain bursts or load drops (depending on the straining mode - load or displacement controlled) where dislocations are ejected from the pillar through free surfaces and therefore do not contribute to hardening. In very small pillars (<1 μm), the probability to find sources or even a dislocation becomes very small, and the multiplication of dislocation requires enormous amount of stresses. Figure 26 (b) also contains results obtained on Cu nanowires and whiskers [41]. Their yield stress exceeds the yield stress for FIB milled pillars and almost reaches the theoretical limit of the yield stress for copper. Note that also the preparation method may play an important role since FIB milling damages the surface, implants Ga ions in a near surface region, and induces small dislocation loops as well as vacancies. In consequence, FIB damage may affect the mechanical properties of micrometre sized structures [42]. For instance, molybdenum alloy micro-pillars prepared from a eutectic did not show any size effects in mechanical compression tests but exhibited a yield strength close to the theoretical limit of the material [43].

Besides the increase of the yield strength with decreasing specimen size, recent studies about the onset of plasticity of semiconductors at the nanoscale have revealed a ductile behavior at room temperature in small pillars. While bulk semiconductors are usually brittle at low temperatures because of deep Peierls valleys, they become ductile at elevated temperatures. In particular in silicon, considered as the model semiconductor, it has been shown experimentally that the brittle-ductile transition (BDT) temperature is of the order of room temperature around few hundreds of nanometers under compressive strain [44, 45]. These new properties of semiconductor at nanoscale allow designing NEMS (Nano-Electro-Mechanical Systems) and electronic devices able to sustain very huge stresses minimizing the failure risk.

In the recent past, several *in situ* mechanical testing methods for micro- and nanometer sized samples have been developed employing μ–Laue diffraction [46, 47, 48, 49]. Raman spectroscopy [50], scanning electron microscopy [40, 45, 51] as well as transmission electron microscopy [52, 53, 54]. These techniques allow for investigating also the elastic properties as well as the plastic ones while *ex situ* studies mainly give access to the plastic regime. Here, we review the combination of *in situ* atomic force microscopy and nanofocused X-ray diffraction to study the mechanical properties of sub-micrometer structures [55, 56, 57]. The *in situ* experiments were performed at the ID01 beamline at ESRF using the specially adapted AFM presented before. The X-ray beam was monochromatized to 9.9 keV using the double bounce channel-cut Si (111) monochromator. For focusing the beam 32 Be compound refractive lenses with a radius of curvature of 200 μm were employed at a distance of 1 m in front of the sample resulting in a spot size of 1.5 and 2.8 μm in vertical and horizontal direction, respectively. The AFM-tip and a particular epitaxial SiGe island ("3D X-Ray Diffraction" section and Figure 7 for sample description) were aligned with respect to the focused X-ray beam by photocurrent imaging and SXDM, respectively. The diffracted X-rays were recorded by a Princeton PI-SCX-1300 2D CCD detector. For mechanical compression studies the AFM tip was electrochemically blunted to a radius of curvature in the range of 1-2 μm to get as close as possible to a homogeneous pressure application like for a flat punch. For pressure application the AFM-tip is lowered towards the sample surface and eventually brought into contact with the selected structure. During mechanical testing the AFM-tip is moved stepwise against the object under investigation and at each step X-ray diffraction data as well as the resonance frequency of the tuning fork are recorded simultaneously. X-ray diffraction patterns recorded in the vicinity of the Si (004) Bragg peak are presented in Figure 27 (a). With increasing applied pressure the CTRs of the SiGe island side facets collapse (II and III) and the Bragg signal of the SiGe island moves to larger q_z values (IV and V) until it is superimposed with the Bragg reflection of the Si substrate (VI). The resonance frequency of the tuning fork is shifted to higher frequencies from about 29 kHz for the free tuning fork to 40 kHz for an applied pressure, where the two diffraction signal from the Si substrate and the SiGe islands are superimposed (see Figure 27 (b)).

In Figure 28, the lattice parameter of the SiGe island a_{SiGe} as well as tuning fork resonance frequency f_{res} are shown as a function of the travel Δz of the sample against the AFM tip which resembles the applied pressure. The island mean lattice parameter was inferred from the center of mass of the diffuse scattering stemming only from the island itself. As can be clearly seen, three regimes exist: In the beginning (up to $f = 31$ kHz) the tip only slightly touches the island top surface establishing a firm contact while leading to small frequency shifts and minor changes in the island lattice parameter. In an intermediate frequency range (31 – 39 kHz) the tip is in well-established contact with the island and the whole structure is compressed. As the frequency reaches 40 kHz, the lattice parameter reaches the value of silicon and the diffraction signal cannot be separated anymore from the substrate. After retracting the AFM-tip and thus releasing the pressure from the island, the 2D XRD image becomes exactly the same as before pressure application (i.e. image I of Figure 27(a)) revealing that the island was deformed purely elastically.

In order to correlate the lattice parameter change with the applied pressure, a complete analysis of the mechanical setup is necessary. When the pressure applied on the investigated island is gradually increased, the oscillating system is no longer the tuning fork alone but rather a coupled system consisting of the tuning fork with tip assembly on one hand and the island on the other. Strong coupling through direct contact of these two elastic systems with comparable stiffness results in the observed frequency shift as well as a broadening of the resonance curves compared to the free tuning fork (see Figure 27 (b)). However, the resonances still exhibit well defined maxima which enable us to observe shifts as small as 1-2 Hz. The stiffness S of the contact area between the AFM tip and the compressed island can be directly inferred from the frequency shift of the tuning fork. The contact area itself can then be deduced from S and the Young modulus of the two materials in contact using the Hertz contact model [58, 59]. This approach allowed us to determine the Young's modulus for a single SiGe island amounting to (108 ± 12) GPa which is very close to the literature value of 105 GPa.

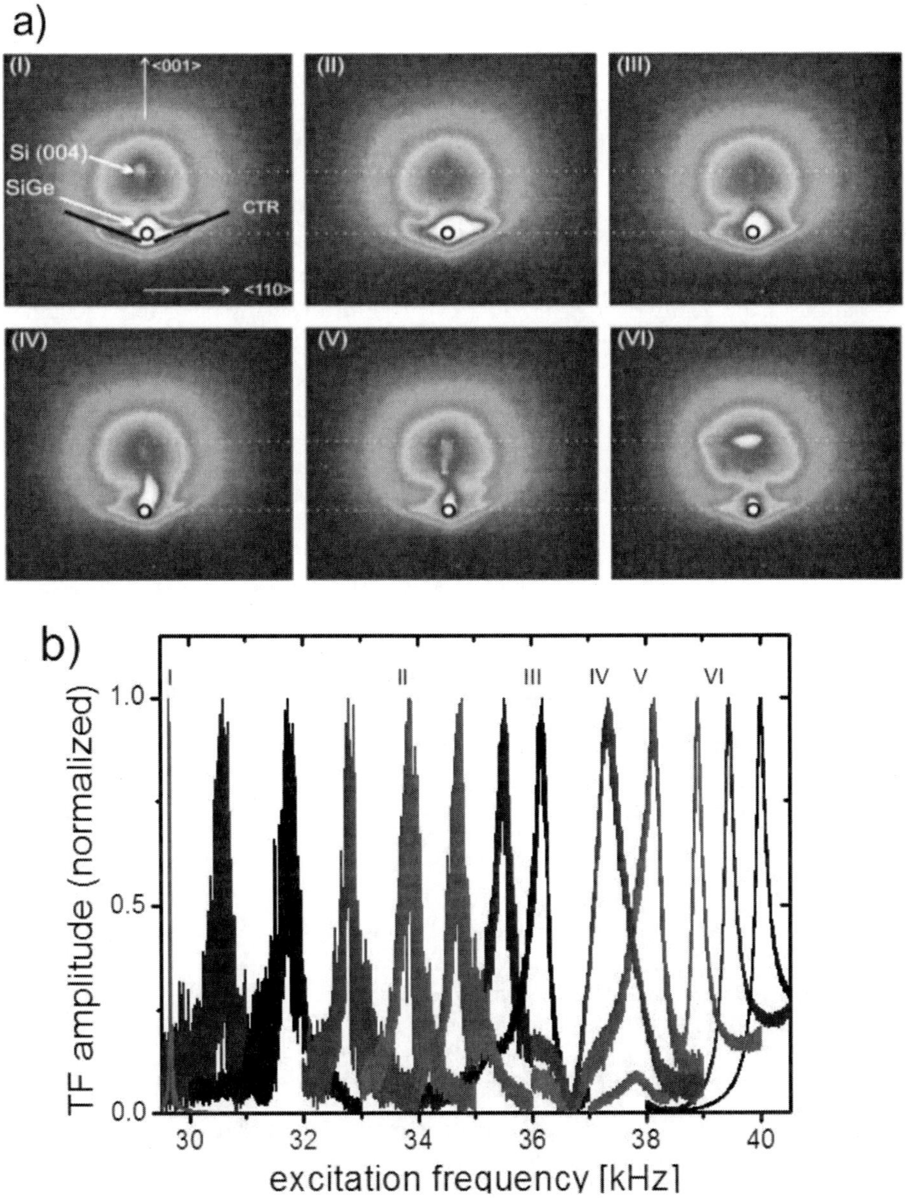

Figure 27. a) X-ray diffraction maps for 6 consecutive pressures applied on an individual SiGe island. The Si (004) and SiGe Bragg reflections are marked by arrows. The CTRs originating from the island side facets are highlighted by black lines, and the substrate CTR is encircled. b) Normalized resonance frequencies of the AFM tuning fork. Note the clear shift of the resonance frequency to higher values occurs with the shift of the SiGe signal towards the Si substrate peak in (a). The resonance curves corresponding to the diffraction maps from panel (a) are indicated by Roman numerals I to VI. Reprinted with permission from [56]. Copyright [2009], American Insitute of Physics.

However, as mentioned above 2D X-ray diffraction images represent only one cut through reciprocal space while Bragg reflections of sub-micrometer structures are three-dimensional. Thus, the three-dimensional intensity distribution has to be recorded *in situ*. As mentioned above 3D-RSMs can be obtained by three different approaches. Every motor movement of the diffractometer induces vibrations which may lead to the damage or the destruction of the AFM-tip and/or the sub-micrometer structure under investigation during *in situ* mechanical testing. Hence, *in situ* 3D-RSMs were recorded by energy scans [57]. This experiment was performed on a SiGe island at the ID01 beamline at ESRF using a Fresnel zone plate for focusing the X-ray beam with energy of 9 keV. The X-rays were focused down to 200 and 300 nm in vertical and horizontal directions, respectively. For increasing the depth of focus the illuminated area on the Fresnel zone plate was reduced to 50 x 50 μm^2 allowing for a variation of the X-ray energy of ~150 eV. The diffracted X-rays were recorded by a MAXIPIX 2D pixel detector with a pixel size of 55 x 55 μm^2 mounted 1.35 m after the sample.

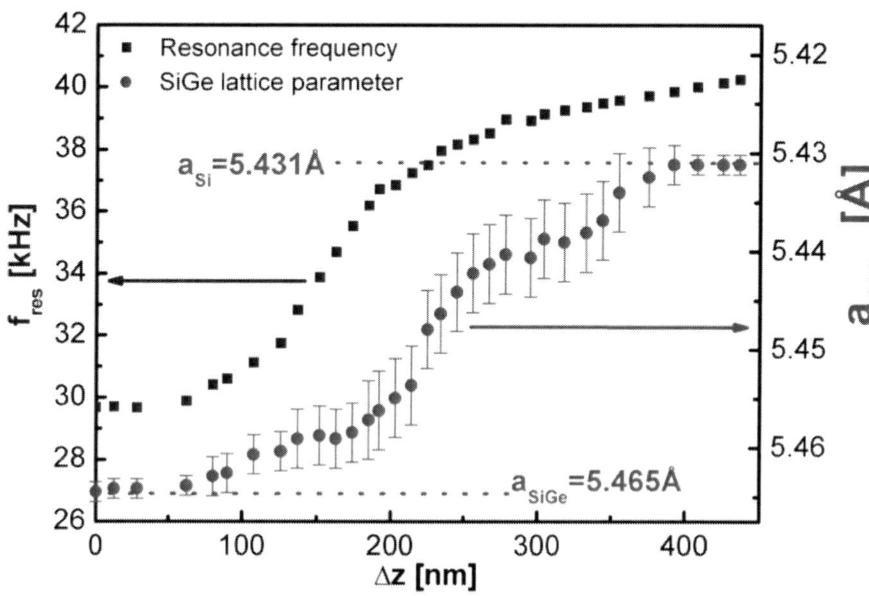

Figure 28. Variation of the SiGe island lattice parameter (red) as a function of the sample displacement Δz. The tuning fork resonance frequency (blue), which varies considerably as the tip touches the island, follows the variation of the lattice parameter. Reprinted with permission from [55]. Copyright [2009], American Insitute of Physics.

To measure 3D RSMs during the deformation energy scans are performed. To this aim, the energy is varied by $\Delta E = \pm$ 100 eV (corresponding to a $\Delta q = \pm 0.516$ nm^{-1} as obtained by a θ-2θ scan in a range $\Delta\theta = \pm 0.37$ °) in steps of 1 eV. The exposure time amounted to 10 s at each energy step. Due to the alignment of the undulator gap at each energy value, a complete energy scan lasted for about one hour. Despite the long measurement time, no drifts, neither of the X-ray beam nor of the AFM, were observed, i.e., two subsequent energy scans at the same compression state revealed the same diffraction signal. This also implies that the radiation damage is negligible for this sample. Figure 29 displays 2D diffraction patterns (first line), the registered 3D reciprocal space maps (middle), and (q_x,q_z) cuts through the 3D-

RSMs (bottom) at $q_y = 0$ for a SiGe island at different deformation stages. The data show the Si (004) Bragg reflection at q_z = 46.3 nm^{-1}, the crystal truncation rod (CTR) of the Si substrate, and the signal of the SiGe island (q_z = 46.05 nm^{-1}), including the CTRs originating from the side facets of the island. The positions of the Si (004) Bragg reflection and the signal of the SiGe island are indicated by dashed lines. In the central vertical cuts the island facet CTRs are highlighted by dotted lines. During pressure application on the island top facet the substrate CTR on the 2D diffraction patterns remains at the same position proving that the whole sample does not tilt during the experiment. Thus, all changes revealed by the *in situ* 3D diffraction mapping originate from the deformation of the specimen. Comparing with the diffraction pattern of the virgin sample (Figure 29(a)), the central part of the diffraction signal of the SiGe island vanishes with increasing mechanical load (see Figure 29 (b) and (c)), while the Si (004) Bragg reflection becomes more diffuse. Hence, a part of the island signal becomes superimposed with the substrate Bragg peak. In addition, the Si (004) Bragg peak develops a sub-structure for increasing load which may be caused by a superposition with the SiGe island signal or by strain induced in the Si substrate. As visible in Figure 29 (b) and (c), the streaks originating from the island side facets turn to steeper angles with increasing load. The load is expressed here by the vertical movement of the AFM-tip against the island top facet after reaching contact, although the applied pressure cannot be determined directly from this movement. Following the direction of enhanced intensity along the contours of the triangular intensity distribution (see dotted lines in Figure 29) the angle of elevation with respect to (110) of the pristine sample is estimated to (36 ± 2)° being in good agreement with the expected 35.26° angle between (110) and (111). The appearance of (111) side facets has been determined for these islands by high-resolution X-ray diffraction [60]. This angle increases to (55 ± 2)° and to (60 ± 2)° when moving the AFM-tip 470 and 750 nm downwards, respectively. The increasing angles of the CTRs indicate that the island shape is being deformed during the mechanical loading. In the stage of highest loading (see Figure 29(d)), when the AFM-tip has been lowered by 1190 nm, the CTR angle returns back to (36 ± 2)° as found for the pristine sample accompanied by a vanishing island signal. In addition, speckle patterns are observed around the position of the SiGe island signal and along the substrate CTR which may indicate the presence of defects. This finding cannot clearly be interpreted. The most probable explanation is a fracture of the central part of the islands is accompanied with strain release of the remaining outer parts, and it brings the side planes back to their initial position before the compression. This hypothesis could not be verified after loading due to a backlash of the AFM-tip during retraction which first increased the load before releasing the pressure.

For a deeper understanding of the experimental results, the displacement field and the shear stress of a SiGe island with a base width of 1 μm and a height of 500 nm epitaxially grown on Si(001) was simulated by the finite-element method (FEM) using COMSOL Multiphysics code. The lattice mismatch between the Si substrate and the $Si_{88}Ge_{12}$ island was simulated in the framework of thermoelasticity. The temperature was taken constant and island and substrate are assigned different thermal-expansion coefficients [61, 62]. For calculations, the complete elastic stiffness tensor was taken into account. The stiffness for the SiGe island was derived employing a law of mixture [10, 56, 63]. For sake of simplification, we treated the islands as homogeneous solid solution neglecting the inhomogeneous Ge composition as determined by high-resolution X-ray diffraction analysis [60]. For computations of the island deformation, a distributed load on the island top facet was used.

The corresponding 3D reciprocal space maps were obtained by Fast Fourier Transformation (FFT) [64]. For this purpose, the phase is calculated which originates from the displacement field in the sample with respect to the Si reference lattice. From this phase, the 3D FFT is computed.

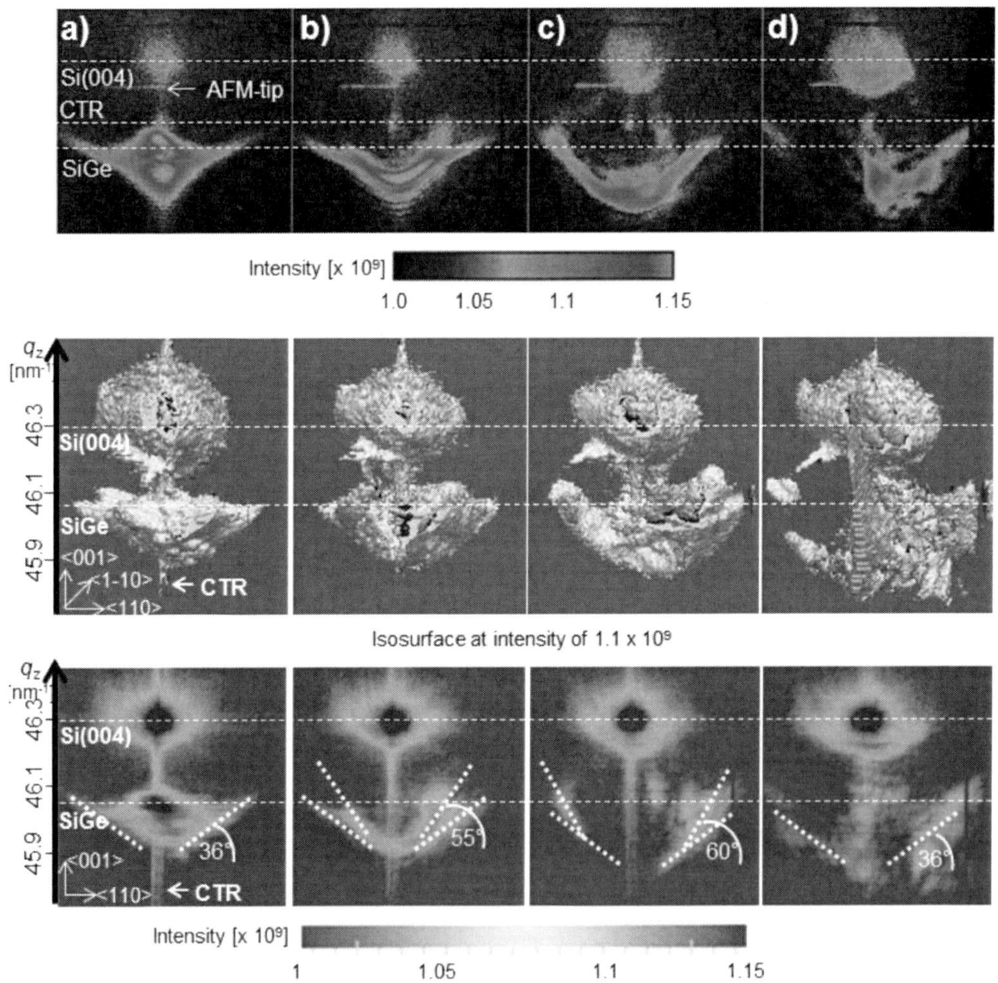

Figure 29. The 2D X-ray diffraction patterns (upper row), reconstructed 3D reciprocal space maps (central row), and (q_x,q_z) cuts through the *in situ* 3D-RSMs at central q_y (lower row) for (a) a pristine SiGe island and (b)-(d) the same SiGe island at different deformation stages. The position of the Si (004) Bragg reflection and the signal of the SiGe island is indicated by dashed lines. The dotted lines highlight the CTRs originating from the island side facets. Reprinted with permission from [57].

In Figure 30, the simulated displacement field, the shear stress, and the shape for a pristine sample (a) consisting of a SiGe island and a Si substrate, as well as for samples with a uniform stress of 1, 3, and 5 GPa on the island top facet, are presented (Figure 30 (b) to (d)). For each of these samples, a 3D reciprocal space map was computed from the FEM simulations by FFT and a vertical central cut through the 3D-RSM was performed corresponding to the presentation of experimental data shown in Figure 29. The calculated

3D-RSM for the pristine sample shows the Si (004) Bragg reflection, the crystal truncation rod of the Si substrate, and the signal of the SiGe island with the CTRs originating from the side facets of the island. The in-plane epitaxial compressive strain induces a 0.8% Poisson out-of-plane expansion in the pristine island, which results in a displacement of more than 4 nm at the top with respect to the Si reference lattice. This out-of-plane displacement within the island is reduced, when an external pressure is applied and becomes negative when the external pressure exceeds 3 GPa, indicating that the lattice parameter of the SiGe island under compression becomes close to the one of Si in this direction. Besides the SiGe island, the substrate below the island exhibits a compressive vertical strain field for large pressures applied on the island top facet resulting in a change of the intensity distribution close to the Si peak. In addition to the displacement field, the shear stress induced by the pressure application is shown in the second row of Figure 30. While a small shear stress exists for the pristine sample, in particular, at the island-substrate-interface, the absolute value of the shear stress at the side facets increases to more than 1 GPa for an applied pressure of 5 GPa. This shear stress causes a shape deformation of the island. In particular, it leads to an outward bending of the island side facets. For a better visualization, the deformation which is shown in the third row of Figure 30 has been increased by a factor of 25. In reciprocal space the bending of the side facets caused by the shear stress leads to a rotation of CTRs to steeper angles and to change from a straight to a curved line as observed in the calculated 3D-RSMs and the central (q_x,q_z) cuts presented in the lower two rows of Figure 30. For visualization the direction of the CTRs are highlighted by dotted lines drawn by eye along the lines of highest intensity. For the pristine sample, the angle between the facet CTR and the horizontal (<110> direction) amounts to 36 ± 2 ° confirms the value found in the *in situ* experiment (Figure 29 (a)). For the ease of comparison, we approximated the curved CTRs for the compressed samples by two parts: One part follows the same angle of 36 ° as found for the pristine island decreasing in length with increasing pressure. It measures the section of the side plane close to the substrate which is not deformed by mechanical load on top of the island. The second part exhibits a steeper angle increasing to 70 ° and 80 ° for an applied pressure of 3 and 5 GPa, respectively. This part clearly measures the deformation of the island top region.

The simulated data are in good qualitative agreement with the experimental ones. According to the calculations, the CTR angle of 60 ± 2 ° found in the experiment corresponds to an external pressure of about 2 to 3 GPa. A detailed analysis of the applied pressure in the experiment was not within the scope of this work. Here, we aimed on the development of a new approach for *in situ* mechanical studies on single sub-micrometer structures using the SiGe islands as a test structure. As mentioned above, the focal spot of the X-ray beam was smaller than the island. Thus, not the complete SiGe island was illuminated and, hence, the measured diffraction signal may not represent the complete deformation of the specimen. The pressure obtained from the CTR angle is a first estimate of the deformation in the top region of the island. The smaller focal spot size may also explain differences between the simulated and the measured 3D-RSMs. This may also be explained by shadowing effects caused by the AFM-tip, or a deviation of the size or shape of the measured island with respect to the island used in the simulation. Further discrepancies between experiment and simulations may be reduced by considering the inhomogeneous Ge distribution within the island in the calculations. The asymmetry of intensity with respect to $q_x = 0$ found in the experimental diffraction patterns but missing in the simulated ones is attributed to a non-central pressure application with the AFM-tip on the island top facet or to a non-central illumination of the island with the X-rays.

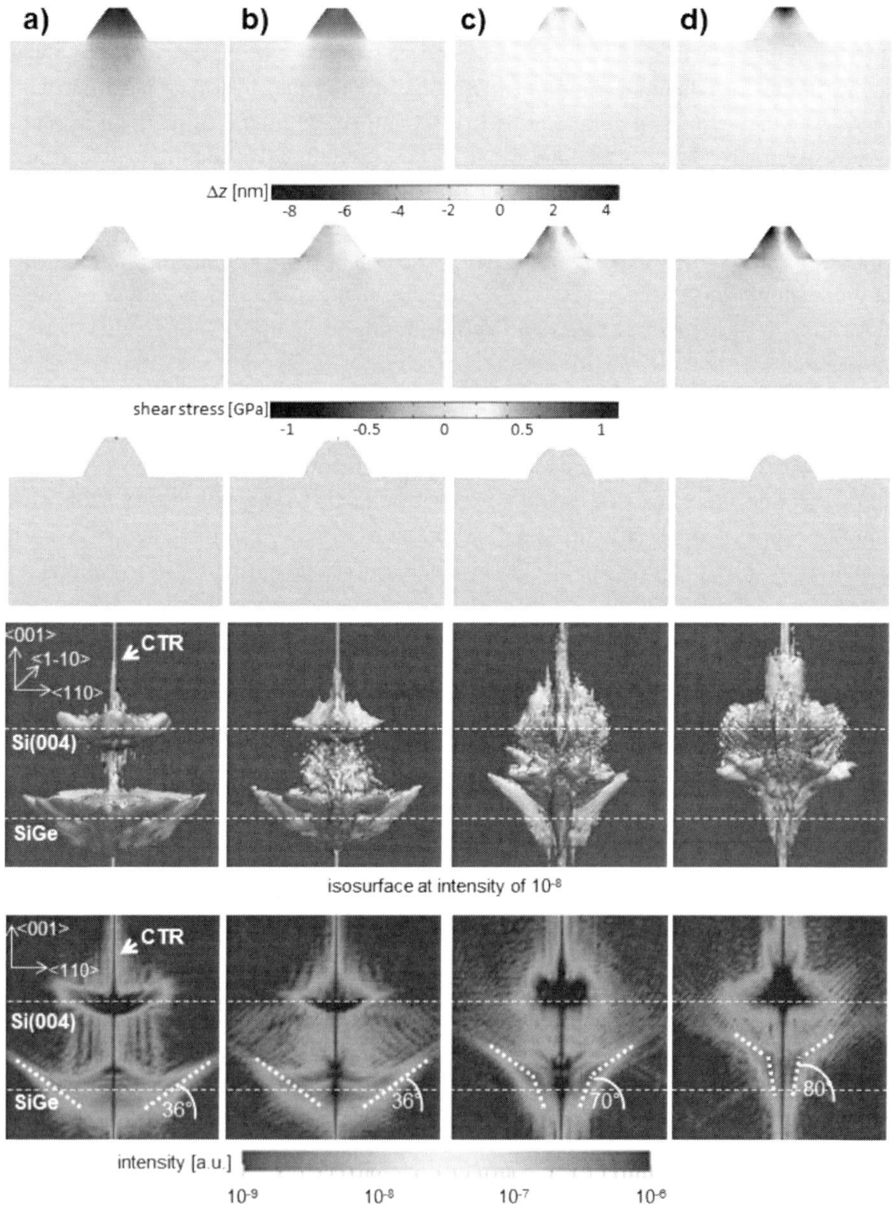

Figure 30. Displacement field (upper row), shear stress (second row), and shape deformation (magnified by a factor of 25) of the island (third row) simulated by the finite-element method using COMSOL Multiphysics. Three-dimensional reciprocal space maps calculated from the displacement fields by fast Fourier transformation (fourth row), and (q_x,q_z) cuts through the 3D-RSMs at central q_y (fifth row) for (a) a SiGe island epitaxially grown on Si(001) and for samples with a distributed load of (b) 1, (c) 3, and (d) 5 GPa applied on the island top facet. The position of the Si(004) Bragg reflection and the signal of the SiGe island are indicated by dashed lines. The dotted lines highlight the CTRs originating from the island side facets. Reprinted with permission from [57].

CONCLUSION

In conclusion, we have shown the advancements which were recently made in the development of specially adapted scanning probe microscopes for *in situ* X-ray studies. In particular, the *in situ* combination of atomic force microscopy and X-ray diffraction opens the door to novel investigations of the mechanical properties of single nanostructures. Despite the great development in nanoscience in recent years regarding the growth of nanosized structures and the characterization of their properties, their mechanical behavior is still not fully understood. X-ray diffraction with its very high strain resolution gives access to the elastic and plastic deformation of mechanically tested nano-objects and the AFM allows for precise force applications and measurements. The existing devices are under constant development for obtaining higher accuracies, in particular, in terms of force measurements. While hard X-rays are nowadays routinely focused down to few hundred nanometers thanks to both 3rd generation synchrotron sources and the advances on focusing optics, several synchrotrons are presently upgraded for providing X-ray beams well below 100 nm in the near future. Within a few years, these developments will allow for studying nanowires, nanotubes, and quantum dots with sizes of few tens of nanometers by X-ray diffraction giving access to their strain originating from the growth process and/or from *in situ* mechanical tests.

REFERENCES

[1] Bragg, W. H.; Bragg, W. L. *Proc. R. Soc. Lond.* A 1913, 88, 428–438.
[2] Ewald, P. *P. Phys. Z.* 1913, 14, 465–472.
[3] Robinson, I. K. *Phys. Rev. B* 1986, 33, 3830–3836.
[4] Feidenhans'l, R. *Surf. Sci. Rep.* 1989, 10, 105-188.
[5] Baudot, S.; Andrieu, F.; Rieutord,; F. Eymery, J. *J. Appl. Phys.* 2009, 105, 114302.
[6] Nam, C.-Y. Jaroenapibal, P. Tham, D. Luzzi, D. E. Evoy, S. Fischer, J. E. *Nano Lett.* 2006, 6, 153–158.
[7] Wang, Z.-J.; Liu, C.; Li, Z.; Zhang, T.-Y. J. *Appl. Phys.* 2010, 108, 083506.
[8] Zhu, Y.; Qin, Q.; Xu, F.; Ding, Y.; Zhang, T.; Wiley, B. J.; Wang, Z. L. *Phys. Rev. B* 2012, 85, 045443.
[9] Dorsch, W.; Strunk, H. P.; Wawra, H.; Wagner, G.; Groenen, J.; Carles, R. *Appl. Phys. Lett.* 1998, 72, 179–181.
[10] Vegard, L. Z. *Phys.* 1921, 5, 17 – 26.
[11] Cornelius, T. W.; Carbone, D.; Jacques, V. L. R.; Schülli, T. U.; Metzger, T. H. *J. Synchrotron Radiat.* 2011, 18, 413–417.
[12] Mimura, H.; Handa, S.; Kimura, T.; Yumoto, H.; Yamakawa, D.; Yokoyama, H.; Matsuyama, S.; Inagaki, K.; Yamamura, K.; Sano, Y.; Tamasaku, K.; Nishino, Y.; Yabashi, M.; Ishikawa, T.; Yamauchi, K. *Nat. Phys.* 2009, 6, 122–125.
[13] Nugent, K. A. *Adv. Phys.* 2010, 59, 1–99.
[14] Diaz, A;, Mocuta, C.; Stangl, J.; Mandl, B.; David, C.; Vila-Comamala, J.; Chamard, V.; Metzger, T. H.; Bauer, G. *Phys. Rev.* B 2009, 79, 125324.
[15] Godard, P.; Carbone, G.; Allain, M.; Mastropietro, F.; Chen, G.; Capello, L.; Diaz, A.; Metzger, T. H.; Stangl, J.; Chamard, V. *Nature Communications* 2011, 2, 568.

[16] Clark, J. N.; Huang, X.; Harder, R.; Robinson, I. K. *Nature Communications* 2012, 3, 993.
[17] Kirkpatrick, P. Baez, A. V. *J. Opt. Soc. Am.* 1948, 38, 766–774.
[18] Snigirev, A.; Kohn, V.; Snigireva, I.; Lengeler, B. *Nature* 1996, 384, 49–51.
[19] Lengeler, B.; Schroer, C.; Tümmler, J.; Benner, B.; Richwin, M.; Snigirev, A.; Snigireva, I.; Drakopoulos, M. *J. Synchrotron Radiat.* 1999, 6, 1153–1167.
[20] http://www.psi.ch/lmn/diffractive-x-ray-lenses
[21] Jefimovs, K.; Bunk, O.; Pfeiffer, F.; Grolimund, D.; van der Veen, J. F.; David, C. *Microelectron. Eng.* 2007, 84, 1467–1470.
[22] Sarkar, S. S.; Solak, H. H.; Raabe, J.; David, C.; van der Veen, J. F. *Microelectron. Eng.* 2010, 87, 854–858.
[23] Mocuta, C.; Stangl, J.; Mundboth, K.; Metzger, T. H.; Bauer, G.; Vartanyants, I. A.; Schmidbauer, M.; Boeck, T. *Phys. Rev.* B. 2008, 77, 245425.
[24] Aristov, V.; Grigoriev, M.; Kiznetsov, S.; Shabelnikov, L.; Yukin, V.; Weitkamp, T.; Rau, C.; Snigireva, I.; Snigirev, A.; Hoffmann, M.; Voges, E. *Appl. Phys. Lett.* 2000, 77, 4058–4060.
[25] Mastropietro, F.; Carbone, D.; Diaz, A.; Eymery, J.; Sentenac, A.; Metzger, T. H.; Chamard, V.; Favre-Nicolin, V. *Optics Express* 2011, 19, 19223–19232.
[26] Larcheri, S.; Rocca, F.; Jandard, F.; Pailharey, D.; Graziola, R.; Kuzmin, A.; Purans, *J. Rev. Sci. Instr.* 2008, 79, 013702.
[27] Pailharey, D.; Mathey, Y.; Jandard, F.; Larcheri, S.; Rocca, F.; Kuzmin, A.; Kalendarev, R.; Purans, J.; Dalba, G.; Graziola, R.; Dhez, O. *J. Phys. Conf. Ser.* 2007, 93, 012038.
[28] Fauquet, C.; Dehlinger, M.; Jandard, F.; Ferrero, S.; Pailharey, D.; Larcheri, S.; Graziola, R.; Purans, J.; Bjeoumikhov, A.; Erko, A.; Zizak, I.; Dahmani, B.; Tonneau D. *Nanoscale Res. Lett.* 2011, 6, 308-308-6.
[29] Chiu, C.-Y.; Chan, Y.-L.; Hsu, Y. J.; Wei, D. H. *Appl. Phys. Lett.* 2008, 92, 103101.
[30] Rose, V.; Freeland, J. W.; Gray, K. E.; Streiffer, S. K. *Appl. Phys. Lett.* 2008, 92, 193510.
[31] Rose, V.; Chien, T. Y.; Hiller, J.; Rosenmann, D.; Winarski, R. P. *Appl. Phys. Lett.* 2011, 99, 173102.
[32] Schmid, I.; Raabe, J.; Quitmann, C.; Vranjkovic, S.; Hug, H. J.; Fink, R. H. *J. Phys. Conf. Ser.* 2009, 186, 012015.
[33] Ishii, M. *Physica B* 2001, 308-310, 1153–1156.
[34] Saito, A.; Maruyama, J.; Manabe, K.; Kitamoto, K.; Takahashi, K.; Takami, K.; Yabashi, M.; Tanaka, Y.; Miwa, D.; Ishii, M.; Takagi, Y.; Akai-Kasaya, M.; Shin, S.; Ishikawa, T.; Kuwahara, Y.; Aono, M. *J. Synchrotron Rad.* 2006, 13, 216–220.
[35] Cummings M. L.; Chien, T.Y.; Preissner, C.; Madhavan, V.; Dising, D.; Bode, M.; Freeland, J. W.; Rose, V. *Ultramicroscopy* 2012, 112, 22-31.
[36] Rodrigues, M. S.; Dhez, O.; Le Denmat, S.; Chevrier, J.; Felici, R.; Comin, F. *J. Instrum.* 2008, 3, 12004.
[37] Uchic, M. D.; Dimiduk, D. M.; Florando, J. N.; Nix, W. D. *Science* 2004, 305, 986–989.
[38] Legros, M. In *Mechanics of Nano-Objects;* Thomas, O.; Prochet, A.; Forest, S.; Ed.; Presses des Mines: Paris; 2011, pp 241 – 269.
[39] Uchic, M. D.; Dimiduk, D. M. *Mat. Sci. Eng. A* 2005, 400-401, 268–278.

[40] Kiener, D.; Grosinger, W.; Dehm, G.; Pippan, R. *Acta Mater.* 2008, 56, 580–592.
[41] Brenner, S. S. *J. Appl. Phys.* 1959, 27, 1484–1491.
[42] Kiener, D.; Motz, C.; Rester, M.; Jenko, M.; Dehm, G. *Mater. Sci. Eng.* A 2007, 459, 262–272.
[43] Bei, H.; Shim, S.; George, E. P.; Miller, M. K.; Herbert, E. G.; Pharr, G. M. *Scripta Mater.* 2007, 57, 397–400.
[44] Östlund, F.; Rzepiejewska-Malyska, K.; Leifer, K.; Hale, L. M.; Tang, Y.; Ballarini, R.; Gerberich, W. W.; Michler, J. *Adv. Funct. Mat.* 2009, 19, 2439–2444.
[45] Michler, J.; Wasmer, K.; Meier, S.; Östlund, F.; Leifer, K. *Appl. Phys. Lett.* 2007, 90, 043123–043123-3.
[46] Kirchlechner, C.; Keckes, J.; Micha, J.-S.; Dehm, G. *Adv. Eng. Mat.* 2011, 13, 837–844
[47] Maaß, R.; van Petegem, S.; Grolimund, D.; van Swygenhoven, H.; Kiener, D.; Dehm, G. *Appl. Phys. Lett.* 2008, 92, 071905–071905-3.
[48] Maaß, R.; van Petegem, S.; Grolimund, D.; van Swygenhoven, H. *Appl. Phys. Lett.* 2007, 91, 131909–131909-3.
[49] Bei, H.; Barabash, R. I.; Ice, G. E.; Liu, W.; Tischler, J.; George, E. P. *Appl. Phys. Lett.* 2008, 93, 071904–071904-3.
[50] Lee, S. W.; Jeong, G.-H.; Campbell, E. E. B. *Nano Lett.* 2007, 7, 2590–2595.
[51] Richter, G.; Hillerich, K.; Gianola, D. S.; Mönig, R.; Kraft, O.; Volkert, C. A. *Nano Lett.* 2009, 9, 3048–3052.
[52] Wang, Z. L.; Poncharal, P.; de Heer, W. A. *J. Phys. Chem. Solids* 2000, 61, 1025–1030.
[53] Minor, A. M.; Lilleodden, E. T.; Stach, E. A.; Morris, J. W. *J. Electron. Matter.* 2002, 31, 958–964.
[54] Oh, S. H. ; Legros, M.; Kiener, D.; Dehm, G. *Nature Materials* 2009, 8, 95–100.
[55] Scheler, T.; Rodrigues, M.; Cornelius, T. W.; Mocuta, C.; Malachias, A.; Magalhães-Paniago, R.; Comin, F.; Chevrier, J.; Metzger, T. H. *Appl. Phys. Lett.* 2009, 94, 023109.
[56] Rodrigues, M. S.; Cornelius, T. W.; Scheler, T.; Mocuta, C.; Malachias, A.; Magalhães-Paniago, R.; Dhez, O.; Comin, F.; Metzger, T. H.; Chevrier, J. *J. Appl. Phys.* 2009, 106, 103525.
[57] Cornelius, T. W.; Davydok, A.; Jacques, V. L. R.; Grifone, R.; Schülli, T.; Richard, M.-I.; Beutier, G.; Verdier, M.; Metzger, T. H.; Pietsch, U.; Thomas, O. *J. Synchrotron Radiat. 2012*, 19, 688–694.
[58] Hertz, H. *J. reine angew. Math.* 1881, 92, 156–171.
[59] Bulychev, S. I.; Alekhin, V. P.; Shorshorov, M. K.; Ternovskii, A. P.; Shnyrev, G. D.; Industrial Lab. 1975, 41, 1409–1412.
[60] Wiebach, T.; Schmidbauer, M.; Hanke, M.; Raidt, H.; Köhler, R.; Wawra, H. *Phys. Rev. B* 2000, 61, 5571–5578.
[61] Christansen, S.; Albrecht, M.; Strunk, H. P.; Maier, H. *J. Appl. Phys. Lett.* 1994, 64, 3617–3619.
[62] Benabbas, T.; Francois, P. ; Androussi, Y.; Lefebvre, A. *J. Appl. Phys.* 1996, 80, 2763–2767.
[63] Wortman, J. J.; Evans, R. A. *J. Appl. Phys.* 1965, 36, 153–156.
[64] Takagi, S. *J. Phys. Soc. Jpn.* 1969, 26, 1239–1253.

In: X-Ray Diffraction: Structure, Principles and Applications ISBN: 978-1-62808-591-4
Editor: Kaimin Shih

Chapter 11

STRUCTURE SOLUTION COMBINING X-RAY AND ELECTRON CRYSTALLOGRAPHY

Junliang Sun*
College of Chemistry and Molecular Engineering,
Peking University, Beijing, China

ABSTRACT

Knowledge of the atomic positions of a new material is crucial for understanding its properties and further modifying the material for better performance. Single crystal X-ray diffraction is an excellent technique for solving this problem, if high quality large single crystals can be obtained. In the case of polycrystalline materials, powder X-ray diffraction is becoming a standard method. For some complicated materials, powder X-ray diffraction may encounter the peak overlapping problem, which makes the structure solution extremely difficult. Electron crystallography can suppy additional information and help to establish the initial model with powder X-ray diffraction data. In this chapter, we will discuss how X-ray diffraction and electron crystallography can be combined for the structure determination using zeolites as examples.

INTRODUCTION

Porous materials have attracted considerable interests because of their diverse industrial applications in the areas of catalysis, separation and ion exchange. Porous materials are normally divided into macro-, meso- and micro-porous materials based on their pore sizes. Among them, crystalline porous materials, such as zeolites (1 Baerlocher and McCsker), aluminoborates (2 Ju et al., 2003), metal-organic-frameworks (3 Phan et al., 2010) etc., are particularly interesting due to their selectivity at a molecular scale. The properties of these crystalline porous materials are closely related to the atomic structure which determines the

* E-mail: junliang.sun@pku.edu.cn.

pore sizes/shapes, positions of active sites and their environments, etc. Thus, to understand their properties and further modify the materials for better performance, it is essential to determine their crystal structures.

X-ray crystallography, including single crystal X-ray diffraction (SXRD) and powder X-ray diffraction (PXRD), has been the major method for their structure determination. The principle of X-ray crystallography can be simply explained by two equations.

$$F_{hkl} = \sum_n f_n \exp(2\pi i(hx_n + ky_n + lz_n)) \tag{1}$$

where f_n is the scattering power for the nth atom; (x_n, y_n, z_n) is its atomic positions; and *hkl* is the index for a certain reflection. Equation-1 calculates structure factors from crystal structures. Structure factors are complex numbers and contain both amplitudes and phases.

$$\rho_{xyz} = \frac{1}{V_{cell}} \sum_{hkl} |F_{hkl}| \cos(2\pi(hx + ky + lz) - \varphi_{hkl}) \tag{2}$$

where x, y and z are the fractional coordinates in the unit cell; V_{cell} is the unit cell volume; $|F_{hkl}|$ is the structure factor amplitude; and φ_{hkl} is the structure factor phase for the reflection *hkl*. Equation-2 calculates the electron densities from structure factors and the peaks of this electron density map correspond to the atomic positions.

In X-ray crystallography, structure factor intensities of all reflections can be observed experimentally and their square roots are structure factor amplitudes. Since structure factor phases are missing, Equation-2 cannot be directly applied to calculate the electron density map. Through different kinds of algorithms, such as Patterson methods (4 Patterson 1934), direct methods (5 Sheldrick 1990), charge-flipping (6 Oszlányi and Sütő 2008), maximum entropy (7 De Vries et al., 1994), genetic algorithm (8 Mitchell 1996), simulated annealing (9 Deem and Newsam 1992) etc., it is possible to recover structure factor phases. In SXRD, all reflections are collected individually with very accurate intensities, so the phases can be readily retrieved by these algorithms. Several new zeolite framework structures, such as SU-15 (SOF) and SU-32 (STW) (10 Tang et al., 2008), PKU-9 (PUN) (11 Su et al. 2009), CoAPO4-CJ40 (JRY) (12 Song et al. 2009), SSZ-57 (*SFV) (13 Baerlocher et al. 2011), Oxonitridophosphate-2 (NPT) (14 Sedlmaier et al. 2011), and GasGeO-CJ63 (JST) (15 Han et al. 2011) have recently been determined by single crystal X-ray diffraction. The limitation for SXRD is that it requires high quality large single crystals (> a few μm) which is not available for many materials, especially for high-silica zeolites. In some cases, even if large crystals are obtained, some materials often have disorders/defects which make SXRD techniques unfeasible.

PXRD are normally used when it is too difficult to synthesize large single crystals. Most of the recently discovered new silicate zeolite framework structures were determined by PXRD, such as IM-16 (UOS) (16 Lorgouilloux et al. 2009), SSZ-56 (SFS) (17 Elomari et al. 2009), LZ-135 (LTF) (18 McCusker et al. 2009), ITQ-34 (ITR) (19 Corma et al. 2008), STA-15 (SAF) (20 Han et al. 2010), MCM-70 (MVY) (21 Xie et al. 2009), IM-20 (UWY) (22 Dodin et al. 2010), ITQ-44 (IRR) (23 Jiang et al. 2010), Linde Type J (LTJ) (24 Broach et al. 2011). In PXRD, millions of small crystals with random orientations are used to generate reflections, and due to the random crystal orientation, each reflection in SXRD becomes a ring in PXRD and its ring size (i.e. its reflection angle, 2θ) is determined by the *d*-spacing

(Figure 1a). By scanning 2θ, we can collect all reflections, and in other words, the 3-dimensional reciprocal space are projected into a one-dimension space, which results in overlapping of reflections with similar or same *d*-values (Figure 1b). The overlapping problem can be partially solved through the fast iterative Patterson squaring (FIPS) (25 Estermann 1992), repartitioning by histogram matching in a charge-flipping algorithm (26 Baerlocher et al. 2007) or other methods with a serial of scans at different conditions. However, when the unit cell parameters become bigger, the overlapped reflections could be up to 80% within 0.3FWHM (full width at half maximum) even with the state-of-the-art high-resolution synchrotron PXRD. The overlapping makes it extremely difficult to get accurate intensities with mathematical algorithms, as well as correct structure factor phases.

In addition to X-ray crystallography, electron crystallography alone can also be used for structure determination of materials, especially for those that are too small to be studied by single crystal X-ray diffraction or too complex to be studied by PXRD. The major differences between electron crystallography and X-ray crystallography are that 1) electrons interact much more strongly with matter than X-rays do; 2) electron beams can be focused to form images by electromagnetic lenses (27 Williams and Carter, 2009). These differences make electron crystallography unique in various applications (28 Zou et al. 2012).

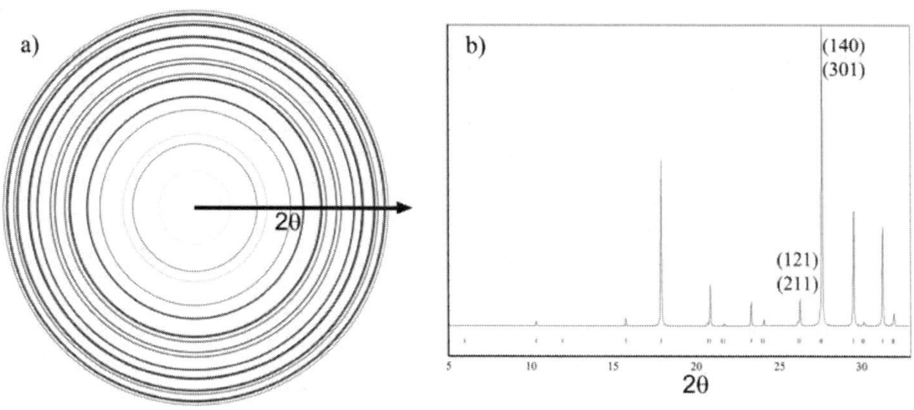

Figure 1. a) Powder X-ray diffraction rings of SrB_4O_7 (Space group: *P*3 with a = 17.145Å, c = 4.2527Å); b) PXRD pattern of SrB_4O_7 where a few reflections have similar or same 2θ values.

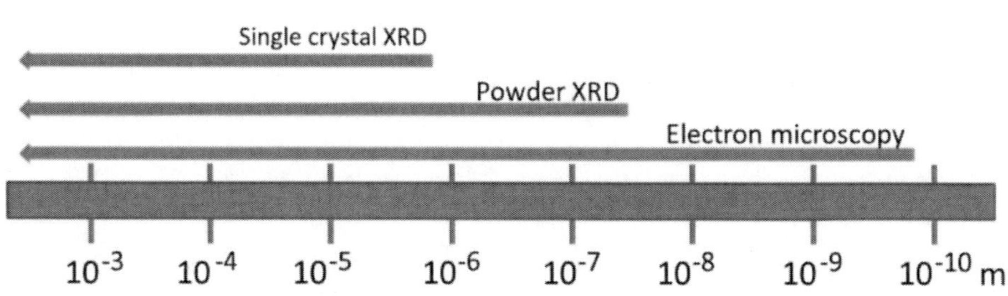

Figure 2. Crystal size limits for the structure determination by different techniques.

The strong interactions between electrons and materials provide both advantages and disadvantages for the structure determination by electron crystallography. As shown in Figure

2, SXRD can be used to determine crystal structures for crystals larger than a few μm with synchrotron X-ray sources. PXRD extends the size limitation for structure determination to about 50nm. When crystal sizes become even smaller, reflection peaks in PXRD are broadened and start to have severe overlapping with a media-sized unit cell. While, with electrons, a crystal with a few unit cells can produce strong enough signals for structure determination, i.e. single-crystal-like electron diffraction data can be obtained from polycrystalline phases. The traditional way to collect electron diffraction data is using manual tilting, which is time-consuming and requires well trained people. A few examples have been solved in such a way, such as MCM-22 (29 Nicolopoulos et al. 1995) and SSZ-48 (30 Wagner et al. 1999). Recently, a few 3D electron diffraction techniques have been developed, such as automated electron diffraction tomography (ADT) (31 Kolb et al. 2011), rotation electron diffraction (RED) (32 Zhang et al. 2010). In these techniques, the data collection is done in an automated manner and takes only about an hour. Various types of materials have been solved by these techniques, such as zeolites, oxides, intermetallic compounds etc.

The main drawbacks of the strong interactions are the dynamical effects and the beam damage. Due to the strong interaction, electrons are diffracted more than once when passing through the material. The electron diffraction is no longer kinematical but dynamical, i.e. the electron diffraction intensities are not simply related to the structure factor intensities, and this makes structure determination from electron diffraction much more difficult than that from single crystal X-ray diffraction. Although the newly developed 3D electron diffraction techniques (ADT, RED) reduce dynamical effects significantly, some structures still have problems for obtaining an initial model. During the interaction between electrons and materials, electrons pass energy to the material, and it may directly break chemical bonds or slowly heat up the sample, which cause some materials losing the crystallinity. This beam damage limits the data resolution as well as the reliability of the diffraction intensities. Thus, for the beam sensitive sample, such as zeolites, metal-organic-frameworks, special care needs to be taken for collecting a high quality electron diffraction data.

The possibility to focus electron beams by electromagnetic fields allows us to bend the diffracted electron beams back to interfere and form images, similar to what we do in optical microscopes. Due to the short wavelength of electrons, this type of images can have extremely high resolution beyond 1-2 Å, i.e. so-called high resolution transmission electron microscopy (HRTEM) images, where atomic columns can be observed directly.

As we discussed above, all diffraction techniques can only obtain structure factor amplitudes, while the information about both the structure factor amplitudes and phases can be obtained experimentally by Fourier transform of the images since they can be roughly considered as a projected potential of the 3-dimensional crystal structure. From a single HRTEM image, a 2-dimensional slice of reciprocal space with amplitudes and phases can be obtained, but it is not enough to obtain the 3-dimensional atomic coordinates. By combining amplitudes and phases from HRTEM images along different zone axes, 3-dimensional reflections can be obtained.

It is not necessary to have all reflections and as long as most of the strong reflections are obtained with both amplitudes and phases, Equation-2 can be used to generate a 3-dimemsional potential map which roughly presents the real crystal potential. This method has been used to determine a number of complicated zeolite structures, such as IM-5 (33 Sun et al. 2010), zeolite beta polymorph-B (34 Corma et al. 2008), ITQ-39 (35 Willhammar 2012) etc. One of the biggest problems for the HRTEM images is the beam damage, which makes it extremely difficult to reach atomic resolution for many types of materials.

The techniques discussed above are all powerful, but have their own limitations. In many complicated cases, one technique is not enough to solve the crystal structure, and different techniques may supply complementary structural information to support each other for tackling more complicated problems. In this chapter, we will use zeolite structures as examples to show how we can combine these techniques.

1. Combine HRTEM Images with X-Ray Diffraction

As we discussed above, the X-ray diffraction methods can only obtain structure factor amplitudes and all phases are missing. Although these phases can be retrieved through different algorithms, it fails quite often for powder X-ray diffraction when structures are too complicated.

HRTEM images contain some information about structure factor phases. Although these phases suffer a lot from various distortions in electron microscopes and may have some errors from the real structure factor phases, they can be used to facilitate the recovering of structure factor phases from powder X-ray diffraction. Here, we can use the examples below to demonstrate how the phases from HRTEM images can be used to help the structure solution together with X-ray diffraction.

Combine with PXRD

In PXRD, the major problem is the peak overlapping which makes the phase recovering extremely difficult. Since phases can be obtained experimentally in HRTEM images with certain accuracy, they can be used to help solve the structure by PXRD. TNU-9 is one of the most complicated zeolites with a monoclinic unit cell (a = 28.2219 Å, b = 20.0123 Å, c = 19.4926 Å, β = 92.331°) (36 Gramm et al. 2006). The sample is in a polycrystalline form and the large unit cell causes severe peak overlapping.

Additionally, the undefined impurity and similar lattice parameters (a ~ $\sqrt{2}b$ and $b \sim c$) make the peaks even more difficult to be separated. With a high-resolution synchrotron X-ray source, the peak overlapping is still above 85% within 0.3FWHM. A direct trial to solve the structure from the PXRD failed. Through three HRTEM images, phases of 258 reflections can be obtained using *CRISP* (37 Hovmöller, 1992).

Using the common reflections, reflections from different images can be merged into one dataset. To combine these phases with PXRD data, the program *Focus* was used (38 Grosse-Kunstleve et al. 1997). *Focus* is a structure solution program that uses the chemical information (e.g. SiO_4 in tetrahedra) to facilitate the structure solution from less reliable data, for example the PXRD data with severe overlapping. In the input, some phases from HRTEM images were input instead of using random phases. After 16 days of CPU time, the correct model was found by *Focus* and it contains 3-dimensional 10-ring channel system with 24 unique Si atomic positions.

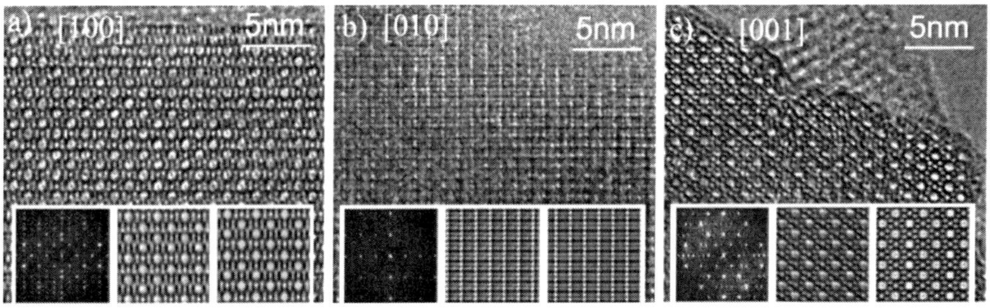

Figure 3. HRTEM images of IM-5 along [100] (a), [010] (b) and [001] (c) directions, from which structure factor amplitudes and phases can be retrieved through Fourier transform. The inserts are their Fourier transform, averaged potential map in *P*1 and averaged potential map in *Cmcm* (39 Baerlocher et al. 2007).

IM-5 is polycrystalline in an orthorhombic cell with about 85% overlapped PXRD peaks, and its structure was solved by combining high-resolution PXRD with HRTEM images in a charge-flipping algorithm by the program *superflip* (39 Baerlocher et al. 2007; 40 Palatinus and Chapuis 2007). Three HRTEM images were taken along three main zone axes shown in Figure 3, and the phases of 95 reflections were derived. When the phases from HRTEM images were directly used as initial phases in charge-flipping iterations, the obtained electron density map was difficult to interpret for locating Si atoms. Thus, a rough structure model was first established with the space group *C*2*cm* by 3-dimensional reconstruction of HRTEM images. This initial model was used to calculate phases such as the starting phases in charge-flipping iterations. Since there are some errors in the initial model, 25% of the phases were input in a random manner but others were used as they were calculated. With 1000 different sets of random phases, the five best ones were combined to identify Si and O positions. The final Rietveld refinement against PXRD shows that IM-5 has a higher symmetry (*Cmcm*) with a 2-dimensional channel system.

Combine with SXRD

Not just structure solution from PXRD needs the help from HRTEM images, and in some special cases, the structure determination from single crystal is not straightforward either. For example, there is severe disordering for SU-78 (41 Yu et al. 2012). Although synchrotron SXRD data were collected from a 30μm-sized crystal of SU-78, it was impossible to obtain the unit cell directly due to both twinning and disordering. From selected area electron diffraction patterns of a small area, the twinning and disordering problems were identified and then the unit cell can be determined as monoclinic with a = 12.6 Å, b = 12.8 Å, c =13.8 Å, β = 108°. Using this unit cell, the SXRD data can be indexed into four twin components. From this indexing, only reflections *hkl* with $h = 3n$ are sharp while others are elongated or form streaks due to the disorder. Since the completeness of this *hkl* file is only about 33%, it was not possible to solve the structure directly using SXRD methods. Thus, reflections 0*kl* were taken out to determine the projected structure along the [100] direction (Figure 4b). The project structure along the [010] direction is disordered and can be determined from HRTEM images (Figure 4c).

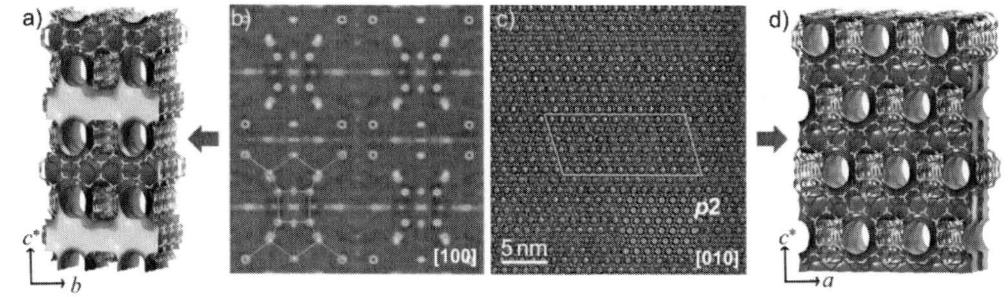

Figure 4. SU-78 structure determined by combining SXRD (b) and HRTEM images (c). As shown in (a) and (d), SU-78 is an intergrowth of two new zeolite beta polymorphs. Viewing along the *a*-axis (a), it has a projection similar to zeolite beta polymorph C. Along with the *b*-axis (d), it is an intergrowth of polymorph A and B. (41 Yu et al. 2012).

Combing the projected structure along [100] and [010] directions, the final structure can be determined as two new polymorphs of zeolite beta with a 3-dimensional 12-ring channel system. This model can be refined with all sharp reflections in SXRD to a reasonable *R*-value ($R_1 = 11.4\%$).

2. COMBINED ED WITH POWDER X-RAY DIFFRACTION

Collecting HRTEM images is normally very time-consuming and requires both good instrument and well-trained operators, especially when the sample is electron beam sensitive. Electron diffraction (ED) patterns can be taken with a lower electron dose, so samples can be stable for longer time which allows less skilful operators to obtain structural information (42 McCusker and Baerlocher 2009).

There are a few different techniques to collect electron diffraction patterns such as selected area electron diffraction (SAED), precession electron diffraction (PED), and 3-dimensional electron diffraction (ADT and RED). The electron diffraction intensities from all these techniques have been used to help structure solutions of crystal structures, which can mainly be divided into two different ways.

Use ED to Obtain the Part/Whole Structure Model

Since it is possible to solve structures from ED directly, one way of using it is to use these intensities to obtain an initial structure model or some structure factor phases, which can then be used for structure solution from PXRD as we discussed above. In the structure determination of ITQ-26 (IWS framework type) (43 Dorset et al. 2008), Dorset and co-workers used the program MICE (44 Gilmore and Bricogne 1997) based on the maximum entropy and the likelihood to obtain the projected potential from the ED pattern along the [001] and [100] directions.

As shown in Figure 5, the channel system from the [001] direction can be clearly seen, indicating that the obtained phases were reasonable for the reflections *hk*0. These phases were then used as starting phases to facilitate the whole structure solution using the program *Focus* from PXRD (38 Grosse-Kunstleve et al. 1997).

The recently developed 3-dimensional electron diffraction techniques make the establishment of structure model from electron diffraction intensities much easier. The aluminophosphate, ITQ-51, has relative low crystallinity and the reflections above 2.5 Å have very low intensities (Figure 6) (45 Martínez-Franco et al. 2013).

It was difficult to solve the structure directly from PXRD, while with RED methods, more than 80% of unique reflections can be collected in about one hour (Figure 6a), from which all Al/P positions and O positions can be obtained immediately through direct methods by the *SHELX* program package (46 Sheldrick 2008).

Surprisingly, from the T-O distances (T = Al/P), it was possible to identify which one is Al and which one is P. The refinement of RED data have a relatively high R_1 (0.37) due to dynamical effects etc.

To confirm the structure model, a Rietveld refinement against the PXRD data was done with soft restraints for T-O distances (Figure 6b). The final refinement confirms the initial model from RED with average shifts by 0.11 Å for T atoms and 0.13 Å for oxygen atoms.

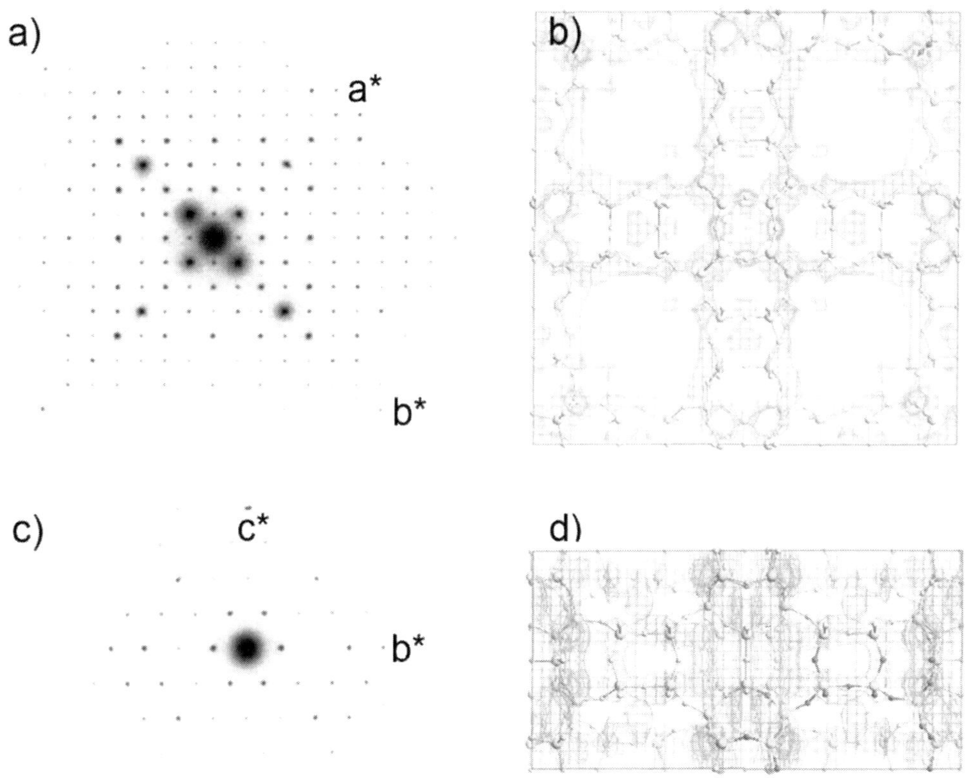

Figure 5. a) The ED pattern of ITQ-26 along the *c*-axis; b) the potential map derived from the ED pattern through a maximum entropy and the likelihood algorithm, which shows a clear channel system (43 Dorset et al. 2008).

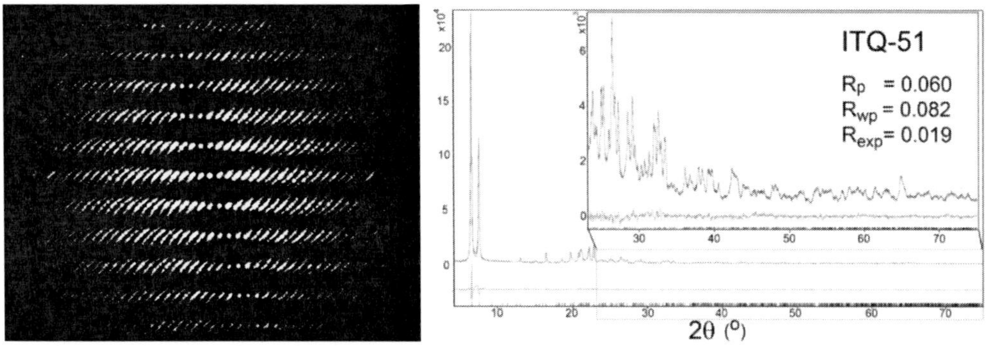

Figure 6. a) Reconstructed 3-dimentional ED data of ITQ-51 from RED methods; b) Rietveld refinement of the ITQ-51 structure model obtained from RED data (45 Martínez-Franco et al. 2013).

Use ED Intensities to Help the Structure Solution from PXRD

Using ED to solve structures is not always feasible, especially when the data quality is not high due to dynamical effects or beam sensitivity. Since the structure factors of X-rays and electrons have some similarity; electron diffraction intensities can be used to partially solve the overlapping problems.

The germanosilicate ITQ-37 was extremely electron beam sensitive (47 Sun et al. 2009), and its HRTEM images have only a resolution of about 6 Å. The electron diffraction extended to a much higher resolution, corresponding to about 3.3 Å, but it is still difficult to obtain complete dataset due to the beam sensitivity. Since all reflections are separated in ED, the SAED intensities were used for pre-partitioning overlapped reflections in PXRD. This pre-partition can be expressed in Figure 7a and b.

Before using ED intensities, all three reflections are equally partitioned. With the knowledge of ED intensities, a single-crystal-like diffraction data could be obtained in such a way (Figure 7b). The pre-partitioning significantly increased the possibilities to find the correct structure model by a charge-flipping algorithm. The final structure refined by PXRD showed that ITQ-37 is a chiral zeolite containing an extra-large 3-dimensional 30-ring channel system.

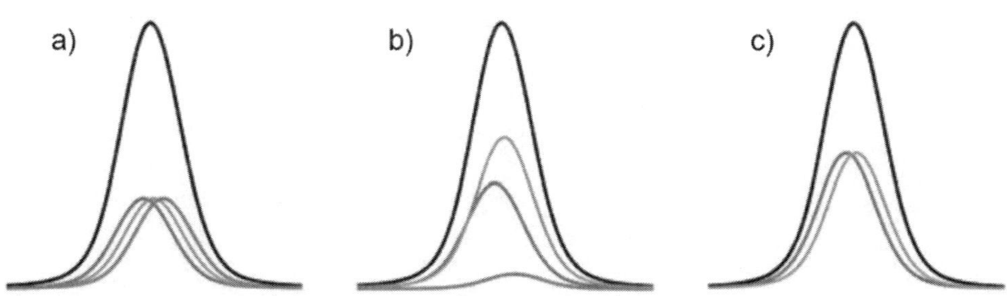

Figure 7. Partitioning of overlapped reflections in powder X-ray diffraction with a) equal partitioning in a traditional way, b) pre-partioning using electron diffraction intensities, and c) equal partitioning after removing one weak reflection (42 McCusker and Baerlocher 2009).

In fact, it is not necessary to use the exact ED intensities to partition overlapped reflections in PXRD and the intensity order is already enough to facilitate the structure solution from PXRD. The weak reflection elimination method shows that we only need to know which peaks are weak from ED patterns. Based on this information, the weak reflections are removed from the overlapped peaks. As shown in Figure 7c, after removing one weak reflection, the overlapped intensity was equally partitioned into two strong reflections, which has a smaller error than the equal partitioning of three peaks. Moreover, this elimination reduces the overlapping significantly and speeds up the structure solution by charge-flipping iterations. The test on the zeolite ZSM-5 decreases the overlapping ratio from 55% to 25%.

Conclusion

In summary, X-ray diffraction is a powerful tool to determine crystal structures, but both single crystal X-ray diffraction and powder X-ray diffraction have their limitations. Electron crystallography is complementary with X-ray crystallography in many ways, and the information obtained from electron crystallography can be used to facilitate the structure solution from powder X-ray diffraction in different approaches. The phases from HRTEM images can be combined with powder X-ray diffraction through dual-space structure solution algorithms, and they can also be used directly to give information about atomic positions. The electron diffraction doesn't have the overlapping problem which is the major problem for powder X-ray diffraction. Therefore, the electron diffraction intensities can be used to give structure information directly or help to pre-partition the overlapped reflections in powder X-ray diffraction.

References

[1] Baerlocher, Ch.; McCusker, L.B. Database of Zeolite Structures: http://www.iza-structure.

[2] Ju, J.; Lin, J.H.; Li, G.B.; Yang, T.; Li, H.M.; Liao, F.H.; Loong, Ch.-K.; You, L.P. Aluminoborate-based molecular sieves with 18-octahedral-atom tunnels Angew. *Chem. Int*. Ed. 2003, 42, 5607-5610.

[3] Phan, A.; Doonan, C.J.; Uribe-Romo, F.J.; Knobler, C.B.; O'Keeffe, M.; Yaghi, O.M. *Acc. Chem. Res*. 2010, 43, 58.

[4] Patterson, A. L. *Phys. Rev*.1934, 46, 372-376.

[5] Sheldrick, G. M. *Acta Crystallogr*. 1990, A46, 467-473.

[6] Oszlányi, G.; Sütő, A. *Acta Cryst*. 2008, A64, 123-134.

[7] De Vries, R. Y.; Briels, W. J.; Feil, D. *Acta Crystallogr*. 1994 A50, 383–391.

[8] Mitchell, M. *An Introduction to Genetic Algorithms*. 1996, Cambridge, MA: MIT Press.

[9] Deem, M.W.; Newsam, J.M. Framework crystal structure solution by simulated annealing: test application to known zeolite structures *J. Am. Chem. Soc*. 1992, 114, 7189-7198.

[10] Tang, L.Q.; Shi, L.; Bonneau, C.; Sun, J.L.; Yue, H.J.; Ojuva, A.; Lee, B.L.; Kritikos, M.; Bell, R.G.; Bacsik, Z.; Mink J.; Zou, X.D. *Nature Mater.*, 2008, 7, 381.
[11] Su, J.; Wang, Y.X.; Wang, Z.M.; Lin, J.H. J. *Am. Chem. Soc.*, 2009, 131, 6080.
[12] Song, X.W.; Li, Y.; Gan, L.; Wang, Z.P.; Yu, J.H.; Xu, R.R. Angew. *Chem. Int.* Ed., 2009, 48, 314.
[13] Baerlocher, Ch.; Weber, T.; McCusker, L.B.; Palatinus, L.; Zones, S.I. *Science* 2011, 333, 1134-1137.
[14] Sedlmaier, S.J.; Döblinger, M.; Oeckler, O.; Weber, J.; Schmedt auf der Günne, J.; Schnick, W. *J. Am. Chem. Soc.* 2011 133, 12069-12078.
[15] Han, Y.; Li, Y.; Yu, J.; Xu, R. Angew. *Chem. Int.* Ed. 2011, 50, 3003-3005.
[16] Lorgouilloux, Y.; Dodin, M.; Paillaud, J.L.; Caullet, P.; Michelin, L.; Josien, L.; Ersen, O.; Bats, N. *J. Solid State Chem.*, 2009, 182, 622.
[17] Elomari, S.; Burton, A.; Medrud, R.C.; Grosse-Kunstleve, R. *Microporous Mesoporous Mater.*, 2009, 118, 325.
[18] McCusker, L.B.; Baerlocher, Ch.; Wilson, S.T.; Broach, R.W. *J. Phys. Chem.* C, 2009, 113, 9839.
[19] Corma, A.; Diaz-Cabanas, M.J.; Jorda, J.L.; Rey, F.; Sastre, G.; Strohmaier, K.G. *J. Am. Chem. Soc.*, 2008, 130, 16482.
[20] Han, Z.X.; Picone, A.L.; Slawin, A.M.Z.; Seymour, V.R.; Ashbrook, S.E.; Zhou, W.Z.; Thompson, S.P.; Parker, J.E.; Wright, P.A. *Chem. Mater.*, 2010, 22, 338.
[21] Xie, D.; McCusker, L.B.; Baerlocher, Ch.; L. Gibson, A.W. Burton and S.J. Hwang, *J. Phys. Chem.* C, 2009, 113, 9845.
[22] Dodin, M., Paillaud, J.-L., Lorgouilloux, Y., Caullet, P., Elkaïm, E. and Bats, N. *J. Am. Chem. Soc.* 2010. 132, 10221-10223.
[23] Jiang, J., Jorda, J.L., Diaz-Cabañas, M.J., Yu, J. and Corma, A. Angew. *Chem. Int.* Ed. 2010, 49, 4986-4988.
[24] Broach, R.W.; Kirchner, R.M. *Microporous Mesoporous Mater.* 2011, 143, 398-400.
[25] Estermann, M.A.; McCusker, L.B.; Baerlocher, Ch. *J. Appl. Cryst.,* 1992, 25, 539.
[26] Baerlocher, Ch.; McCusker, L.B.; Palatinus, L. Z. Kristallogr. 2007, 222, 47.
[27] Williams, D.B.; Carter, C.B. *Transmission Electron Microscopy*, 2009, Springer, New York.
[28] Zou, XD; Hovmöller, S.; Oleynikov, P. Electron Crystallography - Electron microscopy and Electron Diffraction. IUCr Texts on Crystallography 16, Oxford University press 2011.
[29] Nicolopoulos, S.; González-Calbet, J.M.; Vallet-Regí, M.; Corma, A.; Corell, C.; Guil, J.M.; Pérez-Pariente, J. *J. Am. Chem. Soc.*, 1995, 117, 8947.
[30] Wagner, P.; Terasaki, O.; Ritsch, S.; Nery, J.G.; Zones, S.I.; Davis, M.E.; Hiraga, K. *J. Phys. Chem.,* 1999, B103, 8245.
[31] Kolb, U.; Mugnaioli, E.; Gorelik, T.E. Automated electron diffraction tomography – a new tool for nano crystal structure analysis. *Cryst. Res. Technol.* 2011, 46, 542-554.
[32] Zhang, D.; Oleynikov, P.; Hovmo¨ller, S.; Zou, X. D. Collecting 3D electron diffraction data by the rotation method. Z. Kristallogr. 225, 94–102 (2010).
[33] Sun, J.L.; He, Z.B.; Hovmöller, S.; Zou, X.D.; Gramm, F.; Baerlocher, Ch.; McCusker, L.B. Z. Kristallogr., 2010, 225, 77.
[34] Corma, A.; Moliner, M.; Ángel Cantín, María J. Díaz-Cabañas, José L. Jordá, Daliang Zhang, Junliang Sun, Kjell Jansson, Sven Hovmöller & Xiaodong Zou "A rational

synthesis of a Beta zeolite highly enriched in polymorph B and the direct structure determination of polymorph B" *Chem. Mater.* 2008, 20, 3218-3223.

[35] Willhammar, T.; Sun, J.L.; Wan, W.; Oleynikov, P.; Zhang, D.L.; Zou, X.D.; Moliner, M.; Gonzalez, J.; Martínez, C.; Rey, F.; Corma; A. Structure and catalytic properties of the most complex intergrown zeolite ITQ-39 determined by electron crystallography Nature Chem. 2012, 188-194.

[36] Gramm, F.; Baerlocher, Ch.; McCusker, L.B.; Warrender, S.J.; Wright, P.A.; Han, B.; Hong, S.B.; Liu, Z.; Ohsuna, T.; Terasaki, O. *Nature*, 2006, 444, 79.

[37] Hovmöller, S.: CRISP - Crystallographic Image Processing on a Personal Computer. *Ultramicroscopy*, 1992, 41, 121-135.

[38] Grosse-Kunstleve, R.W.; McCusker, L.B.; Baerlocher, *Ch. J. Appl. Crystallogr.* 1997, 30, 985–995.

[39] Baerlocher, Ch.; Gramm, F.; Massüger, L.; McCusker, L. B.; He, Z. B.; Hovmöller, S.; Zou, X. D.: Structure of the polycrystalline zeolite catalyst IM-5 solved by enhanced charge flipping. *Science* 2007, 315 1113-1116.

[40] Palatinus, L.; Chapuis, G. *J. Appl. Crystallogr.* 2007, 40, 786–790.

[41] Yu, Z.B.; Han, Y.; Zhao, L.; Huang, S.L.; Zheng, Q.-Y.; Lin, S.Z.; Córdova, A.; Zou, X.D.; Sun, J.L. Intergrown New Zeolite Beta Polymorphs with Interconnected 12-Ring Channels Solved by Combining Electron Crystallography and Single-Crystal X-ray *Diffraction Chem. Mater.* 2012, 24, 3701-3706.

[42] McCusker, L.B.; Baerlocher, *Ch. Chem. Comm.* 2009, 1439.

[43] Dorset, D.L.; Strohmaier, K.G.; Kliewer, C.E.; Corma, A.; Díaz-Cabañas, M.J.; Rey, F.; Gilmore, C.*J. Chem. Mater.* 2008, 20, 5325.

[44] Gilmore, C.J.; Bricogne, G. *Methods Enzymol.* 1997, 277, 65.

[45] Martínez-Franco R.; Moliner, M. Yun, Y.F.; Sun, J.L.; Wan, W.; Zou, X.D.; Corma, A. Synthesis of a new extra-large molecular sieve using proton sponges as organic structure directing agents, PNAS 2013, 110, 3749-3754.

[46] Sheldrick, G.M. A short history of SHELX. *Acta Cryst.* 2008, A64, 112-122.

[47] Sun, J.L.; Bonneau, C.; Cartín, Á.; Corma, A.; Díaz-Cabañas, M.J.; Moliner, M.; Zhang, D.L.; Li, M.R.; Zou, X.D. *Nature*, 2009, 458, 1154.

INDEX

C

D

E

K

L

M

N

O

P

Q

R

S

T

U

V

W

X

Y

Z